Carlos A. Cingolani
Editor

Pre-Carboniferous Evolution of the San Rafael Block, Argentina

Implications in the SW Gondwana Margin

Editor
Carlos A. Cingolani
Centro de Investigaciones Geológicas
  (UNLP-CONICET)
National University of La Plata
La Plata
Argentina

and

División Geología
Museo de La Plata
La Plata
Argentina

ISSN 2197-9596   ISSN 2197-960X (electronic)
Springer Earth System Sciences
ISBN 978-3-319-50151-2   ISBN 978-3-319-50153-6 (eBook)
DOI 10.1007/978-3-319-50153-6

Library of Congress Control Number: 2016958498

© Springer International Publishing AG 2017
This work is subject to copyright. All rights are reserved by the Publisher, whether the whole or part of the material is concerned, specifically the rights of translation, reprinting, reuse of illustrations, recitation, broadcasting, reproduction on microfilms or in any other physical way, and transmission or information storage and retrieval, electronic adaptation, computer software, or by similar or dissimilar methodology now known or hereafter developed.
The use of general descriptive names, registered names, trademarks, service marks, etc. in this publication does not imply, even in the absence of a specific statement, that such names are exempt from the relevant protective laws and regulations and therefore free for general use.
The publisher, the authors and the editors are safe to assume that the advice and information in this book are believed to be true and accurate at the date of publication. Neither the publisher nor the authors or the editors give a warranty, express or implied, with respect to the material contained herein or for any errors or omissions that may have been made.

Printed on acid-free paper

This Springer imprint is published by Springer Nature
The registered company is Springer International Publishing AG
The registered company address is: Gewerbestrasse 11, 6330 Cham, Switzerland

Springer Earth System Sciences

Series editors

Philippe Blondel, Bath, UK
Eric Guilyardi, Paris, France
Jorge Rabassa, Ushuaia, Argentina
Clive Horwood, Chichester, UK

More information about this series at http://www.springer.com/series/10178

(1940–2013)

This book is dedicated to the memory of Maria Adela Montalvo, my wife. She was a key technical assistant for several works done by the geological group from the Department of Geology-La Plata Museum and National Research Council, on the San Rafael Block and other argentine regions for more than 40 years. She crossed successfully technical secretary 'paradigms': the old and the electric typewriter and finally the computer systems, always with enthusiastic expertise.

Carlos A. Cingolani

Preface

The aim of the book is to offer all available information, largely presented at scientific conferences and meetings about the pre-Carboniferous units of the San Rafael Block, compiled after over 15 years of research in the area carried out by the working group. The opportunity to present data in the Springer Publishing enables greater visibility of all data available to an international audience. A first proposal was made in 2013, with the support and enthusiasm of Dr. Jorge Rabassa as regional editor, for the series of Springer Briefs, but we soon realized that this would not be enough in order to publish the full dataset. It was therefore requested to Springer to change to Earth Sciences Series of Books and the deadline was extended until 2016 that finally constitutes the work presented here to be available to the geological community. We hope that this book to be a review of the state of the art in the field that stimulates more discussions about the San Rafael Block's geological evolution, in the context of the proto-Andean Gondwana margin.

La Plata, Argentina
August 2016

Carlos A. Cingolani

Acknowledgements

As the editor I would like to express my gratitude to all contributors to this book and appreciation for the support received from the reviewers and especially from the Springer regional editor Jorge Rabassa. Thanks to Profs. Florencio G. Aceñolaza, Victor A. Ramos, Héctor A. Leanza, Peter Königshof, Luis Spalletti, Carlos Rapela, Silvio Peralta, Graciela Vujovich, Alejandro Ribot, Eugenio Aragón, Joaquín García-Sansegundo, Pedro Farias, Ana María Sato, Umberto Cordani, Koji Kawashita, Farid Chemale Jr., Miguel Basei, Eduardo Llambías, Maximiliano Naipauer, Gustavo W. Bertotto, Pablo González, Emilio González Díaz, Daniel Poiré, Robert Pankhurst, for several comments and discussions about the geological evolution of the San Rafael and other regions. Many thanks to Springer Books editors and publication team for their encouragement and technical assistance. A special recognition is for Norberto Uriz (University of La Plata), for his tireless collaboration to improve and even to 'create' the illustrations that are presented thorough the book. Thanks to Paulina Abre for helping us in language correction of the early draft chapters. The actual working group of the Department of Geology at La Plata Museum: Andrea Bidone, Mario Campaña, Arón Siccardi, Miguel Cricenti is deeply acknowledged. A remembrance to Humberto Lagiglia, since during his days as director of the Museo Municipal de Historia Natural of San Rafael, illustrated us with his knowledge of archeology of the area and kindly offered logistic support during several fieldworks carried out by the working research group. I also deeply thank Sofía de Occhipinti and her family from Villa 25 de Mayo for their loving hospitality. A profound remembrance is to the late Professors at the University of La Plata Alfredo J. Cuerda, Osvaldo C. Schauer and Luis Dalla Salda, who had participated in these investigations. I have received important laboratory support from the CIG-Geological Research Centre since many years and I greatly acknowledge its entire scientific and technical staff. Thanks to the University of La Plata, specially the Department of Geology of the La Plata Museum where I learnt, worked and taught during 50 years. The National Research Council (CONICET) and the National Research Agency (ANPCyT) have provided financial support for field and laboratory works throughout the grants during several years since 1999. The international cooperation within FAPESP (Brazil) and CONICET (Argentina) was very fruitful for isotope laboratory works at the University of São Paulo. Finally, I would like to express sincerely acknowledge my family for creating and maintaining the home that helped develop my work.

Contents

**Pre-Carboniferous Evolution of the San Rafael Block, Argentina.
Implications in the SW Gondwana Margin: An Introduction**   1
Carlos A. Cingolani

**The Mesoproterozoic Basement at the San Rafael Block,
Mendoza Province (Argentina): Geochemical and Isotopic
Age Constraints.** 19
Carlos A. Cingolani, Miguel A.S. Basei, Ricardo Varela,
Eduardo Jorge Llambías, Farid Chemale, Jr., Paulina Abre,
Norberto Javier Uriz and Juliana Marques

**Sedimentary Provenance Analysis of the Ordovician Ponón
Trehué Formation, San Rafael Block, Mendoza-Argentina** 59
Paulina Abre, Carlos A. Cingolani, Norberto Javier Uriz and Aron Siccardi

**Ordovician Conodont Biostratigraphy of the Ponón Trehué
Formation, San Rafael Block, Mendoza, Argentina** 75
Susana Heredia and Ana Mestre

**The Pavón Formation as the Upper Ordovician Unit Developed
in a Turbidite Sand-Rich Ramp. San Rafael Block, Mendoza,
Argentina.** 87
Paulina Abre, Carlos A. Cingolani and Marcelo J. Manassero

**Lower Paleozoic 'El Nihuil Dolerites': Geochemical and Isotopic
Constraints of Mafic Magmatism in an Extensional Setting
of the San Rafael Block, Mendoza, Argentina** 105
Carlos A. Cingolani, Eduardo Jorge Llambías, Farid Chemale, Jr.,
Paulina Abre and Norberto Javier Uriz

**Magnetic Fabrics and Paleomagnetism of the El Nihuil
Mafic Unit, San Rafael Block, Mendoza, Argentina** 127
Augusto E. Rapalini, Carlos A. Cingolani and Ana María Walther

Low-Grade Metamorphic Conditions and Isotopic Age Constraints of the La Horqueta Pre-Carboniferous Sequence, Argentinian San Rafael Block 137
Hugo Tickyj, Carlos A. Cingolani, Ricardo Varela and Farid Chemale, Jr.

La Horqueta Formation: Geochemistry, Isotopic Data, and Provenance Analysis 161
Paulina Abre, Carlos A. Cingolani, Farid Chemale, Jr. and Norberto Javier Uriz

Silurian-Devonian Land–Sea Interaction within the San Rafael Block, Argentina: Provenance of the Río Seco de los Castaños Formation 183
Carlos A. Cingolani, Norberto Javier Uriz, Paulina Abre, Marcelo J. Manassero and Miguel A.S. Basei

Primitive Vascular Plants and Microfossils from the Río Seco de los Castaños Formation, San Rafael Block, Mendoza Province, Argentina 209
Eduardo M. Morel, Carlos A. Cingolani, Daniel Ganuza, Norberto Javier Uriz and Josefina Bodnar

The Rodeo de la Bordalesa Tonalite Dykes as a Lower Devonian Magmatic Event: Geochemical and Isotopic Age Constraints 221
Carlos A. Cingolani, Eduardo Jorge Llambías, Miguel A.S. Basei, Norberto Javier Uriz, Farid Chemale, Jr. and Paulina Abre

Pre-Carboniferous Tectonic Evolution of the San Rafael Block, Mendoza Province 239
Carlos A. Cingolani and Victor A. Ramos

San Rafael Block Geological Map Compilation 257
Carlos A. Cingolani

Index 265

Contributors

Paulina Abre Centro Universitario de la Región Este, Universidad de la República (CURE-UDELAR), Ruta 8 km 282, Treinta y Tres, Uruguay

Miguel A.S. Basei Centro de Pesquisas Geocronológicas (CPGeo), Instituto de Geociencias, Universidade de São Paulo, São Paulo, Brazil

Josefina Bodnar División Paleobotánica, Museo de La Plata, UNLP, La Plata, Argentina; Consejo Nacional de Investigaciones Científicas y Técnicas (CONICET), La Plata, Argentina

Carlos A. Cingolani Centro de Investigaciones Geológicas-CONICET, Universidad Nacional de La Plata, La Plata, Argentina; División Geología del Museo de La Plata, La Plata, Argentina; Consejo Nacional de Investigaciones Científicas y Técnicas (CONICET), La Plata, Argentina

Farid Chemale Jr. Programa de Pós-Graduação em Geologia, Universidade do Vale do Rio dos Sinos (UNISINOS), São Leopoldo, Brazil

Daniel Ganuza División Paleobotánica, Museo de La Plata, UNLP, La Plata, Argentina

Susana Heredia CONICET—Instituto de Investigaciones Mineras, Universidad Nacional de San Juan, San Juan, Argentina

Eduardo Jorge Llambías Centro de Investigaciones Geológicas (CONICET-UNLP), Universidad Nacional de La Plata, La Plata, Argentina

Marcelo J. Manassero Centro de Investigaciones Geológicas-CONICET, Universidad Nacional de La Plata, La Plata, Argentina

Juliana Marques Laboratorio de Geologia Isotópica, Universidade Federal do Río Grande do Sul, Porto Alegre, Brazil

Ana Mestre CONICET—Instituto de Investigaciones Mineras, Universidad Nacional de San Juan, San Juan, Argentina

Eduardo M. Morel División Paleobotánica, Museo de La Plata, UNLP, La Plata, Argentina; Comisión de Investigaciones Científicas de la Provincia de Buenos Aires (CIC), La Plata, Argentina

Victor A. Ramos Instituto de Estudios Andinos Don Pablo Groeber (IDEAN), Universidad de Buenos Aires–CONICET, Ciudad Universitaria, Buenos Aires, Argentina

Augusto E. Rapalini Departamento de Ciencias Geológicas, Facultad de Ciencias Exactas y Naturales, Instituto de Geociencias Básicas, Aplicadas y Ambientales de Buenos Aires (IGEBA), Universidad de Buenos Aires CONICET, Buenos Aires, Argentina

Aron Siccardi División Geología del Museo de La Plata, Centro de Investigaciones Geológicas (CONICET-UNLP), La Plata, Argentina

Hugo Tickyj Facultad de Ciencias Exactas y Naturales, Universidad Nacional de La Pampa, Santa Rosa, La Pampa, Argentina

Norberto Javier Uriz División Geología, Museo de La Plata, UNLP, La Plata, Argentina

Ricardo Varela Centro de Investigaciones Geológicas-CONICET, Universidad Nacional de La Plata, La Plata, Argentina

Ana María Walther Departamento de Ciencias Geológicas, Facultad de Ciencias Exactas y Naturales, Instituto de Geociencias Básicas, Aplicadas y Ambientales de Buenos Aires (IGEBA), Universidad de Buenos Aires CONICET, Buenos Aires, Argentina

Pre-Carboniferous Evolution of the San Rafael Block, Argentina. Implications in the SW Gondwana Margin: An Introduction

Carlos A. Cingolani

Abstract The San Rafael Block as part of the Cuyania terrane, lies eastwards of the present-day Andean Cordillera in Mendoza province and it develops south of the Nazca flat-slab subduction zone. It is a geographical region constituted by a set of rather convex elevation oriented NW–SE from Sierra de las Peñas to Cerro Nevado as the eastern Neogene volcanic arc and ending at the transitional zone known as the La Escondida mining district; as a geological province was also cited as 'Sierra Pintada'. The knowledge about this geological region started on 1891 and was continued during the twentieth century with intense regional mapping projects carried out by different Argentine institutions through which the geological background was founded. This book is dedicated to the tectonic evolution of the pre-Carboniferous units and was organized by chronological stages, in order to know the implications in the proto-Andean SW Gondwana margin, as follows: The Mesoproterozoic basement of Laurentian affinity; tectonic extension, passive margin and Cuyania terrane collisional event during the Lower Paleozoic; Silurian-Lower Devonian orogenic sedimentation, Chanic compressional phase during the Chilenia terrane accretion in the Upper Devonian-Lower Carboniferous and finally the tectonic evolution synthesis. An updated geological map compilation is also subdivided in three regions: Sierra de las Peñas, Sierra Pintada-Cerro Nevado and La Escondida transitional zone.

Keywords Gondwana · Proto-Andean margin · Paleozoic accreted terranes · Chanic tectonic phase

C.A. Cingolani (✉)
División Geología del Museo de La Plata, Centro de Investigaciones Geológicas-CONICET, Universidad Nacional de La Plata, Diag. 113 N. 275, 1904 La Plata, Argentina
e-mail: ccingola@cig.museo.unlp.edu.ar; carloscingolani@yahoo.com

1 The San Rafael Block in the Context of Paleozoic Gondwana Margin

The southern South America Pacific Paleozoic Gondwana margin is characterized by the presence of orogenic belts oriented in a north–south direction (Ramos et al. 1986). They had been accreted to the cratonic areas during Upper Precambrian–Lower Cambrian (Pampean cycle), Cambrian–Devonian (Famatinian cycle) and Upper Devonian–Lower Carboniferous (Gondwanian cycle). The San Rafael Block is part of the Cuyania composite terrane (Ramos 2004 and references therein), it is linked to the Famatinian cycle and lies eastwards of the present-day Andean Belt (Fig. 1).

The Cuyania composite terrane comprises four sectors, which from north to south, are: Precordillera thin-skinned (partially thick-skinned) fold and thrust belt that was generated by shallow east-dipping flat-slab subduction of the Nazca plate; the Pie de Palo area with Grenvillian-age basement; the San Rafael Block and Las Matras Block (Fig. 1). The subsurface lithofacies along the Triassic 'Cuyo basin' linked the Precordillera with the San Rafael Block (Rolleri and Criado Roqué 1970; Criado Roqué and Ibáñez 1979; Ramos 2004) with an extensional subsidence sedimentary regime.

The Cuyania terrane has been considered from a stratigraphical standpoint unique to South America due to the presence of Lower Paleozoic carbonate and

Fig. 1 Location of the San Rafael Block within the Cuyania terrane in the SW Gondwana margin

siliciclastic deposits overlying an igneous-metamorphic crust of Grenville-age (Ramos 2004; Sato et al. 2000, 2004; Varela et al. 2011); and it has been the objective of several lines of research during recent years, attempting to constrain its allochthonous or para-autochthonous origin with respect to Gondwana (Rapela et al. 1998). Some tectonic interpretations had been proposed: one explains that the terrane was detached from the southern Appalachians of Laurentia in Cambrian times, was transferred to western Gondwana during the Early to Middle Ordovician, and was amalgamated to the early proto-Andean margin by the Mid-Late Ordovician (Thomas and Astini 2003; Ramos 2004 and references therein). Other studies have claimed a continent-continent collision called 'Occidentalia terrane hyphotesis' (Dalla Salda et al. 1992), and even a para-autochthonous-to-Gondwana origin based on strike-slip displacements from the South Africa-Antarctica regions was proposed (Aceñolaza et al. 2002; Finney 2007). Geological and paleontological evidence constrain the time for the docking of Cuyania to the late Middle to Late Ordovician. The terrane deformation linked to the collision started in the Ordovician, and continued until the time of approach of the Chilenia terrane during Late Devonian, against the Pacific side (Fig. 2).

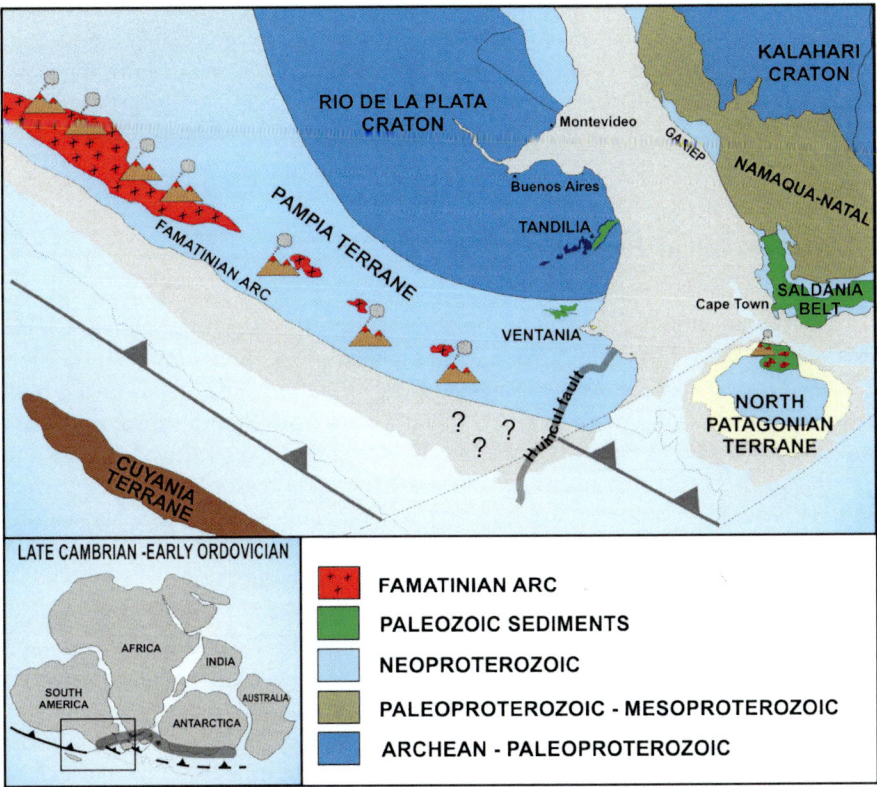

Fig. 2 Paleogeographic reconstruction of SW Gondwana region after Pankhurst et al. (2006), Gregori et al. (2008), Uriz et al. (2011) and Uriz (2014)

The San Rafael Block, as a part of Cuyania terrane, is developed south of the Andean flat-slab subduction zone as shown on Fig. 1.

2 Some Geomorphological Aspects

The variety of colours of the Upper Paleozoic volcanic and volcaniclastic rocks inspired the first denomination of the region: 'Sierra Pintada' (Burckhardt and Wehrli 1900). It is a geographical unit constituted by a set of rather convex elevations oriented approximately NW–SE (34° 14′S to 36° 10′S—68° 06′W to 69° 06′W); it rises east of the Andes from Sierra de las Peñas to Cerro Nevado (3980 m a.s.l.) as the eastern Neogene volcanic arc (Fig. 3). Two main streams cross the block in west to east direction, the Diamante and Atuel Rivers. To the north of the Río Diamante, the highest hill is Cerro de la Chilena (1775 m a.s.l.). These rivers cut across ranges and erode canyons that can reach in some places over 300 m deep. Polanski (1949, 1954) studied the geomorphological aspects of 'Mendocino territory' giving the name of '**San Rafael block**' to the mountains located southwest of the eponymous city that closely resembles the westward Frontal Cordillera for the expanded Upper Paleozoic volcaniclastic rocks. From north to south, the main positive regions are: Sierra de las Peñas, Sierra Pintada to Cerro Nevado and Agua Escondida mining district (Fig. 3). The San Rafael Block has a rich hydroelectric potential, uranium mining and tourist attractions.

3 A Geological Province

As it has been aforementioned, at the stage of the first scientific pioneers (since 1890–1945), the geological province was as named as the 'Sierra Pintada' (or Sierra Pintada System) and later as the 'San Rafael block'; afterwards, it was linked to the so-called 'Mendocino-Pampeano mobile belt' (see Criado Roqué 1969a, b, 1972) to form the geologic province 'Sanrafaelino-Pampeana' as exposed by Criado Roqué and Ibáñez (1979) which integrated the La Escondida region located southwards (Fig. 3). The San Rafael region exposed several faults and lineaments within a seismic area demonstrated by the massive earthquake of 30 May 1929, suggesting that the region is still submitted to shortening (Costa et al. 2006).

Following this criterion in the geological map compilation presented in Chapter "San Rafel Block Geological Map Compilation" it has been divided into three parts: the northern sector corresponding to the Sierra de las Peñas, the central one which comprises the Sierra Pintada-Cerro Nevado, and the southern sector which includes the Agua Escondida (or La Escondida) district. In each of these sectors we have identified places where studies on pre-Carboniferous units were done and that are

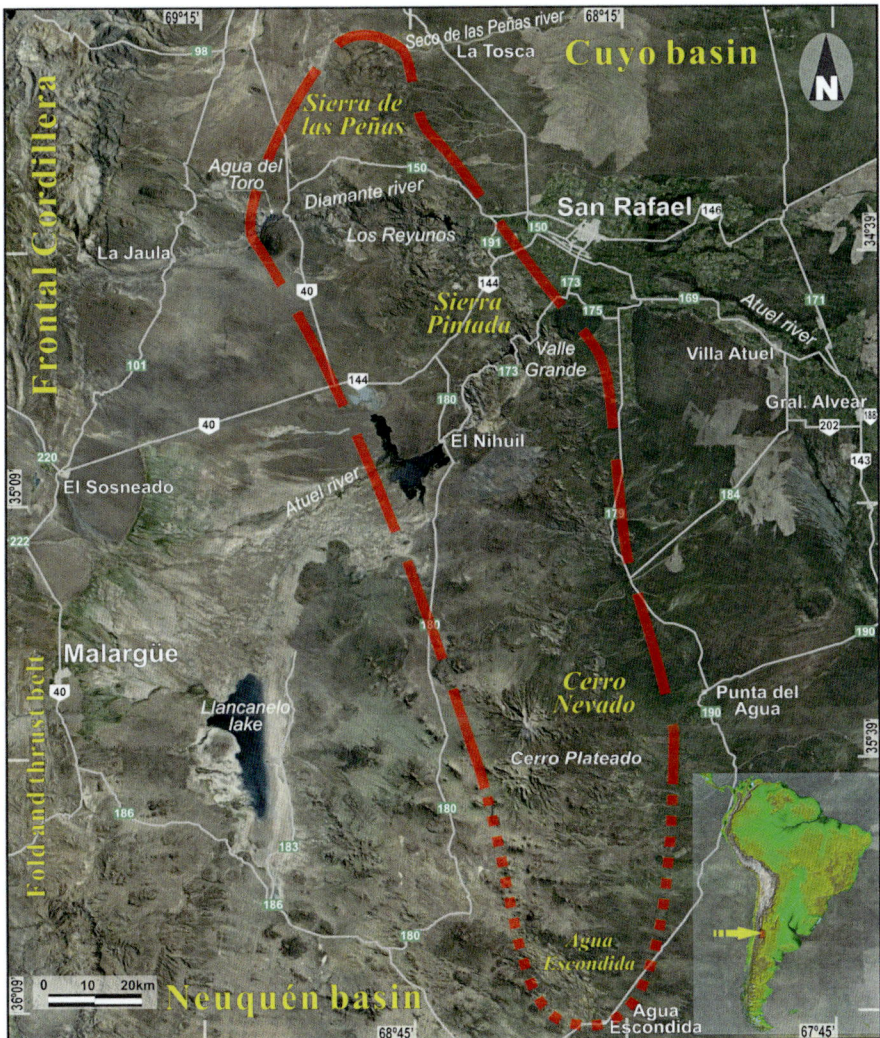

Fig. 3 Satellite image showing in *dotted red lines* the location of the San Rafael Block and the main geographic aspects as the Sierra de las Peñas, Sierra Pintada-Cerro Nevado and towards the south the transition to the Agua Escondida mining district

the aim of the herein presented book. It is important to note that throughout the text, the term San Rafael Block has been used to denote the geological province enclosing all outcrops within the Las Peñas River towards the north to the region covered by the Cenozoic volcanism of Cerro Nevado, and as a transitional zone the Agua Escondida district to the south. The Cuyo and Neuquén basins bound the San Rafael Block to the north-east and south respectively (Figs. 1 and 3).

4 Main Aspects of the Pioneer Geological Knowledge

The first geological published reference of this geological unit is due to Bodenbender (1891) who noted the presence of pre-Jurassic 'porphyry' rocks. Then Hauthal et al. (1895–96) exposed the first geographic and geological description of the 'Sierra Pintada' where Pb and Ag ore mines are present. This scientific program was prepared by Francisco P. Moreno as Director of the La Plata Museum. The topographer Lange (Lange et al. 1895) mentioned that the highest point is the Cerro Nevado (3810 m a.s.l.). Wehrli and Burkhardt (1898) and Lange, Burkhardt and Wehrli (1900) crossed the Sierra Pintada from San Rafael to Malargüe and described it as a 'small mountain range' elongated from NW to SE. Stappenbeck (1913) in an extensive travel from the Sierra Pintada to Cerro Nevado, described the main rock types and observed their relationships. In the La Estrechura (south of Río Leones, Ponón Trehué sector) he mentioned the presence of basement gneisses that supposedly were brought up to the surface by modern basalts. Later in 1934, Stappenbeck (unpublished report to the YPF Oil Company) distinguished the greywacke, mica-schist and other low-grade metamorphic rocks assigned to the Precambrian, Middle Paleozoic and what he called 'Estratos de Paganzo' which included the metasedimentary rocks of the Cerro Bola and Arroyo Pavón. He mentioned the presence of a granitic pluton related to the Las Picazas mine district near the Río Diamante, with sphalerite and galena mineralization. Wichmann (1928) mentioned basement rocks such as gneisses, granites, pegmatites, amphibolites among others as well as limestones in the Ponón Trehué area that are considered similar to those of Cerro de la Cal and Salagasta (near the city of Mendoza); he also described Carboniferous shales and sandstones Finally, he stressed the presence of an intrusive called 'diabase' near Nihuil town. Storni (1933) studied the physiographic aspects of the region west of the Sierra Pintada and north of the Río Diamante, describing an intrusive as 'quartz diorite' in schists to the NW of Agua de la Chilena. He named the region as the 'Sierra Pintada block'. Groeber (1929, 1939) worked on the volcanic and modern series of Rincón del Atuel. Keidel (1947) gave an overview of the regional geology of the San Rafael region. Padula (1949, 1951) when working for the YPF Oil Company distinguished several formations as the local 'basement' of the area. Holmberg (1948a, b) provided details from the Cerro Bola area and described the sequences as 'Estratos del Arroyo Pavón' assigned to the Upper Paleozoic. Feruglio (1946), in his work on 'Orographic Systems of Argentina', reviewed the main morphological and geological features of the region, and described the pre-Paleozoic-Paleozoic greywacke-schists and metamorphic rocks from the Sierra Pintada System and Cerro Nevado. He expressed 'this mountain can be considered as an eroded gigantic block'. Dessanti (1945) discovered the fossils within the Carboniferous El Imperial unit. After intense regional mapping in the area of the Sierra Pintada, Dessanti (1954, 1956) published the 'Hoja Geológica 27c Cerro Diamante', as a result of previous detailed and thorough field investigation. Along the present book we have used the basic stratigraphy done by Dessanti. Polanski ('Hoja Geológica 26c La

Tosca' 1964) mentioned the positive structures of the Frontal Cordillera and the San Rafael Block and describes the pre-Jurassic metasediments especially in the Los Gateados creek. To the south, the author confirms that this unit called La Horqueta is intruded by granitic stock of the Agua de la Chilena. Based on several studies González Díaz (1964) recognized the region as a morphostructural unit and named it as San Rafael Block, following Feruglio (1946) and Polanski (1964) and extended the southern boundary beyond Cerro Nevado Cenozoic volcano, to the region known as La Escondida (González Díaz 1972). He proposed the subdivision of the original 'La Horqueta Formation' into two units, a metamorphic one known as La Horqueta and the metasedimentary section called the Río Seco de los Castaños Formation (González Díaz 1981). Núñez (1976) mapped the 'Hoja Geológica 28c Nihuil', which remains as an unpublished report of the Secretaría de Minería (Buenos Aires), describing the outcrops of the Río Seco de los Castaños Formation that is composed of sandstones, greywackes and siltstones outcropping in the Lomitas Negras and around Agua del Blanco. The Loma Alta gabbros, known as 'pre-Ordovician mafic rocks' are also mentioned by the author. Then González Díaz (1964, 1972), in the geological descriptions of the San Rafael and Agua Escondida maps, distinguished a probably Devonian, Upper Paleozoic, continental Mesozoic rocks and several Cenozoic units. The region of the Cerro Nevado volcano ('Hoja Geológica 29d') was mapped by Holmberg (1973) who recognized the Cerro Las Pacas Formation as part of the metamorphic basement rocks. It was Núñez (1979) who mapped the region of Ponón-Trehué in the 'Hoja Geológica 28d Estación Soitué' and recognized the basement rocks and the carbonate Lower Paleozoic sedimentary cover with Ordovician fossils. After these pioneers works, the National Geological Survey from Argentina (SEGEMAR) organized new compilations of the geology of the San Rafael Block (see the Mendoza province Geological Map 1:500,000; Sepúlveda et al. 2001, 2007).

5 Comments on Heritage of Human History Along the San Rafael Region

The primitive inhabitants of southern Mendoza province formed several tribes spread across the plains and foothills before and after the arrival of the Spanish conquerors. Huarpes, Puelches, Pehuenches and Mapuches were bellicose and combative people, and skilled riders that knew very well the mountain passes across the Andes. Pictographs drawings on the rocks are signs that remain of those peoples. After Lagiglia (2002) they became from the 'Old Pre-Ceramic period' as hunters that lived between 15,000 and 12,000 years BP. Evidence are found near the Atuel and Diamante rivers where very rustic stone tools were found. The beginning of the 'Christian era' brought changes that influenced the 'agro-pottery' cultures, such as the addition of growing vegetables, as well as social reorganizations. During the 'Inca period', between 1450 and 1550 AD, the metallurgy acquired relevance as well as agriculture and irrigation methods. Finally, the

'Spanish Colonial period' developed after 1550: the arrival of Europeans broke the social organization of the Incas. Abundant water with two main rivers (Diamante and Atuel) permitted the settlement of several groups of original populations. Under these circumstances, the 'Fortín San Rafael del Diamante' was founded in 1805 by the Spanish Viceroy, Marqués Rafael de Sobremonte. This fortress was located near the Río Diamante, where is now the Villa 25 de Mayo. For over 70 years, this locality became a bastion of civilization, protecting those who came to the Diamante valley. Since 1875 the Italian, French and Spanish immigration dedicated to agriculture colonized the area. The Villa 25 de Mayo gradually became the 'old villa' since 1903, where the Mendoza government declared the communal authorities near the French settlement as the city of San Rafael which is one of the most important of western Argentina. The arrival of the railway from Mendoza changed the history of San Rafael. In 1903 the journal 'Los Andes' communicated the official opening of the new branch of the railroad. During 1944, it was extended southwards to connect the towns of San Rafael and Malargüe. This branch was called as 'black gold' conceived mainly for transport carbonate rocks, coal and even oil. Unfortunately the railway is unused since many years ago. Nowadays, the main activities of the region are related to its numerous wineries and olive oil production founded mainly by the French, Italian and Spanish communities.

6 Relevance of the pre-Carboniferous Geological History

The San Rafael Block shows similarities and differences compared to Cordillera Frontal and Precordillera of western Argentina; it records an interesting history of terrane accretions (Cuyania-Chilenia) during the Paleozoic in the proto-Andean Gondwana margin. As it can be seen in Fig. 4, pre-Carboniferous units of the San Rafael Block comprise: the Mesoproterozoic basement which can be correlated to other basement rocks within the Cuyania terrane; a fossiliferous Ordovician carbonate platform which unconformably overlies the basement; Ordovician siliciclastic rocks bearing interesting graptolite biozones; MORB dolerites that resemble ocean floor mafic rocks from western Precordillera; Silurian-Devonian siliciclastic units forming distinguishable sedimentary facies that record the Chanic deformation as evidence of the Chilenia terrane collision.

After more than thirty years of work experience mainly in the Lower and Middle Paleozoic geology of the Precordillera in western Argentina, we started the biostratigraphic study of the Ordovician sedimentary rocks outcropping in the Cerro Bola region (Cuerda and Cingolani 1998). In parallel, we conducted a sampling of basement outcrops of the Río Leones type-section, and their Rb-Sr results allowed us to recognize a Mesoproterozoic basement (1 Ga) similar to that found elsewhere of the Cuyania terrane (Cingolani and Varela 1999). These results and the findings of Ordovician graptolite biozones around the Cerro Bola encouraged further studies in the area. Thus, with financial support from CONICET (National Research Council) grants, sedimentological investigations (Manassero et al. 1999) and

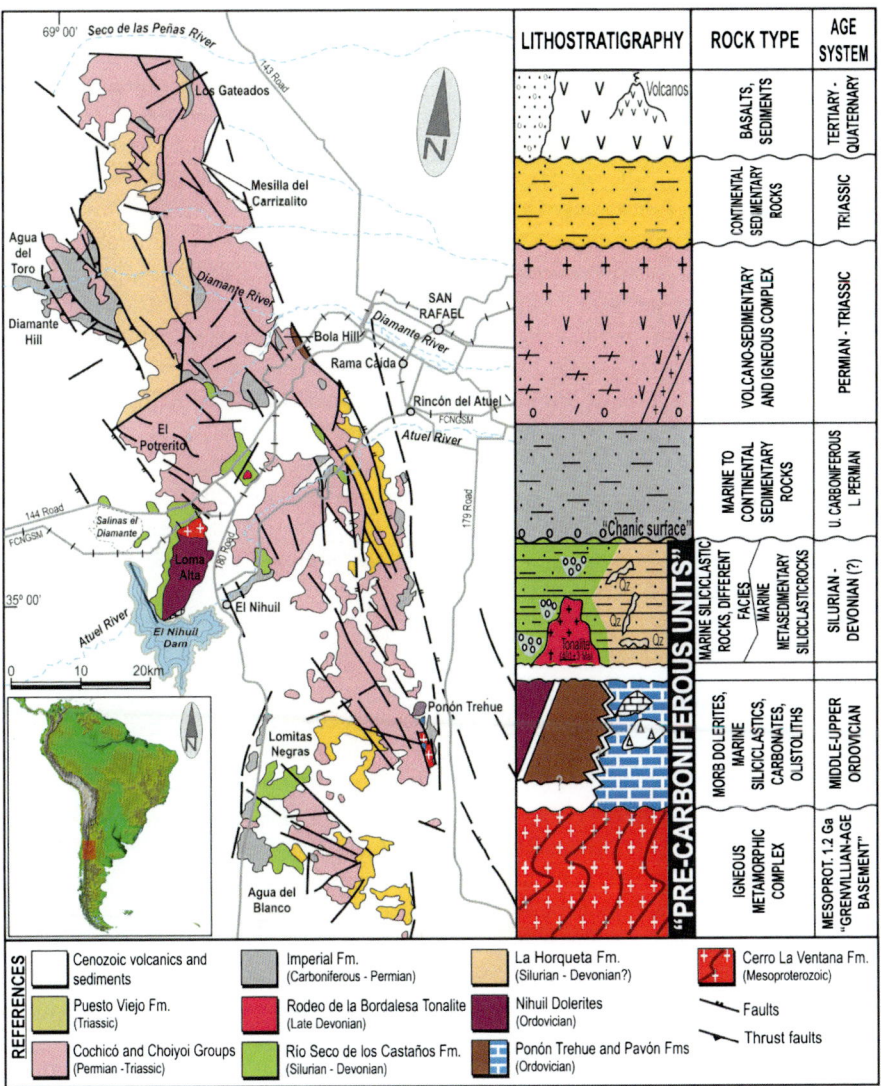

Fig. 4 Geological sketch map and stratigraphic column of the San Rafael Block showing the pre-Carboniferous units

sedimentary provenance analysis based on petrographical, geochemical and isotopic Sm–Nd data (Cingolani et al. 2003; Manassero et al. 2009) were performed. At that time, the possibility of developing a doctoral thesis on the provenance of the Ordovician sedimentary units of the Precordillera and San Rafael Block was planned. This resulted in the first detrital zircon ages presented by Abre (2007) in her Ph.D. thesis defended at the University of Johannesburg, South Africa. Interesting facts found and the geological characteristics of the region that somehow linked a transition to the eastern sector of the Frontal Cordillera encouraged the group to apply for a grant to the Argentine National Research Agency (ANPCyT) with the aim of extending the studies to all the 'pre-Carboniferous' units of the San Rafael Block. Therefore, studies on geochemical and isotope analyses of the mafic rocks of the El Nihuil region (Cingolani et al. 2000) began with international cooperation with two laboratories: the Isotope Geology of the Federal University of Río Grande do Sul, Porto Alegre and the Geochronology Center at the University of São Paulo, Brazil; the first one under the direction of Prof. Farid Chemale Jr. and Prof. Koji Kawashita and the second with the lead of Prof. Miguel A. Stipp Basei. Several papers were published in peer-reviewed international journals as a result of these studies. Rapalini and Cingolani (2004) carried out paleomagnetic analyses of the Pavón Formation (Upper Ordovician) and defined the first Ordovician paleopole of Cuyania. Abre et al. (2009, 2011, 2012) published papers related to the Ordovician sedimentary provenance of Ponón Trehué Formation and other units from Precordillera. From 2002, the group has published partial results in several Geological Congresses, as well as in the South American Symposium on Isotope Geology and Gondwana Symposium, because interesting facts about geochemistry and isotopic geology were achieved. A complete bibliographical list can be found within each of the following chapters.

7 Book Contents

This book is mainly conceived for Earth Science professionals working in scientific research and industry, as well as university students and general interest readers. The first chapter corresponds to the Introduction and Contents of this book, whereas the following chapters deal with the pre-Carboniferous units from Mesoproterozoic basement to Silurian–Devonian sedimentary and igneous rocks. All chapters are illustrated with maps, schemes and cross-sections that constitute up-to-date proposals and several original datasets. Different sources are cited at the end of each chapter that can be used to expand particular points discussed in the book. As it is shown on Fig. 5, the book was organized by several stages as follows:

Basement of Laurentian affinity: in the second chapter C. Cingolani, M. Basei, R. Varela, E. J. Llambías, F. Chemale Jr., P. Abre, N. J. Uriz and J. Marques, provide new petro-geochemical and isotopic information to constrain the crustal evolution of the Mesoproterozoic basement. The isotopic data obtained as well as their petrological and geochemical features are reported for two main outcrops (the type-section known as Río Leones-Ponón Trehué region and a part of the El Nihuil Mafic Unit). These data are useful to discuss relationships with equivalent Mesoproterozoic units located along the Cuyania terrane in the proto-Andean Gondwana margin.

Tectonic extension, passive margin stage and collisional event: Chapter "Sedimentary Provenance Analysis of the Ordovician Ponón Trehué Formation, San Rafael Block, Mendoza-Argentina", by P. Abre, C. Cingolani and N. Uriz,

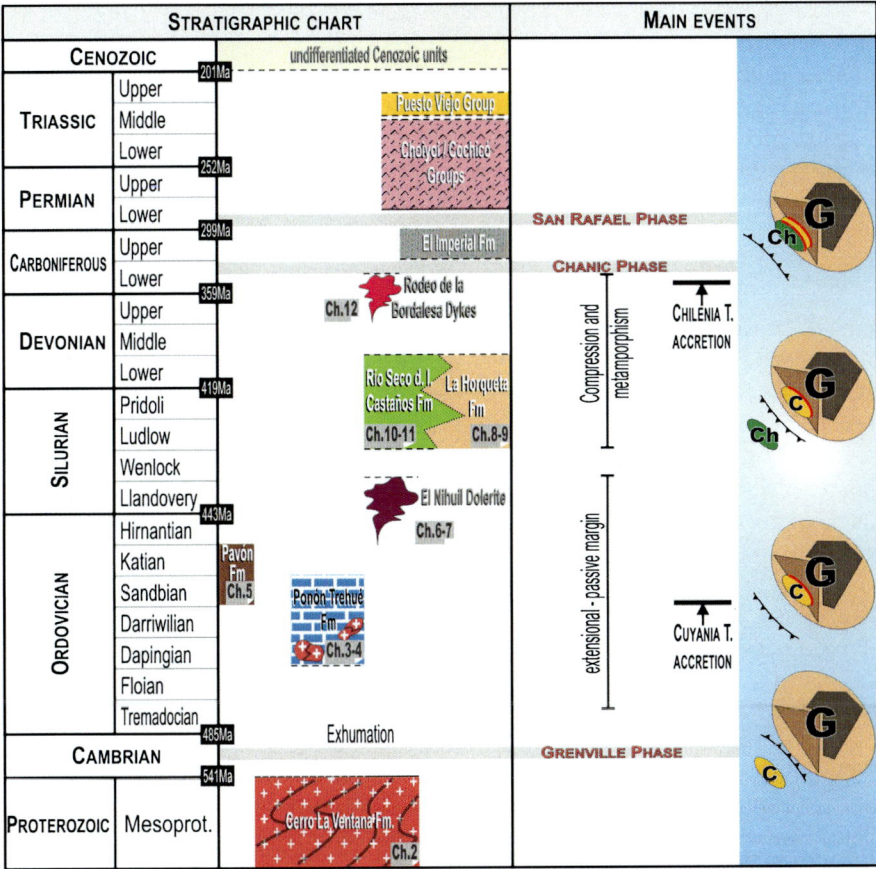

Fig. 5 The San Rafael Block studied units organized in chronostratigraphic order where the main tectonic events are represented in each chapter of the book (e.g. Chapter "The Mesoproterozoic Basement at the San Rafael Block, Mendoza Province (Argentina): Geochemical and Isotopic Age Constraints"). At the *right*, *G* SW Gondwana, *C* Cuyania terrane, *Ch* Chilenia terrane. The polarity of subduction during Paleozoic paleogeographic evolution in the proto-Andean margin is also exposed

deals with provenance analyses of the carbonate-siliciclastic Ordovician sedimentary Ponón Trehué Formation (Darriwilian to Sandbian). This is the only sequence which exhibits a direct contact with the Mesoproterozoic basement through an unconformity, not only within the San Rafael Block, but rather for the entire Cuyania terrane. On Chapter "Ordovician Conodont Biostratigraphy of the Ponón Trehué Formation, San Rafael Block, Mendoza, Argentina", by S. Heredia and A. Mestre, a review of the Middle Ordovician conodont fauna of the Ponón Trehué Formation is offered. Different genera and species of conodonts were recovered from clastic-carbonate beds and correlated to the Precordilleran units. Chapter "The Pavón Formation as the Upper Ordovician Unit Developed in a Turbidite Sand-

Rich Ramp. San Rafael Block, Mendoza, Argentina", by P. Abre, C. Cingolani and M. Manassero is dedicated to the deformed siliciclastic Pavón Formation outcropped at cerro Bola region. The graptolite fauna, in particular the presence of Climacograptus bicornis Biozone indicate a Sandbian age (Upper Ordovician). The complete provenance dataset suggests the basement of the San Rafael Block (Cerro La Ventana Formation) as the main source of debris. The siliciclastic sequence was deposited in a basin at a latitude of around 26°S, and linked to the accretion of the Cuyania terrane towards west of Gondwana; this accretion caused uplift by thrusting of the Mesoproterozoic crust to the east at ca. 460 Ma. Chapter "Lower Paleozoic 'El Nihuil Dolerites': Geochemical and Isotopic Constraints of Mafic Magmatism in an Extensional Setting of the San Rafael Block, Mendoza, Argentina", by C. Cingolani, E. J. Llambías, F. Chemale Jr., P. Abre and N. Uriz, describe new data from the 'El Nihuil Mafic Unit' that is exposed at the Loma Alta region. The authors present the petrology, geochemistry, isotope data, and determinations of emplacement conditions of the undeformed Lower Paleozoic dolerites (El Nihuil Dolerites) with a tholeiitic ocean floor basalt geochemical signature (Cingolani et al. 2000) that could represent a sliver of Cuyania-Chilenia terranes suture after the passive margin stage (Fig. 5).

Chapter "Magnetic Fabrics and Paleomagnetism of the El Nihuil Mafic Unit, San Rafael Block, Mendoza, Argentina", A. Rapalini, C. Cingolani and A. Walther offer the reconnaissance of magnetic fabric and paleomagnetic studies carried out on the El Nihuil Mafic Unit both in gabbros and dolerites. Most samples showed well-defined but scattered remanence directions which indicate that no significant paleomagnetic data can be obtained from El Nihuil Mafic Unit.

Orogenic sedimentation, Chanic compressional phase during Chilenia terrane accretion: Chapter "Low-Grade Metamorphic Conditions and Isotopic Age Constraints of the La Horqueta pre-Carboniferous Sequence, Argentinian San Rafael Block" by H. Tickyj, C. Cingolani, R. Varela and F. Chemale Jr, is dedicated to the siliciclastic sedimentary La Horqueta Formation of Middle Paleozoic age. It is characterized by asymmetric open to similar folds, with southeast vergence that underwent very low-grade (high anchizonal) to low-grade (epizonal) metamorphic conditions that slightly increase from south to north in an intermediate pressure regime. Whole-rock Rb-Sr isochronic ages were obtained on metapelites from the key outcrops indicating that metamorphism and deformation occurred during the Devonian Chanic Orogenic phase, probably related to Chilenia terrane collision (see Fig. 5). U-Pb (LA-MC-ICP-MS) detrital zircon age patterns suggest that the La Horqueta Formation received a dominant sedimentary input from Mesoproterozoic sources. In Chapter "La Horqueta Formation: Geochemistry, Isotopic Data and Provenance Analysis", P. Abre, C. Cingolani, F. Chemale Jr., and N. Uriz expose the provenance analyses based on whole-rock geochemistry and isotope data for the La Horqueta Formation. Whole-rock geochemical data point to a derivation from a source slightly less evolved than the average upper continental crust. The εNd values are within the range of variation of data from the Mesoproterozoic Cerro La Ventana Formation, which is part of the basement of the Cuyania terrane outcropping within the San Rafael Block. Despite geochemical similarities this unit display different proportions of detrital zircon ages, when compared to the Río Seco de los Castaños Formation. Chapter "Silurian-Devonian Land-Sea Interaction Within the San Rafael Block, Argentina: Provenance of the Río Seco de los Castaños Formation" by C. Cingolani, N. Uriz, P. Abre, M. Manassero, M. A. S. Basei deals with the Río Seco de los Castaños Formation as one of the most relevant 'pre-Carboniferous units' outcropping within the San Rafael Block assigned to Upper Silurian-Lower Devonian age. The authors review the provenance data obtained by petrography and geochemical-isotope analyses as well as the U-Pb detrital zircon ages. Comparison with the La Horqueta Formation is also discussed. These data suggest an Early Carboniferous (Mississipian) low-metamorphic (anchizone) event for the unit correlated with the Chanic tectonic phase that affected the Cuyania terrane and also linked to the collision of the Chilenia terrane in the western pre-Andean Gondwana margin (Fig. 5). The authors comment that the studied Río Seco de los Castaños samples show dominant source derivation from Famatinian (Late Cambrian-Devonian) and Pampean-Brasiliano (Neoproterozoic-Early Cambrian) cycles. Detritus derived from the Mesoproterozoic basement are scarce. U-Pb data constrain the maximum sedimentation age to the Silurian-Early Devonian. In Chapter "Primitive Vascular Plants and Microfossils from the Río Seco de los Castaños Formation, San Rafael Block, Mendoza Province, Argentina", by E. Morel, C. Cingolani, D. Ganuza, N. Uriz and J. Bodnar, the authors describe fossil plant remains that comprise non-forked and forked axes without or with delicate lateral expansions. They refer them to primitive land plants and discuss about their systematic affiliation. Diverse acritarch assemblages are present in the same unit. On the basis of the taxonomical information and stratigraphic correlation, they could infer that the Río Seco de los Castaños Formation has an Early Devonian age. The taphonomical conditions of this fossil association would indicate that the plants were transported some distance from their presumed coastal and riverbank habitats in a warm to cool temperate paleoclimatic conditions. Chapter "The Rodeo de la Bordalesa Tonalite Dykes as a Lower Devonian Magmatic Event: Geochemical and Isotopic Age Constraints" by C. Cingolani, E. Llambías, M. A. S. Basei, N. Uriz, F. Chemale Jr. and P. Abre is dedicated to describe small intrusive bodies in the Río Seco de los Castaños unit composed of tonalite, lamprophyre ('spessartite-kersantite') and aplite dykes. Geochemical and isotopic data from the grey tonalitic rocks were characterized by high to medium potassium concentration, with metaluminous composition and I-type calc-alkaline signature. The 401 ± 4 Ma U-Pb zircon age corresponds to the emplacement time. The Rb-Sr whole-rocks and biotite isochronic age of 374 ± 4 Ma could be related to deformation during the Chanic tectonic phase. The crystallization age corresponds to a Lower Devonian time and suggests that part of the Late Famatinian magmatic event is present in the San Rafael Block. The geochemical and geochronological data allowed us to differentiate the Rodeo de la Bordalesa tonalite from the mafic rocks exposed at the El Nihuil area.

Tectonic evolution synthesis: In Chapter "Pre-Carboniferous Tectonic Evolution of the San Rafel Block, Mendoza Province" by C. Cingolani and V.

A. Ramos, the pre-Carboniferous evolution of the San Rafael Block is described in different stages (Fig. 5). The first one is referred to the Mesoproterozoic basement. The signature of this basement indicates a common origin with the present eastern part of Laurentia. The carbonate platform of Cuyania terrane has been drifted away during Early Cambrian to Early Ordovician times. The Ordovician silico-carbonate sequences of the San Rafael Block are unconformably deposited over the basement near the present eastern slope of the Cuyania terrane. The El Nihuil dolerites with a tholeiitic ocean floor signature considered the southern end of the Famatinian ophiolites were interpreted as a Late Ordovician–Early Silurian extensional event. The collision of Cuyania produced a new west polarity subduction and a magmatic arc, represented by the Devonian Rodeo de la Bordalesa tonalite and some granitoids of the Agua Escondida Mining District. The Late Silurian–Early Devonian sequences of La Horqueta and Río Seco de los Castaños formations were deformed during the collision and accretion of the Chilenia terrane against the proto-Andean margin, and recorded an east vergent cleavage developed on the previous deformed rocks (Fig. 5). This collision produced the strong angular unconformity between the La Horqueta/Río Seco de los Castaños Formations and the El Imperial Formation (Upper Paleozoic). The new subduction with east polarity characterized the beginning of the Gondwanian cycle. The new magmatic arc was interrupted by the intense Lower Permian deformation of the San Rafael tectonic phase (Fig. 5).

Compilation of the San Rafael geological map: to conclude the book, a compilation of the San Rafael Geological Map updated with the new information is offered. It is subdivided into three parts: the northern sector corresponds to Sierra de las Peñas, the central to Sierra Pintada-Cerro Nevado and the southern to La Escondida mining district (34° 14′S to 36° 10′S—68° 06′W to 69° 06′W). Also as supplementary material, we present a complete geological map of the San Rafael Block that will allow readers to obtain more details of the studied pre-Carboniferous units. These compilations were done with the collaboration of Mario Campaña and Norberto Uriz from the Department of Geology of the La Plata Museum.

Acknowledgements I would like to express many thanks to Jorge Rabassa, Víctor Ramos and Paulina Abre for comments and suggestions.

References

Abre P (2007) Provenance of Ordovician to Silurian clastic rocks of the Argentinean Precordillera and its geotectonic implications. Ph.D. Thesis University of Johannesburg, South Africa. UJ web free access

Abre P, Cingolani CA, Zimmermann U, Cairncross B (2009) Detrital chromian spinels from Upper Ordovician deposits in the Precordillera terrane, Argentina: a mafic crust input. J S Am Earth Sci 28:407–418

Abre P, Cingolani CA, Zimmermann U, Cairncross B, Chemale Jr F (2011) Provenance of Ordovician clastic sequences of the San Rafael Block (Central Argentina), with emphasis on the Ponón Trehué Formation. Gondwana Res 19(1):275–290

Abre P, Cingolani C, Cairncross B, Chemale Jr F (2012) Siliciclastic Ordovician to Silurian units of the Argentine Precordillera: constraints on Provenance and tectonic setting in the Proto-Andean margin of Gondwana. J S Am Earth Sci 40:1–22

Aceñolaza FG, Miller H, Toselli AJ (2002) Proterozoic-Early Paleozoic evolution in western South America - a discussion. Tectonophysics 354:121–137

Bodenbender G (1891) Apuntes sobre rocas eruptivas de la pendiente oriental de los Andes entre el río Diamante y el río Negro. Revista Argent de Hist Nat I (3):177–191, Buenos Aires

Burkhardt C, Wehrli H (1900) Profils géologiques transversaux de la Cordillere Argentino-Chilienne. Anales Museo de La Plata, Secc Geología y Minas II:1–136, La Plata

Cingolani CA, Varela R (1999) The San Rafael Block, Mendoza (Argentina): Rb-Sr isotopic age of basement rocks. II South Am Symp Isotope Geol, Actas An SEGEMAR, 23–26, Córdoba

Cingolani CA, Llambías EJ, Ortiz LR (2000) Magmatismo básico pre-Carbónico del Nihuil, Bloque de San Rafael, Provincia de Mendoza, Argentina. 9° Congreso Geológico Chileno, 2:717–721, Puerto Varas

Cingolani CA, Manassero M, Abre P (2003) Composition, provenance and tectonic setting of Ordovician siliciclastic rocks in the San Rafael Block: Southern extension of the Precordillera crustal fragment, Argentina. J S Am Earth Sci 16:91–106

Costa C, Cisneros H, Salvarredi J, Gallucci A (2006) La neotectónica del margen oriental del bloque de San Rafael: Nuevas consideraciones. Asociación Geol Argent Serie D: Publi Espec 6:33–40

Criado Roqué P (1969a) El Bloque de San Rafael. In: Leanza AF (ed) Geología regional Argentina. Academia Nacional de Ciencias, Córdoba, pp 237–287

Criado Roqué P (1969b) Cinturón Móvil Mendocino Pampeano. In: Leanza AF (ed) Geología Regional Argentina. Academia Nacional de Ciencias, Córdoba, pp 297–303

Criado Roqué P (1972) Bloque de San Rafael. In: Leanza AF (ed) Geología Regional Argentina. Academia Nacional de Ciencias, Córdoba, pp 283–295

Criado Roqué P, Ibáñez G (1979) Provincia geológica Sanrafaelino-Pampeana. Geol Reg Argent, Acad Nac de Cienc de Córdoba 1:837–869

Cuerda AJ, Cingolani CA (1998) El Ordovícico de la región del Cerro Bola en el Bloque de San Rafael, Mendoza: sus faunas graptolíticas. Ameghiniana 35(4):427–448

Dalla Salda L, Cingolani C, Varela R (1992) Early Paleozoic orogenic belt of the Andes in southwestern South America: result of Laurentia-Gondwana collision? Geology 20:617–620

Dessanti RN (1945) Sobre el hallazgo del Carbónico Marino en el Arroyo El Imperial de la Sierra Pintada. Notas del Museo de La Plata Geol X 42:205–220

Dessanti RN (1954) La estructura geológica de la Sierra Pintada (Departamento de San Rafael, Provincia de Mendoza). Revista de la Asoc Geol Argent 9(4):246–252

Dessanti R (1956) Descripción geológica de la Hoja 27c, Cerro Diamante (Provincia de Mendoza). Dir Nac de Geol y Min, Bol 85, 79p, Buenos Aires

Feruglio E (1946) Los Sistemas Orográficos de la Argentina. In: Geografía de la República Argentina. GAEA. Sociedad Argentina de Estudios Geográficos, Buenos Aires, Vol 4, pp 1–225

Finney S (2007) The parautochthonous Gondwanan origin of the Cuyania (greater Precordillera) terrane of Argentina: A re-evaluation of evidence used to support an allochthonous Laurentian origin. Geol Acta, 5(2):127–158

González Díaz E (1964) Rasgos geológicos y evolución geomorfológica de la Hoja 27d (San Rafael) y zona occidental vecina. Rev de la Asoc Geol Argent, 19(3):151–188, Buenos Aires

González Díaz EF (1972) Descripción geológica de la Hoja 27d San Rafael, Mendoza. Serv Min Geol Bol 132:127, Buenos Aires

González Díaz E (1981) Nuevos argumentos a favor del desdoblamiento de la denominada Serie de La Horqueta del Bloque de San Rafael, Provincia de Mendoza. 8° Congreso Geol Argent 3:241–256

Gregori DA, Kostadinoff J, Strazzere L, Raniolo A (2008) Tectonic significance and consequences of the Gondwanide orogeny in northern Patagonia, Argentina. Gondwana Res 14(3):429–450

Groeber P (1929) Líneas fundamentales de la Geología del Neuquén, sur de Mendoza y regiones adyacentes. Direc Gen de Min Geol e Hidrología, Publ. n. 58

Groeber P (1939) Mapa geológico de Mendoza. Physis, 14, Buenos Aires

Hauthal R, Lange G, Wolff E (1895) Examen topográfico y geológico de los Departamentos de San Carlos, San Rafael y Villa Beltrán, provincia de Mendoza. Rev Museo de La Plata, 7:13–96, La Plata

Holmberg E (1948a) Geología del cerro Bola. Contribución al conocimiento de la tectónica de la Sierra Pintada. Secretaría de Industria y Comercio de la Nación. Direc Gen de Ind y Min Bol 69:313–361, Buenos Aires

Holmberg E (1948b) Geología del Cerro Bola. Rev de la Asoc Geol Argent B Aires 3(4):313–360

Holmberg E (1973) Descripción Geológica de la Hoja 29d, Cerro Nevado. Ser Geol Min Argent (SEGEMAR) Bol 144:71, Buenos Aires

Keidel J (1947) El Precámbrico. El Paleozoico. Geografía de la Repúbl Argent GAEA 1:48–126, Buenos Aires

Lagiglia HA (2002) Arqueología prehistórica del sur mendocino y sus relaciones con el centro-oeste argentino. In: Gil A, Neme G (eds) Entre Montañas y Desiertos: Arqueología del sur de Mendoza, pp 43–64. Sociedad Argentina de Antropología, Buenos Aires

Lange G, Wolff E, Hauthal R (1895) Exámen topográfico y geológico de los Departamentos de San Carlos, San Rafael y Villa Beltrán, provincia de Mendoza. Anales del Museo de La Plata, p 13–45

Manassero M, Cingolani CA, Cuerda AJ, Abre P (1999) Sedimentología, Paleoambiente y Procedencia de la Formación Pavón (Ordovícico) del Bloque de San Rafael, Mendoza. Rev de la Asoc Argent de Sedimentología 6(1–2):75–90

Manassero MJ, Cingolani CA, Abre P (2009) A Silurian-Devonian marine platform-deltaic system in the San Rafael Block, Argentine Precordillera-Cuyania terrane: lithofacies and provenance. In: Königshof P (ed) Devonian change: case studies in palaeogeography and palaeoecology. The Geological Society, London, Special Publications Vol 314, pp 215–240

Núñez E (1976) Descripción geológica de la Hoja Nihuil. Informe Inédito. Serv Geol Nac, p 112, Buenos Aires

Núñez E (1979) Descripción geológica de la Hoja 28d, Estación Soitué, Provincia de Mendoza. Serv Geol Nac Bol 166:1–67, Buenos Aires

Padula E (1949) Descripción geológica de la Hoja 28-C "El Nihuil", Provincia de Mendoza. YPF (unpublished report)

Padula E (1951) Contribución al conocimiento geológico del ambiente de la Cordillera Frontal, Sierra Pintada, San Rafael (Mendoza). Rev de la Asoc Geol Argent 6(1):5–13, Buenos Aires

Pankhurst RJ, Rapela CW, Fanning CM, Márquez M (2006) Gondwanide continental collision and the origin of Patagonia. Earth Sci Rev 76:235–257

Polanski J (1949) El bloque de San Rafael. Dirección de Minas de Mendoza (unpublished report), Mendoza

Polanski J (1954) Rasgos geomorfológicos del territorio de la Provincia de Mendoza. In: Instituto de Investigaciones Económicas y Tecnológicas. Cuaderno de Estudios e Investigaciones 4:4–10. Ministerio de Economía, Gobierno de Mendoza. Argentina

Polanski (1964) Descripción geológica de la Hoja 26c La Tosca (prov. de Mendoza). Direc Nac de Geol Min Bol 101:86, Buenos Aires

Ramos VA (2004) Cuyania, an exotic block to Gondwana: review of a historical success and the present problems. Gondwana Res 7:1009–1026

Ramos V, Jordan TE, Allmendinger RW, Mpodozis C, Kay SM, Cortés JM, Palma MA (1986) Paleozoic terranes of the central Argentine-Chilean Andes. Tectonics 5:855–880

Rapalini AE, Cingolani CA (2004) First Late Ordovician Paleomagnetic pole for the Cuyania (Precordillera) terrane of western Argentina: a microcontinent or a Laurentian plateau? Gondwana Res 7:1089–1104

Rapela CW, Pankhurst RJ, Casquet C, Baldo E, Saavedra J, Galindo C (1998) Early evolution of the Proto-Andean margin of South America. Geology 26(8):707–710

Rolleri EO, Criado Roqué P (1970) Geología de la Provincia de Mendoza. IV Jornadas Geológicas Argentinas (Mendoza 1969), II:1–60. Buenos Aires

Sato AM, Tickyj H, Llambías EJ, Sato K (2000) The Las Matras tonalitic-trondhjemitic pluton, central Argentina: Grenvillian-age constraints, geochemical characteristics, and regional implications. J S Am Earth Sci 13:587–610

Sato AM, Tickyj H, Llambías EJ, Basei MAS, González PD (2004) Las Matras Block, Central Argentina (37° S–67° W): the southernmost Cuyania terrane and its relationship with the Famatinian Orogeny. Gondwana Res 7(4):1077–1087

Sepúlveda E, Carpio F, Regairaz M, Zanettini J, Zárate M (2001) Hoja Geológica 3569-II, San Rafael, Provincia de Mendoza. Serv Geol Min Argent Inst de Geol Recursos Min Bol 321:1–77

Sepúlveda EG, Bermúdez A, Bordonaro O, Delpino D (2007) Hoja Geológica 3569-IV, Embalse El Nihuil, Provincia de Mendoza. Serv Geol Min Argent Inst de Geol Recursos Min Bol 268:1–52

Stappenbeck R (1913) Apuntes hidrogeológicos sobre el sudeste de la Provincia de Mendoza. Bol. 6 (Serie B). Dirección General de Minas, Buenos Aires

Storni CD (1933) Rasgos fisiográficos de la región situada al Norte del curso medio del Río Diamante (Departamento San Carlos, Mendoza). Rev Geog Am 1(3):171–180

Thomas WA, Astini RA (2003) Ordovician accretion of the Argentine Precordillera terrane to Gondwana: a review. J S Am Earth Sci 16:67–79

Uriz NJ (2014) Análisis de la procedencia e historia tectónica del Paleozoico inferior sedimentario del Macizo Nordpatagónico: correlaciones e implicancias paleogeográficas. Tesis Doctoral. Facultad de Ciencias Naturales y Museo, Universidad Nacional de La Plata. La Plata. Argentina, p 379. http://hdl.handle.net/10915/38495

Uriz NJ, Cingolani CA, Chemale Jr F, Macambira MJ, Armstrong RA (2011) Isotopic studies on detrital zircons of Silurian-Devonian siliciclastic sequences from Argentinean north Patagonia and Sierra de la Ventana regions: comparative sedimentary provenance. Int J Earth Sci 100:571–589

Varela R, Basei MAS, González PD, Sato AM, Naipauer M, Campos Neto M, Cingolani CA, Meira VT (2011) Accretion of Grenvillian terranes to the south western border of the Río de la Plata craton, western Argentina. Int J Earth Sci 100:243

Wehrli L, Burkhardt C (1898) Rapport preliminaire sur une expedition géologique dans la Cordillere Argentino Chilienne entre le 33° et 36° latitude sud. Rev Museo de La Plata, 8:374–388, La Plata

Wichmann R (1928) Reconocimiento geológico de la región de El Nihuil, especialmente relacionado con el proyectado dique de embalse del Río Atuel. Dirección Nacional de Minería (unpublished report), Buenos Aires

The Mesoproterozoic Basement at the San Rafael Block, Mendoza Province (Argentina): Geochemical and Isotopic Age Constraints

Carlos A. Cingolani, Miguel A.S. Basei, Ricardo Varela, Eduardo Jorge Llambías, Farid Chemale Jr., Paulina Abre, Norberto Javier Uriz and Juliana Marques

Abstract This work provides new petro-geochemical and isotopic information to constrain the crustal evolution of the Precambrian Cerro La Ventana Formation. The Rb–Sr, Sm–Nd, Pb–Pb, and U–Pb isotopic data obtained as well as their

---

Electronic supplementary material The online version of this chapter (doi:10.1007/978-3-319-50153-6_2) contains supplementary material, which is available to authorized users.

C.A. Cingolani (✉) · R. Varela · E.J. Llambías
Centro de Investigaciones Geológicas-CONICET, Universidad Nacional de La Plata, Diag. 113 n. 275, 1904 La Plata, Argentina
e-mail: ccingola@cig.museo.unlp.edu.ar; carloscingolani@yahoo.com

R. Varela
e-mail: ricardovarela4747@gmail.com

E.J. Llambías
e-mail: llambias@cig.museo.unlp.edu.ar

M.A.S. Basei
Centro de Pesquisas Geocronológicas (CPGeo), Universidade de São Paulo, Instituto de Geociencias, São Paulo, Brazil
e-mail: baseimas@usp.br

F. Chemale Jr.
Programa de Pós-Graduação em Geologia, Universidade do Vale do Rio dos Sinos (UNISINOS), 93022-000 São Leopoldo, Brazil
e-mail: faridchemale@gmail.com

P. Abre
CURE-UDELAR Ruta 8 Km 282, 33000 Treinta y Tres, Uruguay
e-mail: pabre@cure.edu.uy

C.A. Cingolani · N.J. Uriz
División Geología del Museo de La Plata, Paseo del Bosque S/N, 1900 La Plata, Argentina
e-mail: norjuz@gmail.com

J. Marques
Laboratorio de Geologia Isotópica, Universidade Federal do Río Grande do Sul, Porto

---

© Springer International Publishing AG 2017
C.A. Cingolani (ed.), *Pre-Carboniferous Evolution of the San Rafael Block, Argentina*, Springer Earth System Sciences,
DOI 10.1007/978-3-319-50153-6_2

petrological and geochemical features are reported. These data are useful to discuss relationships with equivalent Mesoproterozoic units located along the Cuyania terrane in the proto-Andean Gondwana margin. The type section of the basement rocks of the Cerro La Ventana Formation is located in the south-eastern part of the San Rafael Block, Mendoza Province known as Leones-Ponón Trehué-La Estrechura region. Equivalent crustal fragments are also included in this basement, such as ductile-deformed rocks of the El Nihuil Mafic Unit that are intruded by Ordovician undeformed dolerites. The basement exposed along the type section corresponds to a metamorphosed volcano-plutonic complex with hardly any sedimentary protolith. Main rocks are tonalites and foliated gabbros and quartz diorites that pass to amphibolites, and minor granodioritic–dioritic orthogneisses, with abundant angular microgranitoid enclaves now deformed and stretched intruded in mafic to felsic metavolcanics with porphyritic relic textures. The studied samples classified as tonalites and some close to the field of granodiorites following a calc-alkaline trend. Gabbroic samples from the El Nihuil mafic unit show a more tholeiitic signature. The bulk of samples from the Cerro La Ventana Formation plot within the field of metaluminous rocks; although a few are in the peraluminous field. Main groups of samples plot as low-Al TTD field; however, some of them show high Sr/Y ratios which are typical of high-Al TTD. The Mg#/K ratio is higher in the Cerro La Ventana Formation compared with Las Matras TTG series suggesting a minor differentiated grade for the first one. The chondrite-normalized REE diagrams for Leones samples have Eu anomalies rather positive and gabbros from El Nihuil region display patterns with positive Eu anomalies typical of plagioclase-rich igneous rocks. The Rb–Sr data defined an isochron with 1148 ± 83 Ma, initial $^{87}Sr/^{86}Sr = 0.70292 \pm 0.00018$. The low initial ratio is indicative of a slight evolved Mesoproterozoic source. An acceptable isochron was obtained using Sm–Nd methodology indicating an age of 1228 ± 63 Ma. The model ages (T_{DM}) are in the range 1.23–1.64 Ga with $\varepsilon Nd_{(1200)}$ in between -0.94 and $+4.7$ recording a 'depleted' source, less evolved than CHUR for the time of crystallization. In a $^{207}Pb/^{204}Pb$ diagram the samples plot similarly to rocks from the basement of Cuyania Terrane (Pie de Palo Range and crustal xenoliths) showing a distinctive non-radiogenic signature. The tonalitic lithofacies located at the Leones River type section was chosen for zircon U–Pb TIMS dating and the obtained crystallization age was 1214.7 ± 6.5 Ma. The in situ U–Pb (LA-ICP-MS) zircon data done in two different laboratories on samples from El Nihuil mafic unit (tonalitic orthogneisses) plotted in a Concordia diagram, record an intercept at 1256 ± 10 Ma and in Tera-Wasserburg diagram an age of 1222 ± 6.9 Ma. With these isotopic data we confirm the Mesoproterozoic age for the basement of the San Rafael Block. The obtained ca. 1.2 Ga is quite similar to those belonging to the basement of other regions from the Cuyania allochthonous terrane.

Alegre, Brazil
e-mail: juliana.marques@ufrgs.br

Keywords Cuyania terrane · San Rafael Block · Cerro La Ventana Formation · Mesoproterozoic basement · Geochemistry · Geochronology

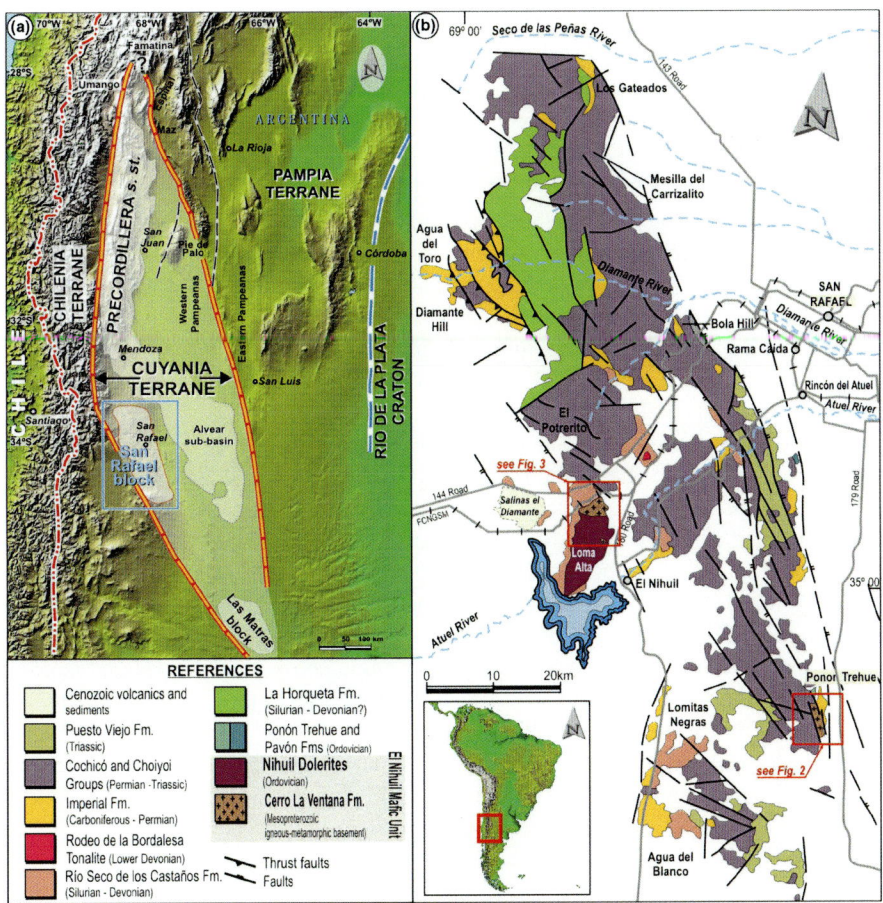

Fig. 1 **a** Location of the San Rafael Block within the Cuyania terrane along the proto-Andean margin of Gondwana; **b** geological sketch map of the San Rafael Block and the position of the studied regions

1 Introduction

The southern South America Paleozoic Gondwana margin (Fig. 1) is characterized by the presence of orogenic belts oriented approximately north-south (Ramos et al. 1986; Dalla Salda et al. 1992). They have been accreted to the cratonic areas during the Cambrian (Pampean), Mid-Ordovician (Famatinian), and Upper Devonian (Gondwanian) tectonic cycles. The Argentine Precordillera or Cuyania composite terrane in the sense of Ramos (2004 and references) is linked to the Famatinian cycle and lies eastward of the present-day Andes. This terrane had been considered from the standpoint of stratigraphy and faunal unique to South America mainly for the Lower Paleozoic carbonate and siliciclastic deposits overlying an igneous–metamorphic crust of Grenville age (Ramos et al. 1998; Sato et al. 2004; Varela et al. 2011). The composite terrane comprises four sectors: The Precordillera (28°–33°S) mainly thin-skinned fold and thrust belt generated by shallow east-dipping flat-slab subduction of the Nazca plate, the Pie de Palo area, and the San Rafael and the Las Matras blocks (Fig. 1).

The San Rafael Block (SRB) lies in west-central Mendoza Province, Argentine (35°S–68° 30′W), and has SSE-NNW structural Cenozoic trend in the pre-Andean region. To the North and South the Cuyo and Neuquén sedimentary basins bound it, respectively. Toward East the outcrops of the SRB vanish under the modern basaltic back arc volcanism and sedimentary cover; the Western boundary is defined by the Andean foothill (Fig. 1). The SRB is interpreted as an extension of the Precordillera region on the light of paleontological and geological evidence (Ramos 1995; Ramos et al. 1999; Keller 1999). Diverse igneous-metamorphic and sedimentary units of Precambrian to Middle Paleozoic age, known as 'pre-Carboniferous units' due to their clear differentiation below a Carboniferous regional unconformity (Dessanti 1956), crop out within the SRB. Basement crustal rocks known as Cerro La Ventana Formation (Criado Roqué 1972) are exposed in the eastern and central part of the SRB (type section: 35° 12′S–68° 17′W), partially covered by calcareous–siliciclastic sedimentary rocks bearing Ordovician macro- and microfossils (Nuñez 1962, 1976; Heredia 2006; Abre et al. 2011; Heredia and Mestre, this volume).

The main purpose of this work is to provide new petro-geochemical and isotopic information to constrain the crustal evolution of the Precambrian Cerro La Ventana Formation. The Rb–Sr, Sm–Nd, Pb–Pb, and U–Pb isotopic data obtained as well as their petrological and geochemical features are reported in this paper. These data are useful to discuss relationships with equivalent Mesoproterozoic units located along the Cuyania terrane in the proto-Andean Gondwana margin.

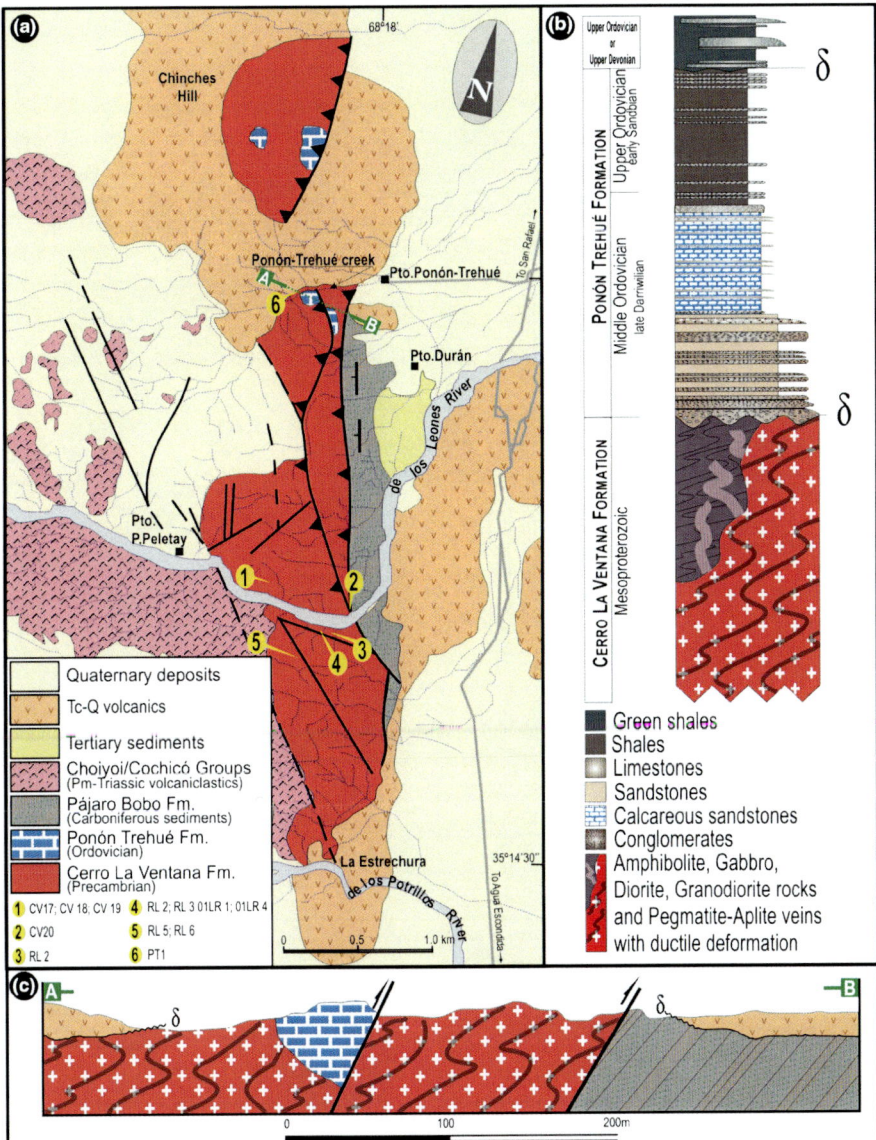

Fig. 2 a Geological sketch map from the Leones-Ponón Trehué creeks toward the south up La Estrechura section (based in Nuñez 1979; Bordonaro et al. 1996; Astini 2002; Cingolani et al. 2005). Locations of the samples are in *yellow*. **b** Stratigraphic column showing the Cerro La Ventana basement and unconformity contact with the Ordovician Ponón Trehué Formation (modified from Heredia 2006). **c** Schematic profile at the Ponón Trehué creek showing the tectonic relationships between basement rocks and other units

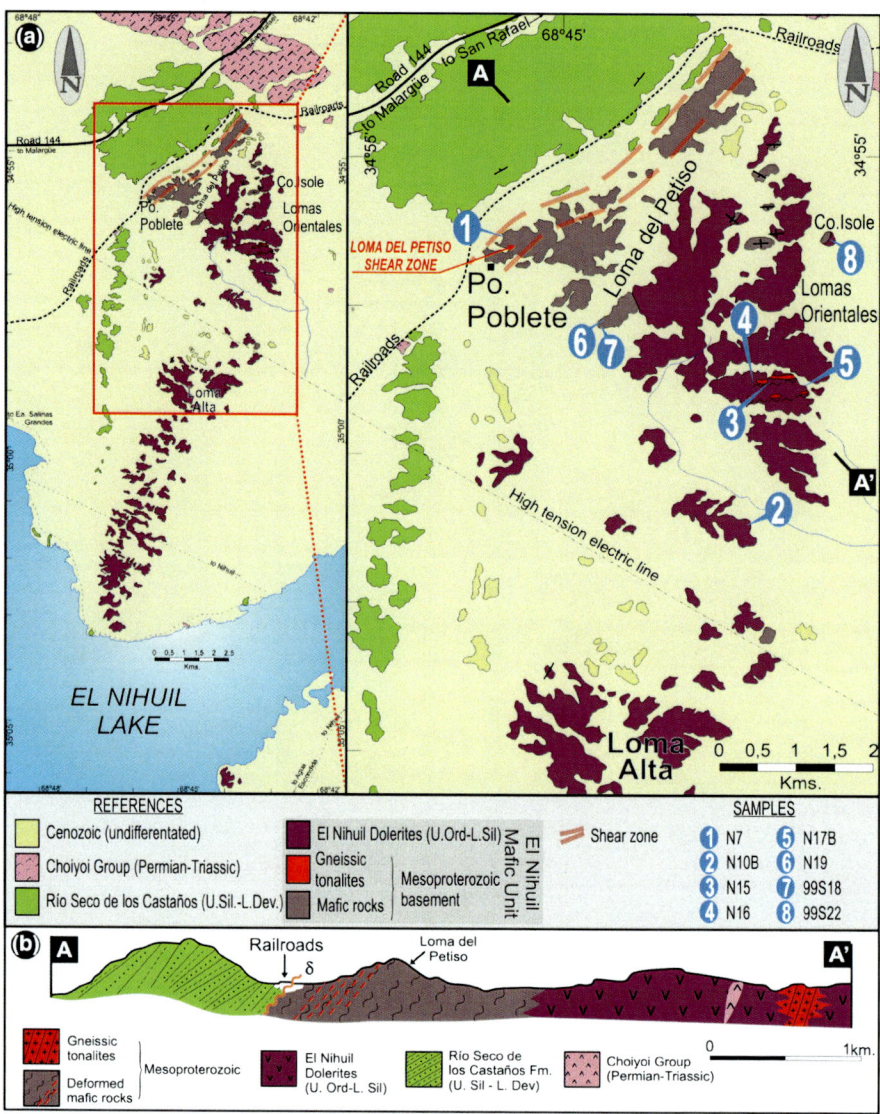

Fig. 3 a Geological sketch map of the El Nihuil mafic unit where the Mesoproterozoic basement is exposed (modified from Cingolani et al. 2000). The locations of the studied samples are in *blue*; b schematic profile showing the relationships between units

2 Geological Setting

The type section of the basement rocks of the Cerro La Ventana Formation (Criado Roqué 1972) or La Ventana Formation (Nuñez 1979) is located in the south-eastern part of the SRB, Mendoza Province (Fig. 2) known as Leones-Ponón Trehué-La Estrechura region (hereafter 'Leones'), in a NNW-SSE trending belt of about 9 km long and 1.5 km wide. It consists of an igneous–metamorphic complex, acidic to intermediate granitoids, as well as pegmatitic and aplitic veins.

Equivalent crustal fragments are also included in this basement, such as ductile-deformed rocks of the El Nihuil mafic unit outcropping at Lomas Orientales and Loma del Petiso (Fig. 3) intruded by undeformed Ordovician dolerites.

Also we mention the 'basement rocks' that were localized in petroleum research wells (Criado Roqué 1979; Criado Roqué and Ibáñez 1979) toward the southeast ('Alvear sub-basin' at Triassic Cuyo Basin). The Cerro Las Pacas Formation (Holmberg 1973) was previously considered as part of the basement; however, new isotopic data done by Cingolani et al. (2014) on mica schists with vertical foliation and intruded by Permian magmatism, record U–Pb detrital zircon ages as young as *ca.* 376 Ma, and therefore it no longer can be considered as part of the Proterozoic basement of the Cuyania terrane. Oil exploration carried out before 1970 in the underground of Triassic Cuyo Basin, 'basement' rocks correlated with the Cerro La Ventana Formation were recognized to the east of the San Rafael Block, at the Alvear sub-basin (Ramos 2004). The borehole IV-D cut garnet–hornblende–biotite schists that recorded a K–Ar age of 605 Ma (Criado Roqué 1979). Close to Corral de Lorca town the exploration well cut metasedimentary rocks (mica schists) from which Cingolani et al. (2012a) obtained U–Pb detrital zircon ages of 395–410 Ma. These rocks are very different to those from the Mesoproterozoic basement and could be correlated to the La Horqueta Formation (Silurian–Devonian age) instead.

It is important to mention that the Leones type section (Ponón Trehué region) constitutes an interesting study area since it records the primary contact between the basement (Cerro La Ventana Formation) and the Precordilleran Lower Paleozoic platform sedimentary rocks (Fig. 2a, b). After Nuñez (1979), Bordonaro et al. (1996), Heredia (1996), and Astini (2002), the two sedimentary members mapped in the area were genetically related to an episode of extension but deposited in separate areas, although both overlie basement rocks. The first one was interpreted as an accumulation of allochthonous (resedimented) carbonate blocks and olistoliths (limestone and igneous blocks) in the proximal reaches of graben. The second unit comprises a lower section composed of sandy carbonates yielding a well-preserved mid-Ordovician conodont and shelly fauna, and resting unconformably on the basement, while the upper section is siliciclastic in composition. Together they provide a record of extensional history (Astini 2002; Abre et al. 2011 and references).

3 History of the Basement Knowledge

Stappenbeck (1913) in his extensive travel from the Sierra Pintada to Cerro Nevado achieved the first reference on basement rocks, and in the La Estrechura (de los Potrillos River) he mentioned the presence of basement gneisses supposedly brought up to the surface by modern basalts. Wichmann (1928) mentioned basement rocks such as gneisses, granites, pegmatites, amphibolites among others, and limestones (outcropping toward west of the Cenozoic volcano Ponón Trehué Hill), which he considered similar to those of the Cerro de la Cal and Salagasta (near the city of Mendoza). Polanski (1949) named the basement as 'Serie de Ponon-Trehue'. Padula (1949, 1951) mapped the region while working for the oil industry and described granitic orthogneisses supporting Ordovician limestones in unconformity and with tectonic contacts with younger Upper Paleozoic sedimentary and volcanic rocks. Criado Roqué (1972) assigned the igneous–metamorphic complex composed mainly of deformed mafic-amphibolite and granitoid rocks, to the Precambrian and named it as Cerro La Ventana Formation. This unit was correlated based on rock similarities to the Pie de Palo outcrops. The basement along the Seco de los Leones River and Ponón Trehué creek was also referred to as La Ventana Formation by Nuñez (1979). He described the petrographical characteristics of the recognized rocks, affected by heterogeneous ductile shear zones, such as amphibolites, mica schists, metaquartzites, gneisses, amphibolitic schists, granites, diorites, and pegmatitic and aplitic veins. Rolleri and Criado Roqué (1970) and Rolleri and Fernández Garrasino (1979) mentioned the presence of high-metamorphic grade 'Pre-Paleozoic Basement' with amphibolites, gabbros, mica schists, amphibolitic schists, and quartzites, intruded by granites and pegmatite–aplite veins. The authors pointed out the presence of intercalated mafic rocks recording pre-orogenic magmatism similar to the Cerros Valdivia and Barbosa (near San Juan city). This complex supports in unconformity an Ordovician sedimentary cover and its exposures are characterized by compressive features such as tectonic contacts acquired during Cenozoic east-vergence thrusting that juxtaposed the basement rocks with both Ordovician and Upper Paleozoic rocks (Fig. 2b, c). Deformation and associated low-grade metamorphism heterogeneously affected this association, originating discrete belts of gneisses. Caminos (1993) correlated the San Rafael Block Precambrian basement to the Western Pampeanas Ranges. Astini et al. (1996) and Thomas et al. (2012) cited an unpublished U–Pb age that yielded a Grenvillian age for the basement rocks. Davicino and Sabalúa (1990) report geochemistry information on mafic rocks and described mylonite and cataclastic rocks along the El Nihuil mafic unit. First Rb–Sr data were obtained by Cingolani and Varela (1999) from seven samples of granodioritic to tonalitic rocks, exposed along the Seco de los Leones and Ponón Trehué creeks; their whole-rock isochron yielded a Mesoproterozoic age of 1063 ± 106 Ma. For more detailed references see also Sato et al. (2000, 2004), Casquet et al. (2006), Vujovich et al. (2004), Ramos (2004), Rapela et al. (2010), and Cingolani et al. (2000, 2012b, 2014). A comprehensive review of the geological and isotopic features of Grenvillian-age

Table 1 GPS coordinates of the studied samples at different outcrop sectors

Sample	GPS	Rock type	Methodology
Leones type section			
CV17	35° 12′ 26″S–68° 18′ 32″W	Tonalite	Pb/Pb; Sm/Nd; geochemistry
CV18	35° 12′ 26″S–68° 18′ 32″W	Gabbro–diorite	Pb/Pb; Sm/Nd; geochemistry
CV19	35° 12′ 26″S–68° 18′ 32″W	Gabbro–diorite (Biot.–Qz)	Pb/Pb; Sm/Nd; geochemistry
CV20	35° 12′ 23″S–68° 17′ 33″W	Deformed tonalite (garnet)	Pb/Pb; Sm/Nd; geochemistry
RL2	35° 12′ 39″S–68° 17′ 53″W	Granitic composition	Rb/Sr
RL3	35° 12′ 40″S–68° 18′ 06″W	Granitic composition	Geochemistry; Rb/Sr
RL4	35° 12′ 40″S–68° 18′ 06″W	Granitic composition	Geochemistry; Rb/Sr
RL5	35° 12′ 53″S–68° 18′ 19″W	Granitic composition	Rb/Sr
RL6	35° 12′ 53″S–68° 18′ 19″W	Tonalite	Geochemistry; Rb/Sr
01RL1	35° 12′ 36″S–68° 17′ 38″W	Gabbro	Geochemistry
01RL2	35° 12′ 36″S–68° 17′ 48″W	Dioritic composition	U–Pb
01RL3	35° 12′ 33″S–68° 18′ 14″W	Granitic composition	U–Pb
01RL4	35° 12′ 33″S–68° 18′ 14″W	Granitic composition	Geochemistry
01RL5	35° 12′ 33″S–68° 18′ 14″W	Dioritic composition	Geochemistry
PT1	35° 10′ 01″S–68° 18′ 53″W	Granitic composition	Rb/Sr
El Nihuil mafic unit			
N7	34° 55′ 35.6″S–68° 45′ 21.6″W	Gabbro	Geochemistry; Rb/Sr
N10B	34° 57′ 37.4″S–68° 43′ 28.1″W	Gabbro	Geochemistry
N19	34° 56′ 07.8″S–68° 44′ 33.7″W	Foliated gabbro	Geochemistry; Rb/Sr
N15	34° 56′ 39.7″S–68° 43′ 18.8″W	Foliated tonalite-granodiorite	Sm/Nd; U–Pb; geochemistry
N16	34° 56′ 36.4″S–68° 43′ 13.7″W	Granitic composition	Geochemistry
N17B	34° 56′ 37.8″S–68° 43′ 04.5″W	Granitic composition	U–Pb; geochemistry
99S22	34° 55′ 40.8″S–68° 42′ 44.0″W	Gabbro	Rb/Sr
99S18	34° 55′ 38.5″S–68° 45′ 25.6″W	Gabbro	Rb/Sr

rocks attached to the southwest of the Río de la Plata craton in Early Paleozoic times was presented by Varela et al. (2011). In this paper, the Mesoproterozoic basement exposures are clearly extended southward of 34°S in the Cuyania terrane, specifically within de San Rafael and Las Matras blocks (Fig. 1). The main difference is the pervasive foliation and subsequent mylonitization that affect the Cerro La Ventana Formation in the San Rafael Block. In this scenario the shallower crustal level is exposed toward the South (Las Matras) as a Grenvillian-age tonalitic-trondhjemitic pluton.

Table 2 Analytical results of major (in w/%) and trace elements (in ppm) in the studied samples (ACTLABS, Canada)

Samples	Leones type section									El Nihuil mafic unit					
	RL3	RL4	RL6	01RL1	01RL2	01RL3	01RL4	01RL5	N7	N10B	N19	N15	N16	N17B
Major elements														
SiO_2	72.65	71.03	68.57	70.92	53.71	55.76	55.37	56.26	48.63	43.24	51.17	65.94	66.04	65.80
Al_2O_3	14.51	14.05	15.12	14.11	16.31	17.96	16.63	15.70	16.01	19.94	23.4	15.06	15.03	15.13
Fe_2O_3	1.62	2.90	3.10	3.64	8.45	7.62	7.91	8.55	9.93	11.74	5.42	4.45	4.14	4.29
MnO	0.02	0.042	0.063	0.07	0.15	0.109	0.125	0.12	0.177	0.127	0.064	0.06	0.07	0.07
MgO	0.50	0.91	1.11	0.79	5.08	3.56	4.11	3.09	6.71	4.34	1.63	2.08	2.11	2.08
CaO	3.25	2.99	3.50	3.65	8.47	7.40	7.08	5.86	10.1	11.3	10.78	4.62	5.22	4.40
Na_2O	4.38	4.18	3.94	3.90	3.61	3.60	3.69	4.02	2.98	2.55	3.76	5.14	3.55	3.68
K_2O	0.90	1.09	1.85	0.81	1.39	1.05	1.37	2.23	0.33	0.56	0.86	0.43	1.23	1.92
TiO_2	0.186	0.341	0.326	0.25	0.52	0.888	0.94	1.29	1.22	2.03	0.71	0.44	0.44	0.45
P_2O_5	0.06	0.12	0.12	0.06	0.11	0.24	0.34	0.47	0.05	0.82	0.12	0.18	0.17	0.18
LOI	1.04	1.35	1.28	1.80	2.20	2.28	2.10	2.40	3.73	2.95	2.54	2.08	2.17	2.14
Total	99.13	99.01	98.98	0.00	0.00	100.46	99.64	99.99	99.86	99.6	100.48	100.49	100.17	100.13
Al_2O_3/SiO_2	0.20	0.20	0.22	0.20	0.30	0.32	0.30	0.28	0.33	0.46	0.46	0.23	0.23	0.23
K_2O/Na_2O	0.21	0.26	0.47	0.21	0.39	0.29	0.37	0.55	0.11	0.22	0.23	0.08	0.35	0.52
Mg#*100	55	55	59	46	70	65	67	59	73	59	54	65	67	66
Trace elements														
Sc	2	3	10			16	18							
V	15	33	43			131	128							
Cr	36	23	34			20	86							
Co	37	33	27			31	34							
Ni	−20	−20	−20			30	75							
Cu	−10	14	15			75	47							

(continued)

Table 2 (continued)

Samples	Leones type section									El Nihuil mafic unit					
	RL3	RL4	RL6	01RL1	01RL2	01RL3	01RL4	01RL5	N7	N10B	N19	N15	N16	N17B
Zn	−30	57	49			79	94							
Ga	14	15	17			18	18		17	19	21	16	16	15
Ge	1.1	0.8	0.9			1.0	1.1							
Rb	13	18	34			17	26		3	13	20	10	23	34
Sr	445	461	407	252.00	370.00	666	482	420.00	588	668	747	378	392	460
Y	2.4	5.1	12.4	12.00	14.00	12.3	18.5	28.00	10.50	14.70	8.90	15.50	15.90	14.20
Zr	75	145	99	75.00	36.00	29	104	193.00	14	23	33	114	114	104
Nb	1.8	2.9	4.6	1.00	0.00	2.4	11.0	9.00	1.00	2.20	2.00	7.3	7.5	6.7
Cs	0.2	0.2	1.0			0.4	0.2		0.4	0.9	0.3	0.70	0.70	1.60
Ba	805	863	918			519	517		398	326	323	397	1160	1400
Hf	1.9	3.9	2.7	3.40	3.60	1.0	2.7	3.20	0.50	0.70	1.00	3.40	3.60	3.20
Ta	2.48	2.09	1.66			0.85	3.51		0.20	0.20	0.70	2.1	3	1.6
Tl	0.08	0.08	0.12			0.11	0.14		0.09	0.11	0.09	0.06	0.13	0.08
Pb	11	8	7			5	5		19	9	9	12	19	6
Th	3.15	6.31	3.25			0.63	1.18		0.12	0.47	0.57	27.08	18.6	15.2
U	0.29	0.60	0.63			0.31	0.71		0.10	0.22	0.26	5.25	4.97	4.49

Table 3 Analytical results obtained by FRX methodology at the University of Río Grande do Sul, Porto Alegre, Brazil

Samples	CV5	CV17	CV18	CV19	CV20
(a)					
SiO_2	56.67	69.86	52.83	50.98	72.11
Al_2O_3	16.27	15.34	17.68	16.72	14.22
Fe_2O_3	6.77	3.28	8.06	7.88	3.93
MnO	0.11	0.05	0.12	0.11	0.07
MgO	3.55	1.13	4.64	6.47	0.66
CaO	6.20	4.19	7.62	6.81	3.77
Na_2O	3.62	3.45	3.09	2.74	3.81
K_2O	2.14	0.88	1.76	0.81	0.14
TiO_2	1.00	0.38	1.04	1.27	0.38
P_2O_5	0.22	0.11	0.22	0.28	0.09
LOI	2.50	1.01	1.98	4.96	0.46
Σ	99.05	99.70	99.04	99.03	99.64
Al_2O_3/SiO_2	0.29	0.22	0.33	0.33	0.20
K_2O/Na_2O	0.59	0.26	0.57	0.30	0.04
Fe_2O_3/MgO	1.91	2.90	1.74	1.22	5.95
(b)					
Cr	25	37	52	195	18
Co	50	bdl	47	46	6
Ni	39	3	28	89	10
Ga	30	15	21	22	16
Rb	96	8	39	bdl	bdl
Sr	963	735	647	656	275
Y	43	4	21	19	20
Zr	139	196	82	121	135
Nb	7	1	4	7	5
Ba	1026	718	794	398	369
Pb	11	9	8	8	6

bdl below detection limit

4 Sampling and Analytical Techniques

The Cerro La Ventana Formation was sampled at the type section (CV17 to CV20, RL2 to RL-6, 01RL1, 01RL4, PT1), along the Ponón Trehué and Leones creeks (simplified as 'Leones'). Another locality was the El Nihuil mafic unit, ductile-deformed rocks within the Loma del Petiso and Lomas Orientales were sampled (N7, N10B, N19, N15, N16, N17 B). Table 1 summarizes the sampling locations and analytical methodologies applied to each sample.

For petrographic studies we use more than 30 samples from different localities covering the whole type basement rocks. Twelve whole-rock samples were analyzed for major, trace, and REE (Table 2) at ACTLABS Canada, by Fusion-ICP and Fusion-ICP MS, except for major and trace elements of four whole-rock samples (CV17 to CV20) that were analyzed by X-ray fluorescence at the Universidade Federal do Rio Grande do Sul, Porto Alegre, Brazil (Table 3). For the **Rb–Sr** systematic, the Rb and Sr concentrations were acquired by XRF and the isotopic composition on natural Sr by mass spectrometry. The sample preparation, chemical attacks, and Sr concentration with cation exchange resin were carried out in the clean laboratory of the Centro de Investigaciones Geológicas (CIG, University of La Plata, Argentina) while FRX and mass spectrometry were done at the Centro de Pesquisas Geocronológicas (CPGeo), São Paulo, Brazil (Cingolani and Varela 1999). Results were plotted on isochron diagram, using the Isoplot model after Ludwig (2001). The technical procedure for **Sm–Nd** methodology performed at the Laboratorio de Geología Isotópica, Universidade Federal do Rio Grande do Sul, Porto Alegre, Brazil started with rock powders spiked with mixed ^{149}Sm–^{150}Nd spike and dissolved using an HF–HNO$_3$ mixture and 6 N HCl in Teflon vials, warmed in hot plate until complete dissolution. The REE were extracted using HDEHP-coated Teflon powder. Isotopic compositions were measured with a VG Sector multicollector mass spectrometer. Sm was loaded on Ta filament and Nd on external Ta triple filament (Ta–Re–Ta) with 0.25 N H$_3$PO$_4$. All analyses are adjusted for variations instrumental bias due to periodic adjustment of collector positions as monitored by measurements of the internal standard; on this basis the analyses of La Jolla Nd average are 0.511859 ± 0.000010. During the course of the analyses Nd and Sm blanks were lesser than 750 and 150 pg, respectively. **Pb–Pb** isotopic analyses were performed at Laboratorio de Geologia Isotópica, Universidade Federal do Río Grande do Sul, Porto Alegre, Brazil; whole-rock samples were completely dissolved in Teflon vial using an HF–HNO$_3$ mixture and 6-N HCl. Ion exchange techniques allowed Pb extraction using AG-1 X 8, 200–400 mesh, anion resin. Each sample was dried to a solid and added a solution of HNO$_3$ with 50 ppb Tl in order to correct the Pb fractionation during the analyses (Tanimizu and Ishikawa 2006). Pb isotopic compositions were measured with a Neptune MC-ICP-MS in static mode, with collecting of 60 ratios of Pb isotopes. The obtained values of NBS 981 common Pb standard during the analyses were in agreement with the NIST values. **U–Pb (ID-TIMS)** procedure of zircon analyses at Centro de Pesquisas Geocronológicas—IGcUSP, Brazil, is as follows: after 10 kg of sample was crushed and reduced to 140–200-mesh grain sizes the portion rich in heavy minerals was treated with bromoform ($d = 2.89$ g/cm^3) and methyl iodide ($d = 3.3$ g/cm^3), and the fraction containing the heavy minerals was processed in the Frantz separator at 1.5 A, and split into several zircon-rich no-magnetic fractions. The final purification of each fraction was achieved by hand picking. Each analyzed fraction represents 30 mg of zircon. Dissolution of zircon crystals was carried out with HF and HNO$_3$ in Teflon micro-bombs in which a mixed ^{205}Pb/^{235}U spike was added. A set of 15 micro-bombs arranged in a metal jacket is left for 3 days in a stove at 200 °C. Then, the HF is evaporated and HCl (6N) added to the micro-bombs, replaced in the stove for 24 h. After the evaporation of HCl 6N, the residue is dissolved in HCl (3N). U and Pb are concentrated and purified by passing the solution in an anionic exchange resin column following the technical procedure after Krogh (1973). The solution enriched in U and Pb is, after addition of phosphoric acid, evaporated until the formation of a micro-drop. The sample is deposited in a rhenium filament and the isotopic composition is determined with multicollector Finnigan MAT 262 solid source mass spectrometer (TIMS). After reduction of the data (PBDAT), the results are plotted in appropriate concordia diagrams using the software ISOPLOT/EX (Ludwig 1999, 2001). During the period of analyses the standard $^{207}Pb/^{206}Pb$ ratios were NBS 983 0.071212 ± 0.00008; NBS 981 0.911479 ± 0.01; and NBS 0.46692 ± 0.01 (Sato and Kawashita 2002). **U–Pb** ages by **LA-ICP-MS** were obtained at Centro de Pesquisas Geocronológicas, São Paulo (N15-N17B), and at Laboratorio de Geología Isotópica, Universidade Federal do Rio Grande do Sul, Porto Alegre (N15), both located in Brazil. Crystal zircons were obtained after crushing and sieving about 3–5 kg of each sample. The fractions retained in less than 140 μm mesh were separated using hydraulic processes to obtain heavy mineral pre-concentrates. These were treated with bromoform to obtain the complete heavy mineral spectra. Methylene iodide was used to achieve a fraction enriched in zircons, followed by an electromagnetic separation with Frantz Isodynamic equipment when necessary. The final selection of crystals, mounted in 2-cm-diameter epoxy resin and polished, was examined under a binocular microscope. The grains were photographed in reflected and transmitted light, and cathodoluminescense (CL) images were produced in order to investigate the internal structures of the zircon crystals as well as to characterize different populations. Choice of spot sites was guided by CL imaging. The U–Pb isotope analyses in both laboratories were performed on zircon grains using a Thermo-Fisher Neptune laser-ablation multicollector inductively coupled plasma mass spectrometer equipped with a 193 Photon laser system. For the operating conditions and instrument settings of the NEPTUNE instrument and laser-ablation system during analytical sessions see Sato et al. (2009) and Bühn et al. (2009), for CPGeo-USP and LGI-UFRGS, respectively. The data are portrayed in Concordia or Tera–Wasserburg diagrams generated with Isoplot/Ex (Ludwig 2001, 2003). U–Pb (LA-ICP-MS) zircon isotopic data are shown in Tables 9 and 10 (in Electronic Supplementary Material).

5 Petrological and Structural Signatures

The studied samples located at the **Leones** type section (Figs. 2 and 4a–g) are classified as basic to mesosilicic gneisses, foliated gabbros, quartz diorites, and tonalites, partially grading to amphibolites and migmatites. This complex is intruded by diorites with structural lineament N10°E/55°SE which yielded migmatite xenoliths.

Fig. 4 Field photographs showing the outcrops of Cerro La Ventana Formation at the Leones type section. **a** Thrust fault with east vergence over the Upper Paleozoic unit; **b–d** quartz diorites and tonalites, partially graded to amphibolites; **e** pegmatite veins; **f, g** angular dioritic enclaves with textural similarity to Las Matras pluton (Varela et al. 2011)

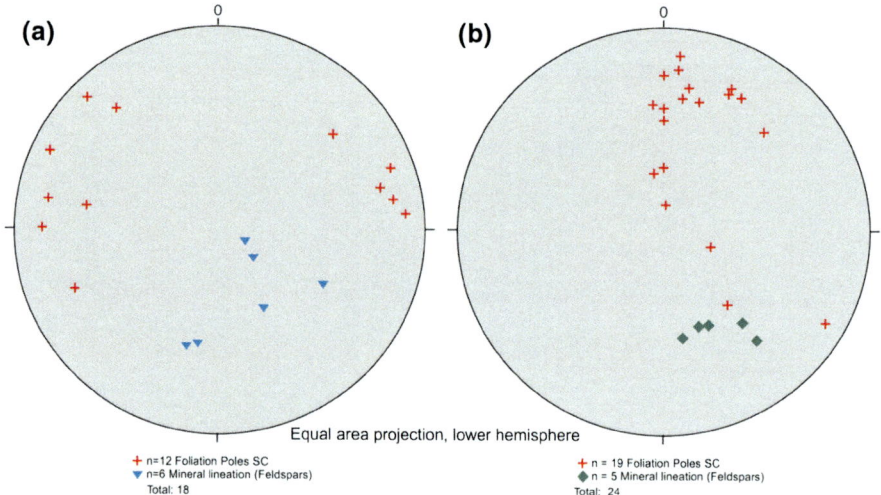

Fig. 5 Measurements of foliation and mineral lineation. **a** Leones type section; **b** El Nihuil mafic unit (tonalitic orthogneisses)

The foliation of some diorites is N20°E/60°NW without any evidence of migmatization. In several places the mentioned rocks are cut by less-deformed pink-granitoids and pegmatites (Fig. 4e). Associated to a vertical faulting with N80°W direction an intense epidotization is present. A ductile deformation is developed with *ca*. N-S shear metamorphic foliation, dipping with high angle toward NW or NE. Linked to the foliation stretching feldspar mineral lineation could be observed, dipping to S or SE. In Fig. 5a, b the structural attitudes of the mylonitic foliation and the mineral lineation are shown, with a main concentration toward N13°W/72°SW. The location of mineral lineation suggests a main transport associated to a shear zone that generated both structures with an important oblique component.

The basement exposed along the type section (Leones River) corresponds to a metamorphosed volcano-plutonic complex with hardly any sedimentary protolith. Main rocks are tonalites and foliated quartz diorites and gabbros that pass to amphibolites, and minor granodioritic–dioritic orthogneisses, with abundant angular microgranitoid enclaves now deformed and stretched (Fig. 4f, g.) intruded in mafic to felsic metavolcanics with porphyritic relic textures. Tabular mafic to felsic dykes cut the country rocks and intrusives and are metamorphosed along with them. In some places from Leones River section the tonalites are rich in garnets (i.e., sample CV20) suggesting that the volcano-plutonic complex could reach the high greenschist–amphibolite facies. Deformation coeval with the metamorphism record the folding of the complex and generated a penetrative foliation S_{x-1} NW-SE dipping 50° to NE and SW. Fold axes have N-S orientation dipping *ca*. 60°S. The foliated quartz-dioritic rocks are composed of less-zoned plagioclase (An_{42}), amphibole, and scarce quartz. Apatite and sphene are most common accessory

Fig. 6 a Lomas Orientales basement outcrops along the El Nihuil mafic unit; **b** studied samples were taken on tonalitic orthogneisses with mafic enclaves

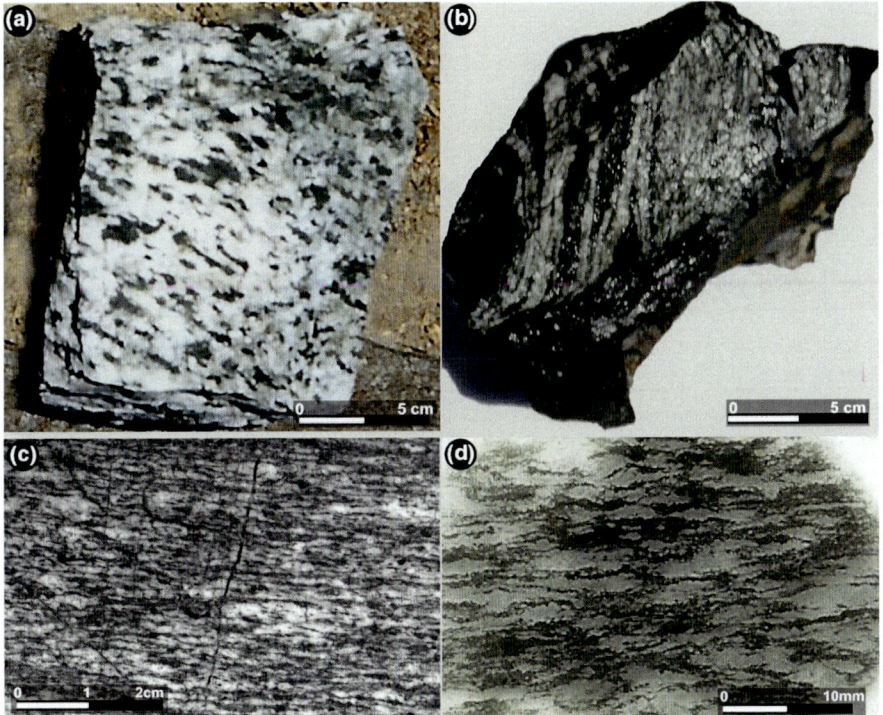

Fig. 7 a, b Hand samples of ductile-deformed gabbros; **c, d** detail of petrographic sections showing mylonitic texture in gabbros, from El Nihuil mafic unit

minerals. The deformed samples show recrystallized amphibole parallel to the foliation. The plagioclase is strongly altered to sericite and epidote. The tonalites are composed of zoned plagioclase (An_{38}), biotite, and quartz. The biotite is mainly

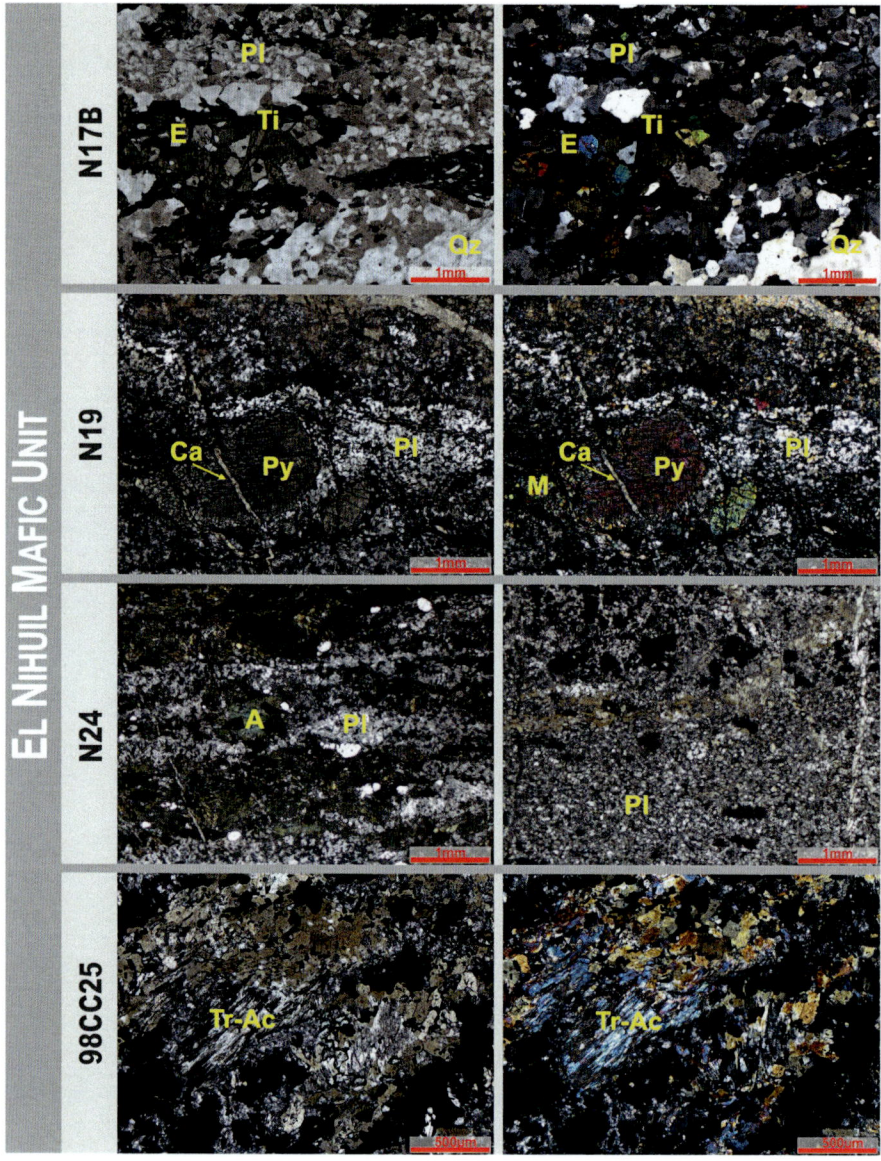

◀**Fig. 8** Photomicrographs (Leica DM 2500P microscope) of samples from the El Nihuil mafic unit ('Loma del Petiso shear zone' and Lomas Orientales). **N17B**: This sample shows well-oriented metamorphic fabric with 8-mm-long hornblende (*A*), with mylonite texture; quartz (*Qz*) and plagioclase (*Pl*) are also present with scarce K-feldspar, showing polygonal mosaic and lensoid texture. The porphyroclasts/porphyroblasts with quartz inclusions lensoidal granoblastic texture within a matrix of plagioclase and quartz. Accessory minerals are sphene (*Ti*) and epidote (*E*). Rock derived from mafic protolith like tonalite, quartz diorite, or quartz amphibolite. (2.5X). **N19**: Sample with pyroxene (*Py*) and plagioclase (*Pl*) as porphyroclasts, partially brecciated, and showing veins of secondary calcite (*Ca*) and with recrystallization of plagioclase. Silicification is also present. Porphyroclasts edges of pyroxene with high-temperature recrystallization forming 'mantle' (2.5X). **N24**: mafic foliated rock with hornblende (*A*), plagioclase (*Pl*) and tremolite—actinolite showing mylonitization in hydrous conditions (2.5X). **98CC25**: Rock sample with granoblastic texture, altered, and brecciated. The amphibole (*A*) was transformed in tremolite–actinolite (*Tr–Ac*). Secondary minerals are apatite and opaques. Other recognized minerals are calcite, quartz, and epidote. Equigranular texture, mylonitizated. Amphibole in mosaics and mafic phenoclasts are replaced by fibrous minerals (2.5X)

altered to chlorite and medium-grained epidote, probably related to a subsolidus crystallization event. Accessory minerals are apatite, sphene, and allanite.

We also include in this chapter, the strongly ductile-deformed gabbros and foliated granodiorite–tonalite orthogneissic samples (Fig. 6a, b) from outcrops preserved at "Loma del Petiso shear zone" and Lomas Orientales within the **El Nihuil Mafic Unit** *ca.* 45–50 km to the NW of the Ponón Trehué region (Cingolani et al. 2000). The petrological characteristics of gneissic rocks are comparable to the tonalites described at Leones River. They have a metamorphic foliation N83°W/54° SW quite similar in all gneissic outcrops suggesting that they correspond to Cerro La Ventana Formation basement relicts (preserved like enclaves or roof pendants) here intruded by Lower Paleozoic dolerites within the El Nihuil mafic unit (Fig. 3; Cingolani et al., this volume).

The basement relicts that crops out at the El Nihuil mafic unit mentioned before are composed of gabbros and tonalites–granodiorites showing mylonitic and cataclastic textures (Figs. 7a–c, 8 and 9).

The relationships with the Lower Paleozoic Ponón Trehué Fm (Fig. 2) and El Nihuil Dolerites (Fig. 3) suggest that the deformation and metamorphism affecting the Cerro La Ventana Formation should have occurred long before its exhumation during Early Ordovician and probably during the late stages of the Mesoproterozoic Grenville orogeny. During the Cenozoic, the final compressive tectonism juxtaposed through east-vergence thrusting basement rocks with both Ordovician and Upper Paleozoic units, as pointed out by Nuñez (1979; Fig. 4a).

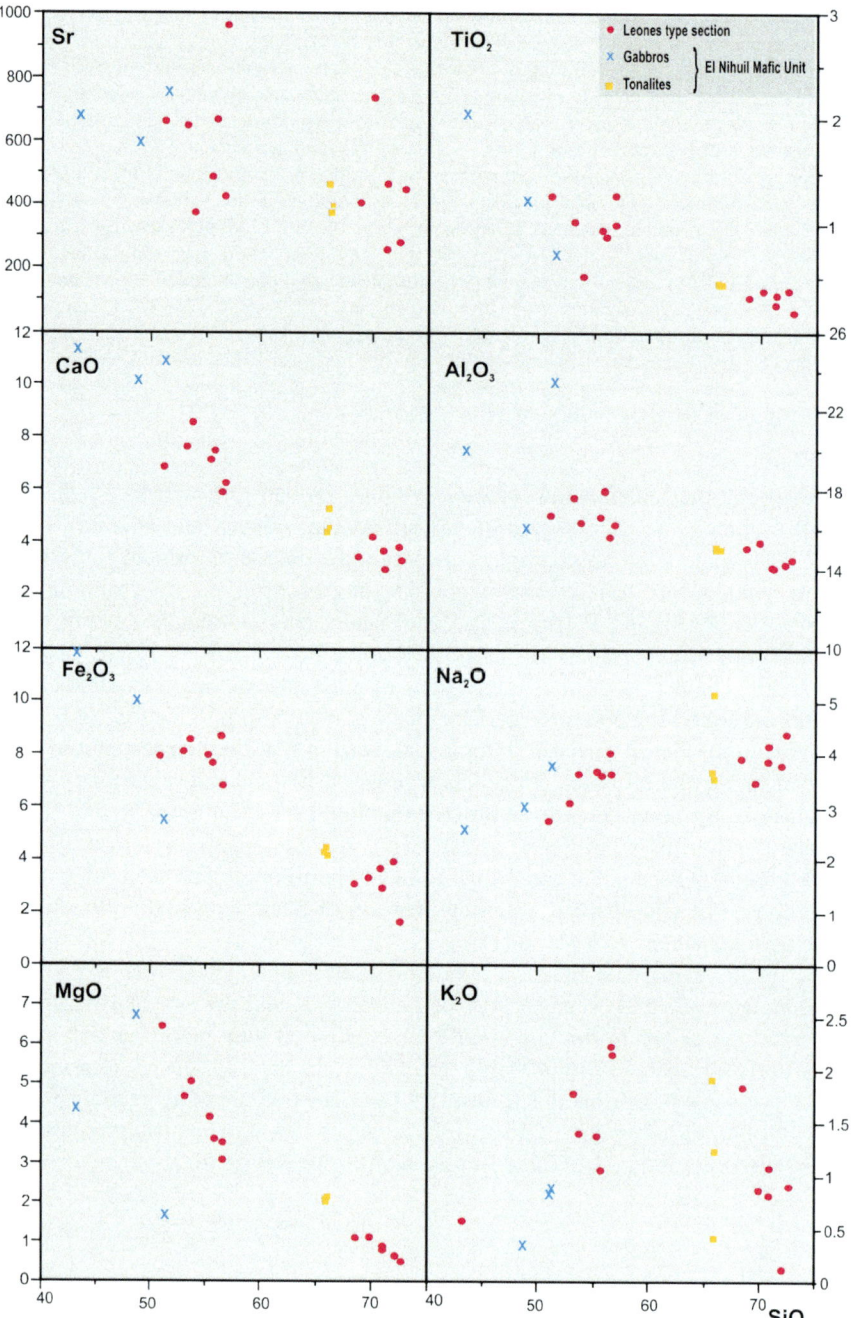

Fig. 9 Harker diagrams for the studied samples for the elements K, Na, Ti, Ca, Fe, Al, Sr, and Mg. A compositional gap can be seen between tonalites and gabbros

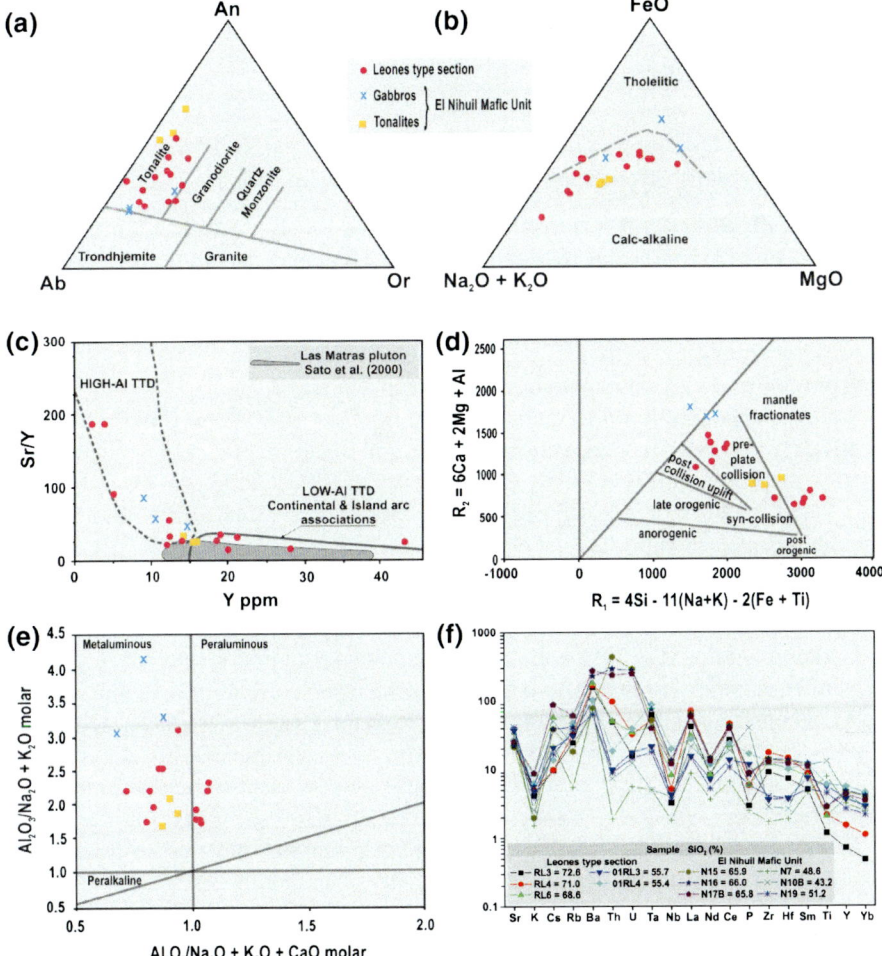

Fig. 10 a Classification of rocks after Barker (1979), based on normative An, Ab, Or; b AFM diagram showing that samples plot mainly in the calc-alkaline field after Batchelor and Bowden (1985); c Sr/Y versus Y diagram after Drummond and Defant (1990) showing that a half of total samples are similar to those from the Las Matras block, interpreted as low-Al TTD (Sato et al. 2000). However, samples from our work show dispersion with high Sr/Y ratios typical of high-Al TTD; d in the R1 versus R2 tectonic discrimination diagram the samples plot mainly in the pre-plate collision to syn-collision fields; e relation of A/CNK (molar) and $Al_2O_3/(Na_2O + K_2O)$ molar, showing the metaluminous/peraluminous signature of studied rocks; f expanded diagram of the trace elements normalized to the primitive mantle (Taylor and McLennan 1985)

6 Lithogeochemistry

6.1 Major Elements

The results of whole-rock samples from the Cerro La Ventana Formation were split into the two groups of outcrops to better describe them (Tables 2 and 3):

1. **El Nihuil Mafic Unit**, (a) tonalites (N15, N16, N17B) with SiO_2 around 66%, Al_2O_3 of 15% and Fe_2O_3 between 4.1 and 4.4%; the concentrations of all major elements are quite similar for the three samples, but a wider spread is found for K_2O and Na_2O; (b) gabbros (N7, N10B, N19) have a SiO_2 concentration varying from 43.2 to 51.1%, Al_2O_3 is in between 16 and 23.4%, Fe_2O_3 ranges from 5.4 to 11.7%, and MgO is in between 1.6 and 6.7%.
2. **Leones type section**, (a) gabbros and tonalites (CV17 to CV20, 01 RL2, 01RL4) with SiO_2 concentrations between 52.8 and 56.6%, Al_2O_3 ranges from 15.7 to 17.9%, Fe_2O_3 contents vary from 6.7 to 8.5%, while MgO is in between 3.1 and 5.1%; (b) the last group here described corresponds to the more acidic component (granodiorites) of the basement and includes samples (RL3, RL4, RL6, 01RL1) showing SiO_2 concentrations above 68.5 and up to 72.6%, Al_2O_3 is in between 14 and 15.3% and Fe_2O_3 contents vary from 1.6 to 3.9%.

Harker diagrams (Fig. 9) clearly show the behavior of major elements and Sr for all studied samples from the two groups: a positive correlation can be seen for Na_2O, whereas a negative correlation is observed for TiO_2, CaO, Fe_2O_3, Al_2O_3, and MgO. Correlations between Sr and K_2O with SiO_2 can be roughly described as negative, despite the dispersion shown. These characteristics could be indicating cogenetic rocks.

In the An-Ab-Or diagram (Fig. 10a) the rock samples plot in the field of the tonalites, although a few are close to the field of granodiorites and two tonalite enclaves from El Nihuil mafic unit plot in the field of trondhjemites close to the tonalite field; in the AFM diagram samples follow a calc-alkaline trend (Fig. 10b). Differences between samples from the type section and gabbroic samples from the El Nihuil mafic unit are evident, since the latter follow a more tholeiitic signature. The bulk of samples from the Cerro La Ventana Formation plot within the field of metaluminous rocks; although a few are in the peraluminous field, their A/CNK ratios are of 1.1 maximum (Fig. 10e). In the R1 versus R2 tectonic discrimination diagram the samples plot in the field of pre-plate collision to syn-collision (Fig. 10d). In the expanded multielement diagram normalized to primitive mantle, it is shown a coherent evolution of the LIL and HFS elements for the studied samples suggesting cogenetic events. Only the presence of the Th and U positive anomaly in the El Nihuil mafic samples show a difference (Fig. 10f).

Table 4 Analytical results of rare earth elements (ppm)

Rare earth elements

Samples	Leones type section					El Nihuil mafic unit					
	RL-3	RL-4	RL-6	01-RL-4	01-RL1	N-7	N10-B	N-19	N15	N16	N17B
La	23.00	39.30	16.70	15.30	61.5	4.76	11.70	8.00	35.3	34.30	32.1
Ce	38.3	63.4	30.9	33.7	6.23	9.3	26.5	15.2	61.5	60.70	56.4
Pr	3.83	6.74	3.47	4.29	23	1.18	3.67	1.84	6.23	6.13	5.79
Nd	13.2	22.5	13.2	18.8	3.99	5.82	18.1	8.41	23	22.90	21
Sm	1.73	2.98	2.66	4.03	0.959	1.56	3.86	1.74	3.99	4.13	3.77
Eu	0.771	0.889	0.751	1.430	3.57	1.44	2.09	1.13	0.959	1.01	0.914
Gd	1.12	2.02	2.37	3.75	0.49	2.03	3.90	1.75	3.57	3.53	3.2
Tb	0.13	0.23	0.41	0.63	2.75	0.31	0.54	0.27	0.49	0.50	0.43
Dy	0.53	1.03	2.26	3.48	0.52	2.05	2.92	1.51	2.75	2.71	2.45
Ho	0.09	0.18	0.42	0.67	1.47	0.41	0.56	0.30	0.52	0.55	0.48
Er	0.22	0.48	1.20	1.89	0.23	1.05	1.40	0.84	1.47	1.45	1.36
Tm	0.027	0.066	0.180	0.277	1.45	0.142	0.179	0.119	0.23	0.23	0.21
Yb	0.18	0.41	1.07	1.67	0.253	0.88	1.01	0.80	1.45	1.51	1.28
Lu	0.026	0.060	0.140	0.233		0.141	0.153	0.117	0.253	0.24	0.208
Sum	83.15	140.29	75.73	90.15	106.41	31.07	76.58	42.03	141.71	139.88	129.59
Sm_n	7.49	12.90	11.52	17.45		6.75	16.71	7.53	17.27	17.88	16.32
Eu_n	8.86	10.22	8.63	16.44		16.55	24.02	12.99	11.02	11.61	10.51
Gd_n	3.66	6.60	7.75	12.25		6.63	12.75	5.72	11.67	11.54	10.46
$s*g_n$	27.41	85.16	89.19	213.80		44.80	212.97	43.08	201.52	206.25	170.67
Eu/Eu*	1.69	1.11	0.91	1.12		2.47	1.65	1.98	0.78	0.81	0.80

$Eu_N/Eu^* = Eu_N/(0.67Sm_N + 0.33Tb_N)$

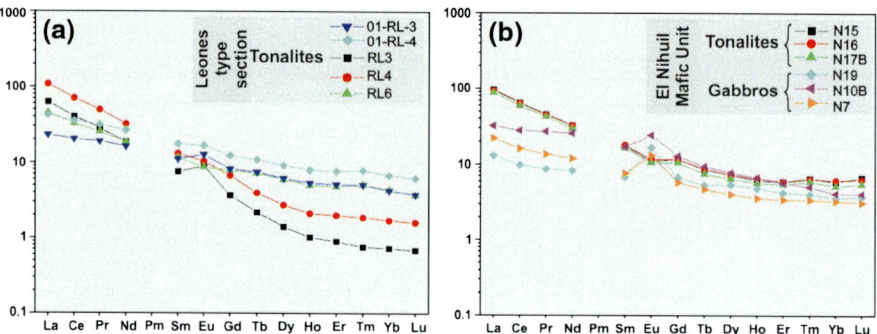

Fig. 11 Chondrite-normalized REE diagrams for, **a** Leones type section; **b** El Nihuil mafic unit. Normalization values from Taylor and McLennan (1985)

6.2 Trace and Rare Earth Elements

Sr/Y versus Y is a useful diagram (Fig. 10c) to discriminate between high and low-Al tonalite–trondhjemite—dacite (TTD) complexes. Main groups of samples plot accordingly to rocks from the Las Matras block (Sato et al. 2000), as low-Al TTD; however, some samples show high Sr/Y ratios which are typical of high-Al TTD in agreement with the diagram of Drummond and Defant (1990).

Samples from the type section of the Cerro La Ventana Formation as well as those from mafic and felsic deformed relicts included in the El Nihuil mafic unit are characterized by high values of Mg# (Mg/Mg + Fe molar ×100; gabbros: 65–67, tonalites 55–70), with respect to rocks with similar silica contents, including those from the Las Matras TTG (Sato et al. 2000, 2004). Furthermore, the Mg# is higher to that produced by slab melts at 1.4 GPa with a residue of gt + cpx ± amph and is superposed with the hybridized slab melts and high magnesia adakites fields, according to Rapp et al. (1999). In order to describe the grade of magnesium enrichment with respect to other differentiated magma, the Mg#/K molar versus silica was plotted. Both the low and high silica rocks studied show similar Mg enrichment respect to K, signature which allows linking both groups of rocks to a same igneous evolution. The Mg#/K ratio is higher for the Cerro La Ventana Formation compared with Las Matras TTG series, suggesting a lower grade of differentiation for the first unit. In the R1 versus R2 diagram the tonalites and diorites from Leones type section plot in two separate groups. The rocks from the crustal enclaves included in El Nihuil mafic unit behave similarly to the tonalites (Fig. 10d). Samples from Leones creek have La/Yb$_N$ ratios in between 5.5 and 86.3 and Eu$_N$/Eu* ranges from 0.9 to 1.7; their chondrite-normalized REE diagram (Table 4; Fig. 11a) shows rather positive Eu anomalies. Samples from the El Nihuil

Table 5 Rb–Sr isotopic data from samples of the Cerro La Ventana type section and El Nihuil mafic unit

	Lab. N°	Field N°	Rb (ppm)	Sr (ppm)	$^{87}Rb/^{86}Sr$	Error	$^{87}Sr/^{86}Sr$	Error
Leones type section	CIG 1003	RL1	16.1	263.7	0.1767	0.0049	0.70598	0.00007
	CIG 1004	RL2	14	288.2	0.1406	0.0039	0.70524	0.00012
	CIG 1005	RL3	13.3	482.6	0.0798	0.0022	0.70409	0.00008
	CIG 1006	RL4	19.3	508.1	0.1099	0.0031	0.70495	0.00009
	CIG 1007	RL5	36.4	449	0.2347	0.0066	0.70667	0.00009
	CIG 1008	RL6	34.3	436.3	0.2276	0.0064	0.70675	0.00009
	CIG 1009	PT1	25.8	265.1	0.2817	0.0079	0.70734	0.00009
	CIG 1015	RL1 D	16.1	263.7	0.1767	0.0049	0.70598	0.00011
	CIG 1016	RL3 D	13.3	482.6	0.0798	0.0022	0.70426	0.00012
El Nihuil Mafic Unit	CIG 1104	N19	23	787.5	0.0845	0.0017	0.704338	0.000028
	CIG 1105	99S22	28.6	492.5	0.1681	0.0034	0.70633	0.000019
	CIG 1197	99S18	7.8	348.6	0.0648	0.0013	0.703779	0.000035
	CIG 1201	N 7	4.3	603.9	0.0206	0.0004	0.703228	0.000063
	CIG 1203	N 19'	21.2	781.1	0.0786	0.0016	0.704282	0.000049
	CIG 1204	99S22'	25.8	480.7	0.1554	0.0031	0.706366	0.000035

mafic unit shows patterns depleted in HREE (Fig. 11b), with La/Yb$_N$ ratios in between 15.3 and 16.9. Therefore, they are dissimilar to diorites from the same unit, which develop REE patterns parallel to N-MORB and have La/Yb$_N$ ratios lower than 1.4. Gabbros display patterns with positive Eu anomalies typical of plagioclase-rich igneous rocks.

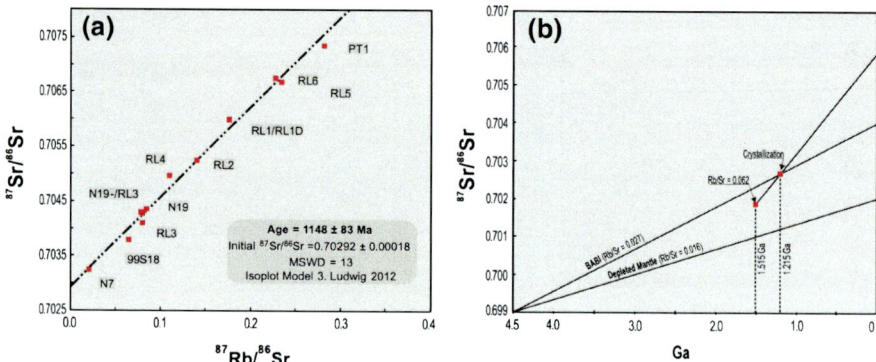

Fig. 12 **a** Rb–Sr isochronic diagram using Isoplot Model 3 by Ludwig (2012) on samples from Leones type section (RL 2, RL 3, RL 5, RL 6, and PT1) and from El Nihuil mafic unit (N7, N19, 99S18). **b** Linear $^{87}Sr/^{86}Sr$ evolution of BABI and of a depleted mantle through time (Faure 1986). The initial $^{87}Sr/^{86}Sr$ found for Cerro La Ventana Fm including samples from El Nihuil mafic unit (0.7029) suggest a derivation from mantle represented by BABI (Rb/Sr = 0.027) considering a 1.2 Ga crystallization age. Another explanation could be refusion of a magma (with Rb/Sr = 0.047) extracted from depleted mantle (Rb/Sr = 0.016) at *ca.* 1.5 Ga

7 Isotopic Data

Rb–Sr: The Rb–Sr systematic was applied by Cingolani and Varela (1999) on 7 whole-rock samples (RL1, RL2, RL3, RL4, RL5, RL6, and PT1), from the granodioritic–tonalitic facies outcropping at the type section of the Cerro La Ventana Formation along Leones and Ponón-Trehué creeks. The Rb concentrations are rather low and vary between 13.3 and 36.4 ppm, while Sr concentrations are high and range from 263.7 to 508.1 ppm; therefore, the $^{87}Rb/^{86}Sr$ ratios are between 0.0798 and 0.2817. According to the model from Williamson (1968), the isochrone age is 1063 ± 106 Ma, and the initial ratio $^{87}Sr/^{86}Sr$ is of 0.7032 ± 0.0003 with MSWD of 1.06. Data recalculated using the Isoplot Model 3 (Ludwig 2012) show an alignment within a low range of $^{87}Rb/^{86}Sr$ (<0.3); the age obtained is 1109 ± 130 Ma (1sigma), initial ratio $^{87}Sr/^{86}Sr$: 0.70304 ± 0.00032, and MSWD of 7.4 (Cingolani and Varela 1999).

Four samples from the El Nihuil mafic unit (N19, 99S22, 99S18 and N7), considered as part of the basement, were recently added (Table 5; Fig. 12a). They show Rb concentrations in between 4.3 and 28.6 ppm, while the Sr concentrations range from 348.6 to 787.5 ppm; the $^{87}Rb/^{86}Sr$ ratios are between 0.0206 and 0.1681. Comparing to data from the type section of the Cerro La Ventana Formation, the El Nihuil mafic unit has lower Rb and higher Sr contents. Using all samples from both regions altogether (excluding 99S22 since it is out of the linear isochrone) following Isoplot Model 3 (Ludwig 2012), an isochron with 1148 ± 83 Ma (1sigma), initial $^{87}Sr/^{86}Sr$ ratio of 0.70292 ± 0.00018, and MSWD = 13 is recorded (Fig. 11a). The low initial $^{87}Sr/^{86}Sr$ ratio is indicative of a

Table 6 Sm-Nd isotopic data for samples taken at Leones type section and El Nihuil Mafic Unit (N15)

Sample	Sm (ppm)	Nd (ppm)	$^{147}Sm/^{144}Nd$	Error	$^{143}Nd/^{144}Nd$	Error (2σ)	$\varepsilon_{Nd(0)}$	$f_{Sm/Nd}$	ε_{Nd} ($t = 1.2$ Ga)	T^a_{DM} (Ma)	ε_{NdTDM}
CV17	2.092	14.104	0.0897	0.0005	0.511996	0.000010	−12.52	−0.54	3.86	1286.8	5.05
CV18	3.146	14.452	0.1316	0.0008	0.512338	0.000012	−5.86	−0.33	4.10	1308.4	5.00
CV19	3.197	14.240	0.1358	0.0008	0.512381	0.000010	−5.00	−0.31	4.32	1291.9	5.04
CV20	3.497	14.221	0.1487	0.0009	0.512503	0.000006	−2.64	−0.24	4.71	1264.1	5.10
N15	3.29	18.88	0.1052	nd	0.511872	nd	−14.94	−0.46	−0.94	1640.0	4.2
RL2	1.46	8.91	0.099344	nd	0.512031	0.000019	−11.84	−0.50	3.1	1345.3	4.91
RL3	1.63	12.75	0.078509	nd	0.511939	0.000012	−13.63	−0.61	4.64	1234.8	5.17
RL4	3.01	22.85	0.079511	nd	0.511863	0.000016	−15.12	−0.60	2.8	1341.9	4.92
RL5	1.85	8.20	0.136597	nd	0.512370	0.000021	−5.22	−0.31	4.0	1325.1	4.96
PT1	2.01	16.17	0.075173	nd	0.511902	0.000027	−14.36	−0.62	4.25	1256.3	5.12

T^a_{DM}: based on De Paolo (1981) ($^{146}Nd/^{144}Nd = 0.7219$); Epsilon$_{Nd(0)} = \{[(^{143}Nd/^{144}Nd)_{sample}/0.512638] - 1\} \times 10^4$, where $^{143}Nd/^{144}Nd_{CHUR} = 0.512638$. $f_{Sm/Nd} = \{[(^{147}Sm/^{144}Nd)_{sample}/0.1967)] - 1\}$, where $^{147}Sm/^{144}Nd_{CHUR} = 0.1967$ (Hamilton et al. 1983). *nd* no data

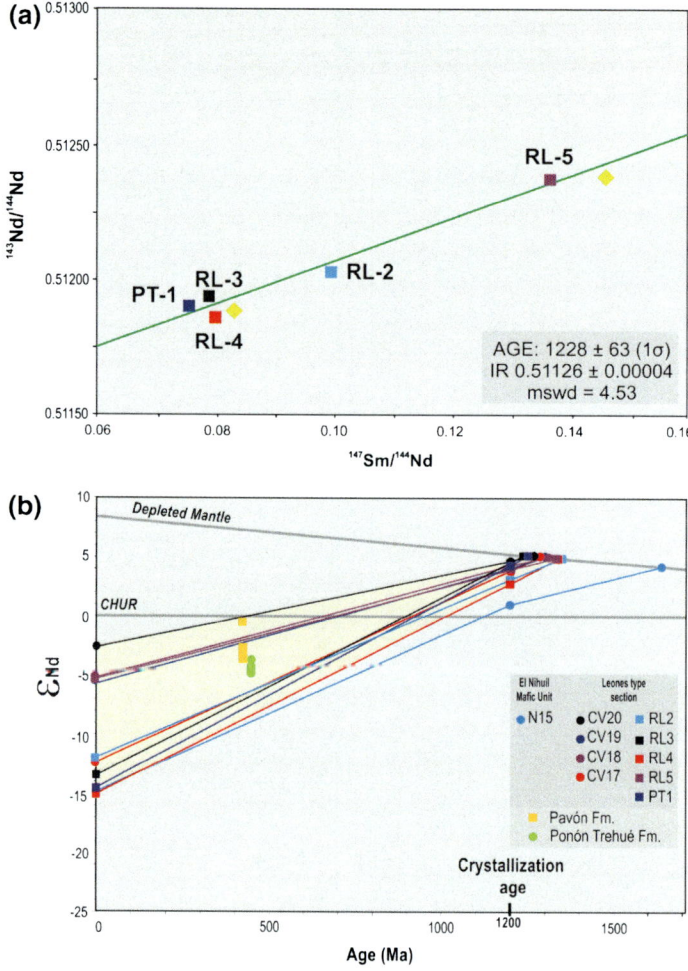

Fig. 13 **a** Sm-Nd isochron diagram from Cerro La Ventana Formation type section samples. **b** ε_{Nd} data for samples from Leones type section and El Nihuil Mafic Unit. Data of Ordovician sequences (Pavón and Ponón Trehué Fms) are also shown (Abre et al. 2011; Abre et al., this volume)

slight evolved Mesoproterozoic source for the tonalitic rocks sampled at Leones and Ponón Trehué creeks as well as for samples from the El Nihuil mafic unit.

If we consider a linear strontium isotopic evolution of BABI through time with $^{87}Sr/^{86}Sr$ from 0.699 to 0.704 (Rb/Sr = 0.027; Faure 1986) during the 4.5 Ga, the $^{87}Sr/^{86}Sr$ value for *ca.*1.2 Ga (crystallization age) is close to the value of 0.70292 ± 0.00018 obtained from the Rb–Sr isochronic diagram (Fig. 12b). This could suggest a direct mantle derivation for the magma at this time. The low $^{87}Sr/^{86}Sr$ ratios as well as the Potassium contents and high Mg# permit to interpret that the studied samples are juvenile rocks (less evolved), similar to the modern

Table 7 Pb–Pb analytical isotopic data of samples from the Cerro La Ventana Fm

Sample	$^{206}Pb/^{204}Pb$	Error[a]	$^{207}Pb/^{204}Pb$	Error[a]	$^{208}Pb/^{204}Pb$	Error[a]
CV17	16.965	0.014	15.398	0.017	37.239	0.020
CV18	17.800	0.013	15.473	0.014	36.685	0.015
CV19	17.580	0.009	15.464	0.011	36.749	0.012
CV20	20.319	0.010	15.684	0.010	38.385	0.010

Isotopic ratios corrected for 0.095%/u.m.a mass fractionation; error[a]: standard relative error (60 measurements in average); total blank during measurements = 119 pg; average isotopic ratios of standard NBS-981 including a deviation of 1sigma (01–08/2010): $^{206}Pb/^{204}Pb$ = 16.898 ± 0.003; $^{207}Pb/^{204}Pb$ = 15.440 ± 0.003; $^{208}Pb/^{204}Pb$ = 36.542 ± 0.010

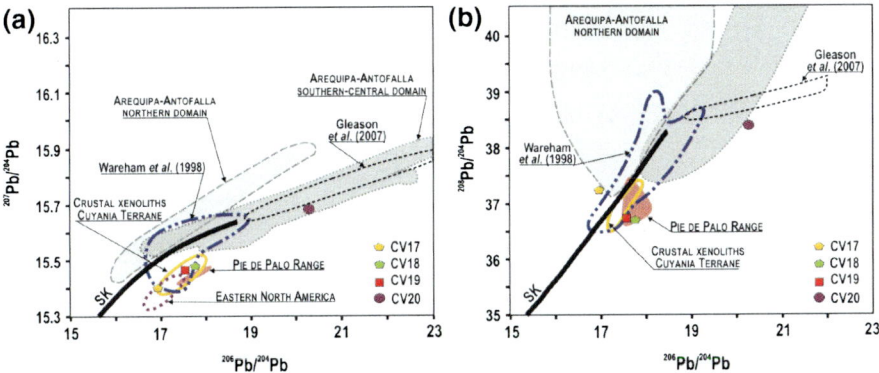

Fig. 14 a, b Pb isotopes of the Cerro La Ventana Fm samples and other basement regions of South America (Ramos 2004). Note that the samples CV17, CV18, and CV19 have distinctive nonradiogenic signature near the field of Pie de Palo complex rocks and crustal xenoliths of Cuyania Terrane; SK = lead isotope curves after Stacey and Kramers (1975)

adakites or Precambrian TTD series. In this sense, are equivalents to the "Grenvillian-age" Las Matras tonalite–trondhjemite pluton (Sato et al. 2000, 2004; Varela et al. 2011).

Sm–Nd: Samples selected for whole-rock Sm–Nd isotopic analyses performed at the Laboratorio de Geología Isotópica, Porto Alegre-Brazil, comprise basement rocks from the Leones (type section of Cerro La Ventana Formation: CV17 to CV20, RL2 to RL5 and PT1) as well as one from the El Nihuil mafic unit (sample N15; Table 6). This isotopic work emphasized the crustal residence (T_{DM}) age determinations and allows us to recognition crustal segments to understand the history of tectonic accretion.

An acceptable isochron was obtained using five Sm–Nd analyses and two duplicates (Fig. 13a), indicating an age of 1228 ± 63 Ma (1 sigma), and MSWD 4.53 and initial $^{143}Nd/^{144}Nd$ 0.51126 ± 0.00004. The model ages (T_{DM}) calculated according to DePaolo (1981) for ten samples are in the range 1.23–1.64 Ga. The $\varepsilon Nd_{(1200)}$ for these samples (Fig. 13b) is in between −0.94 and +4.7 indicating a 'depleted' source, less evolved than CHUR for the time of crystallization, which is

Table 8 U–Pb analytical data and typology of zircons studied by isotope dilution and TIMS methodology at University of São Paulo, Brazil

SPU	Magnetic fraction	Mineral typology	207/235[a]	Error (%)	206/238[a]	Error (%)	Coef.	238/206	Error (%)	207/206	Error (%)	206/204[b]	Pb (ppm)	U (ppm)	Weight (m)	206/238 Age (Ma)	207/235 Age (Ma)	207/206 Age (Ma)
02-RL-103 RIO SECO DE LOS LEONES
2164	D	P(2-4/1), Fr, T, Ambar, cl, Cl, I, F	2.28546	0.49	0.205677	0.484	0.98916	4.86199235	0.484	0.080591	0.0719	5929.2	79.3	385.2	0.037	1206	1208	1212
2165	E	P (2-1.5/1), T, Ot, Pink, Cl, I, F	2.26035	0.481	0.203087	0.477	0.99332	4.92399809	0.477	0.080722	0.0554	4919.5	76.7	348.1	0.028	1192	1200	1215
2161	A	Py(1.5/1), T, Pink, Cl	2.29936	0.487	0.206338	0.487	0.99222	4.84641704	0.484	0.080821	0.0607	3124.1	36.5	176.4	0.037	1209	1212	1217

Typology—Zircon

Shape		Color/Transparency		Internal features	
P(x/y)	Prismatic crystals	T	Transparent crystals	Cl	Crystal with no or few inclusions or fractures
Py	Prismatic crystals with well-developed pyramidal facets	Ot	Opaque or translucent crystal	I	Crystal with frequent inclusions
Fr	Crystals fragments			F	Crystal with frequent fractures

SPU: laboratory number
Magnetic fractions: numbers in parentheses indicated the tilt used on Frantz separator at 1.5 A current
[a] Radiogenic Pb corrected for blank and initial Pb; U corrected for blank;
[b] Not corrected for blank or nonradiogenic Pb
Total U and Pb concentrations corrected for analytical blank
Ages: given in Ma using Ludwig Isoplot/Ex program (1998), decay constants recommended by Steiger and Jäger (1977)

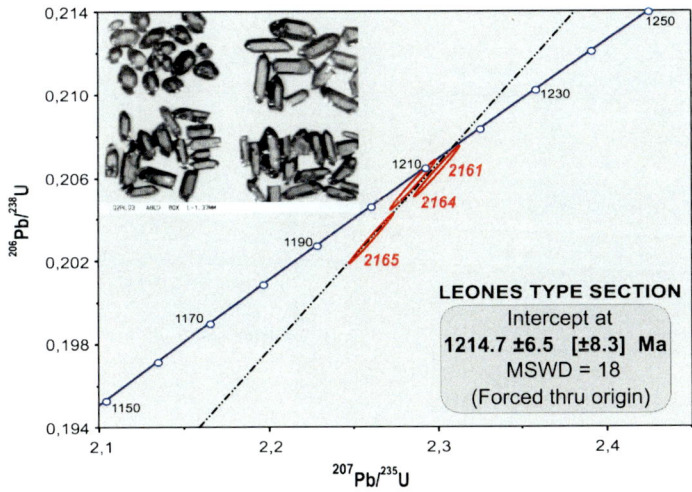

Fig. 15 Concordia plot of zircons from Leones sample analyzed by isotope dilution and TIMS methodology at University of São Paulo, Brazil. In the inset the microscope image of zircon fractions

rather similar to the immature source for the orthogneisses found in Western Pampean Ranges (Sato et al. 2000; Ramos 2004). Dating obtained from both, Sm–Nd and Rb–Sr methods are roughly coincident within errors, indicating that an average age of 1200 Ma can be reasonably proposed as crystallization age for the Cerro La Ventana Formation with contribution of juvenile material.

Pb–Pb: Pb analyses were performed on four whole-rock samples from the Cerro La Ventana Formation (CV17 to CV20); $^{206}Pb/^{204}Pb$ range from 16.965 to 20.319, $^{208}Pb/^{204}Pb$ range from 36.685 to 38.385, whereas the radiogenic lead ($^{207}Pb/^{204}Pb$) is in between 15.398 and 15.684 (Table 7). Samples from the Cerro La Ventana Formation plot in a $^{207}Pb/^{204}Pb$ diagram (Fig. 14a) below the Stacey and Kramers (1975) second-stage Pb evolution curve for average crust, similar to rocks from the basement of Cuyania Terrane (Pie de Palo Range and crustal xenoliths). In a $^{208}Pb/^{204}Pb$ diagram (Fig. 14b) most of the samples also plot below the second-stage Pb evolution curve for average crust as basement rocks from the Cuyania Terrane, except for sample CV17 which is above such a curve. Although sample CV20 (garnet it was described on thin section) remains below SK curve, it shows the higher Pb ratios, particularly regarding $^{206}Pb/^{204}Pb$, and being therefore dissimilar to crustal xenoliths and Pie de Palo rocks from the basement of the Cuyania Terrane.

U–Pb: Two methods were applied: the isotope dilution (ID) with thermal ionization mass spectrometry (TIMS) and in situ LA-ICP-MS as follow:

(a) **ID-TIMS**: A sample of the tonalitic facies located at Leones River type section (01RL03) with 55.78% of SiO_2 was chosen for zircon U–Pb TIMS dating. The analyses were carried out at the Centro de Pesquisas Geocronológicas, São

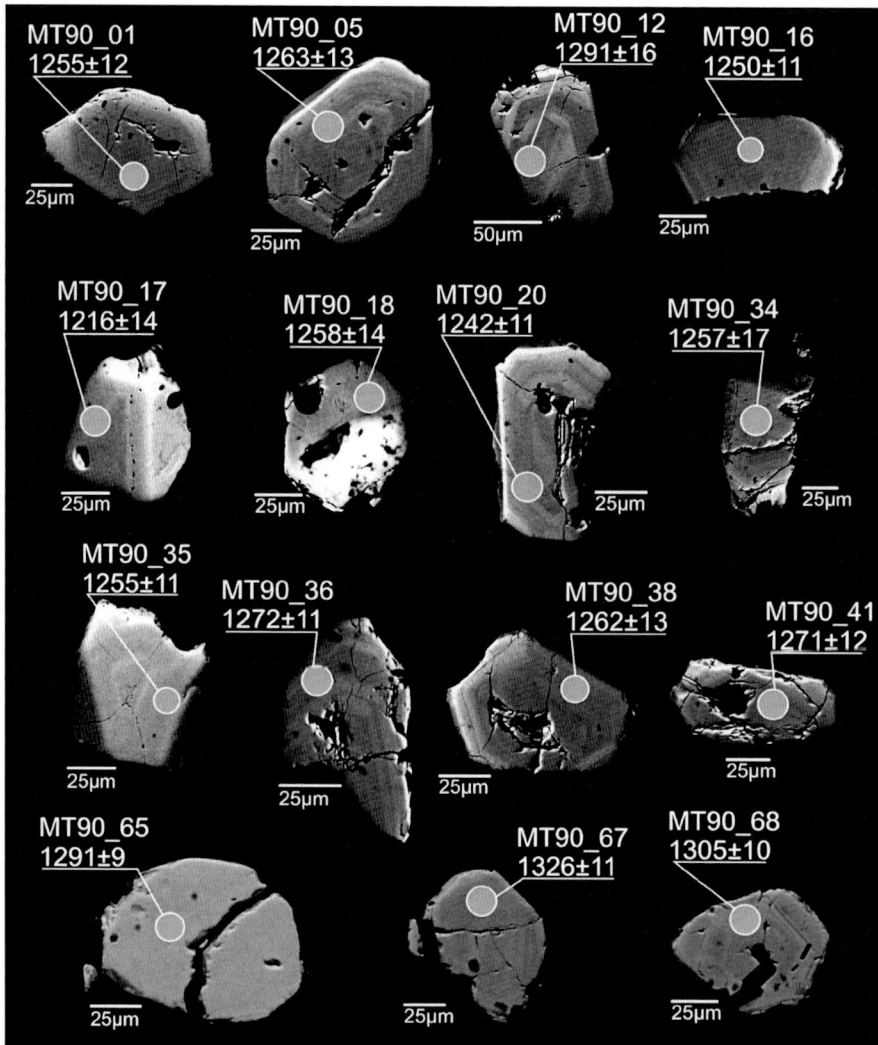

Fig. 16 Backscattering images of selected zircon crystals with spot locations by LA-ICP-MS analyses at LGI-UFRGS, Porto Alegre, Brazil laboratory. Ages in Ma

Paulo, Brazil. Four zircon fractions (Table 8; Fig. 15) were analyzed, with prismatic, euhedral and well-developed crystal faces. All were transparent crystals without inclusions or fractures. The obtained age of 1214.7 ± 6.5 Ma was based on main concordant 2161 and 2164 fractions and 2165 fraction with little Pb lost. The 2162 fractions with probable Pb crustal heritage were not considered in the age determination. The Mesoproterozoic obtained value is

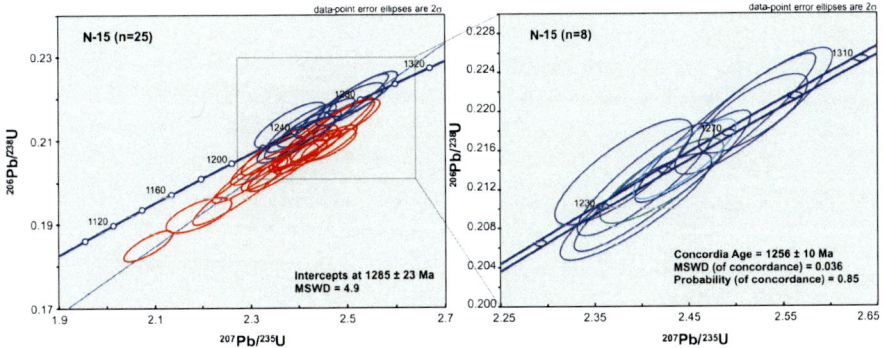

Fig. 17 Concordia plot of zircons from the sample N15 (*n* = 25) from El Nihuil mafic unit; on the right detail of eight concordant results

interpreted as the zircon crystallization age during the emplacement of the tonalitic magmatic rock.

(b) **LA-ICP-MS**: At the region of Lomas Orientales as part of the El Nihuil mafic unit (Cingolani et al. 2012b) crop out a meter-scale relics of tonalite–orthogneissic rocks with intermediate composition (SiO_2 around 66%) bearing a foliation with N83°W/54°SW (Fig. 5). The in situ U–Pb (LA-ICP-MS) zircon data on the sample N15 was obtained at the Laboratorio de Geología Isotópica, Universidade Federal do Río Grande do Sul, Porto Alegre, Brazil. The results yielded a $^{207}Pb/^{208}Pb$ weighted mean age of 1285 ± 23 Ma based on 25 spots (Fig. 16; Table 9). The zircon grains vary from colorless to brown, and are prismatic and subhedral with few inclusions. Some grains show rounded terminations. Backscattering images reveal that most grains display oscillatory zoning (Fig. 16). Some grains exhibit metamictization in their central part. Cracks are relatively common. The data point out a Mesoproterozoic crystallization age when all 25 results are plotted despite some scatter illustrated by the relatively high MSWD (Fig. 17a). Nevertheless, seven concordant results yielded a 1256 ± 10 Ma Concordia age (Fig. 17b) that can be considered the crystallization time.

The sample N15-N17B (tonalitic orthogneiss) taken at the same locality of Lomas Orientales along the El Nihuil mafic unit was analyzed by the in situ U–Pb (LA-ICP-MS) zircon data at the Centro de Pesquisas Geocronológicas, São Paulo, Brazil (Fig. 18). The obtained age plotted in a Tera-Wasserburg diagram is 1222.8 ± 6.9 Ma (*n* = 26) and record MSWD (concord.) = 0.8 (Fig. 19; Table 10) with 2 sigma errors. With these U–Pb isotopic data we confirm Rb–Sr and Sm–Nd age showing the Mesoproterozoic (boundary of Ectasian- Stenian Periods, after the IUGS International Chronostratigraphic Chart 2015) basement rocks of the San Rafael Block. The obtained age *ca*. 1.2 Ga is interpreted as a Grenvillian (Mesoproterozoic) zircon crystallization age for the Cerro La Ventana Formation and it is quite similar to those belonging to the basement of other regions from the

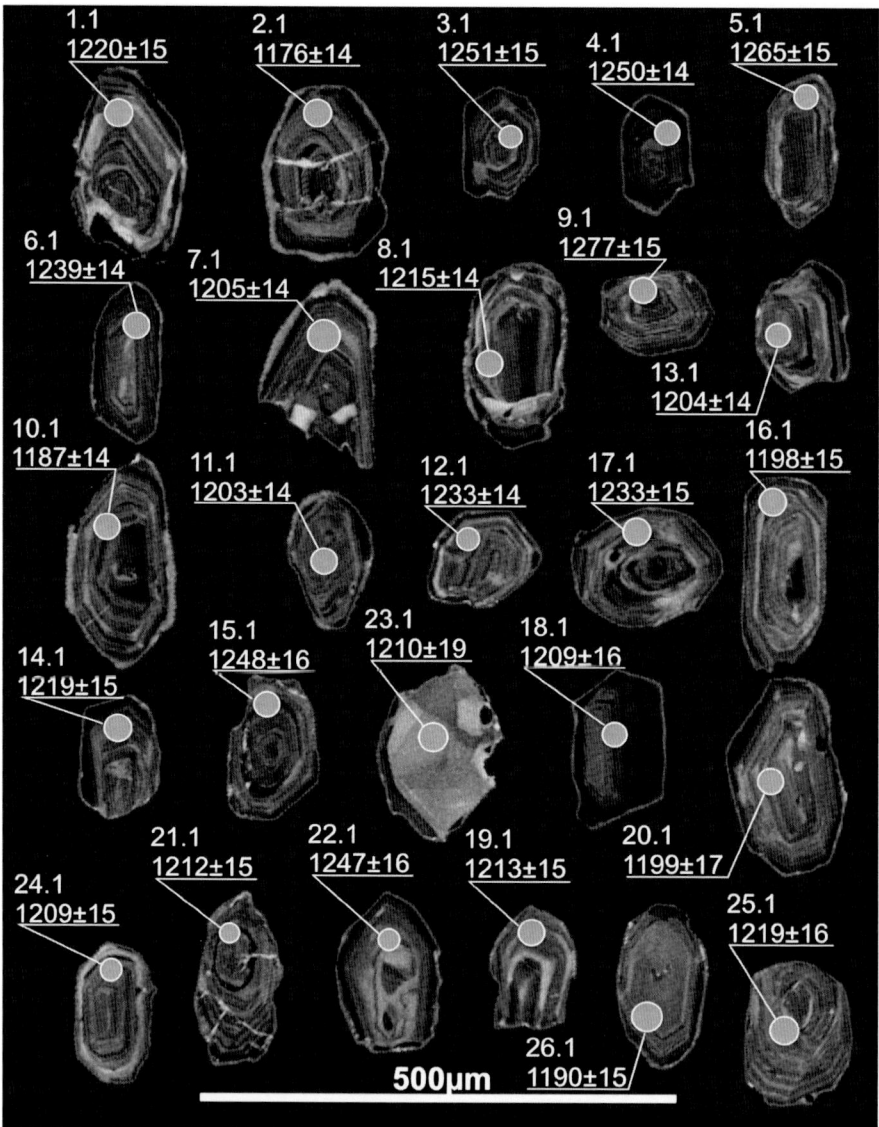

Fig. 18 Cathodoluminiscence (CL) images of selected zircon crystals with spot locations by LA-ICP-MS analyses at CPGeo-USP, São Paulo, Brazil laboratory. Ages in Ma

Cuyania allochthonous terrane (Ramos 2004; Sato et al. 2004; Morata et al. 2010; Varela et al. 2011 and references). Rapela et al. (2010) recognized in Western Pampean Ranges and Precordilleran xenoliths that the Mesoproterozoic geodynamic history evolved during 300 Ma, with a complex orogenic evolution dominated by convergent tectonics and accretion of juvenile oceanic arcs to the continent

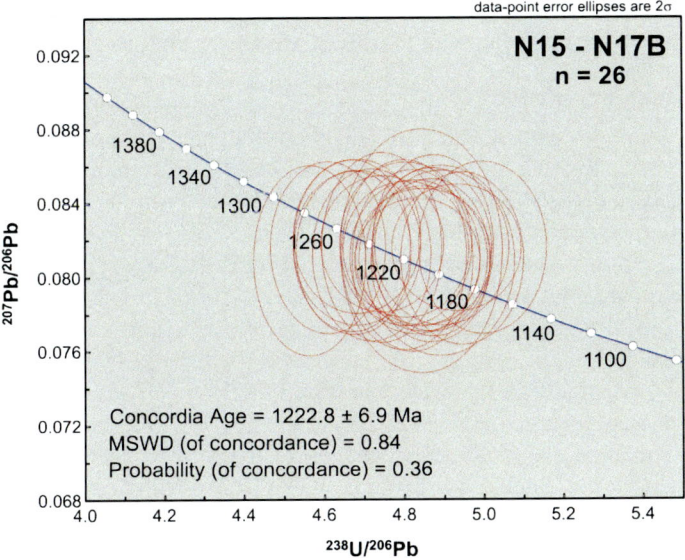

Fig. 19 Tera–Wasserburg diagram showing the plot of zircons from the sample N15-N17B from El Nihuil mafic unit

and intraplate magmatism. On the other hand Kumar et al. (2007) proposed a global thermal event at 1.1 Ga as responsible for the development of widespread global magmatic activity during this time.

8 Final Remarks

In summary, having interpreted all data presented the following statements can be drawn,

(a) The San Rafael Block crustal basement called Cerro La Ventana Formation crops out in two localities, at the Leones (type section) and within El Nihuil mafic unit, both areas showing ductile deformation.
(b) The outcrops at the type section record the primary contact between the basement and Ordovician platform sedimentary rocks, which is unique within the Cuyania terrane.
(c) The Cerro La Ventana Formation corresponds to a metamorphosed volcano-plutonic complex with hardly any sedimentary protolith that reached the high greenschist–amphibolite facies.
(d) The ductile deformation coeval with the metamorphism resulted in the folding of the complex and developed a penetrative foliation, which should have occurred long before its exhumation during Early Ordovician and probably

during the late stages of the Mesoproterozoic Grenville orogeny. This is a clear difference with Las Matras pluton from which no thermal overprint has been reported.

(e) Geochemical data indicate that the Cerro La Ventana Formation comprises cogenetic rocks with a main calc-alkaline and metaluminous character. The bulk of samples have Sr/Y ratios typical of low-Al TTD, although some can be described as high-Al TTD. The Mg#/K ratio is higher in the Cerro La Ventana Formation compared with Las Matras TTG series suggesting less differentiation for the first one. The chondrite-normalized REE diagrams for Leones samples have Eu anomalies rather positive and gabbros from El Nihuil region display patterns with positive Eu anomalies typical of plagioclase-rich igneous rocks.

(f) The Rb–Sr data defined an isochron with 1148 ± 83 Ma, initial $^{87}Sr/^{86}Sr = 0.70292 \pm 0.00018$. The low initial ratio is indicative of a slight evolved Mesoproterozoic source. An acceptable isochron was obtained using Sm–Nd methodology indicating an age of 1228 ± 63 Ma. The model ages (T_{DM}) are in the range 1.23–1.64 Ga with Epsilon $Nd_{(1200)}$ in between -0.94 and $+4.7$ recording a 'depleted' source, less evolved than CHUR for the time of crystallization.

(g) In Pb–Pb diagrams studied samples from the Cerro La Ventana plot similar to the rocks from the basement of Cuyania Terrane (Pie de Palo Range and crustal xenoliths) showing a distinctive nonradiogenic signature such as the Grenvillian belt of eastern Laurentia.

(h) The tonalitic facies located at the Leones River type section was chosen for zircon U–Pb TIMS dating and the obtained crystallization age is 1214.7 ± 6.5 Ma. The in situ U–Pb (LA-ICP-MS) zircon data from El Nihuil Mafic Unit (orthogneisses), analyzed in two different laboratories, plotted in a Concordia diagram record an intercept at 1256 ± 10 Ma, and the second age plotted in a Tera–Wasserburg diagram is 1222.8 ± 6.9 Ma.

(i) The isotopic data confirm the Mesoproterozoic age for the basement of the San Rafael Block; this age can also be found in basement rocks of other regions considered as part of the Laurentia derived Cuyania terrane in the proto-Andean margin of Gondwana.

Acknowledgements Field and laboratory works were financially supported by CONICET (grants PIPs 0647, 199), ANPCyT (grant PICT 07-10829) and University of La Plata (Projects 11/573, 11/704). We thank to colleagues Pablo González, Leandro Ortiz, and Diego Licitra for field work assistance and suggestions on metamorphic structural data. Luis Dalla Salda during 2001–2003 period and Alejandro Ribot, more recently, provide their expertise in discussion on petrographic description of samples. Thanks to Marcio Pimentel and Koji Kawashita for helping us in U–Pb laboratory work and data interpretation. Mario Campaña helps us with technical assistance. We are grateful to Víctor Ramos for his comments and discussions. Finally, we acknowledge to Agapito Aguilera and Domingo Solorza as field guide experts ('baqueanos') near the Leones River and El Nihuil town.

References

Abre P, Cingolani CA, Zimmermann U, Cairncross B, Chemale Jr F (2011) Provenance of Ordovician clastic sequences of the San Rafael Block (Central Argentina), with emphasis on the Ponón Trehué Formation. Gondwana Res 19(1):275–290

Abre P, Cingolani CA, Uriz NJ (this volume) Sedimentary provenance analysis of the Ordovician Ponón Trehué Formation, San Rafael Block, Mendoza-Argentina. In: Cingolani C (ed) Pre-Carboniferous evolution of the San Rafael Block, Argentina. Implications in the SW Gondwana margin. Springer, Berlin

Astini R, Ramos VA, Benedetto JL, Vaccari NE, Cañas FL (1996) La Precordillera: un terreno exótico a Gondwana. 13° Congreso Geológico Argentino y 3° Congreso de Exploración de Hidrocarburos, 5:293–324. Buenos Aires

Astini RA (2002) Los conglomerados basales del Ordovícico de Ponón Trehué (Mendoza) y su significado en la historia sedimentaria del terreno exótico de Precordillera. Revista de la Asociación Geológica Argentina 57:19–34

Barker F (1979) Trondhjemite: definition, environment, and hypotheses of origin, pp 1–12. In Barker F (ed) Trondhjemites, dacites, and related rocks. Elsevier, Amsterdam, 659 p

Batchelor AR, Bowden P (1985) Petrogenetic interpretation of granitoid rock series using multicationic parameter. Chem Geol 48(1985):43–55

Bordonaro OL, Keller M, Lehnert O (1996) El Ordovícico de Ponon-Trehue en la provincia de Mendoza (Argentina): redefiniciones estratigráficas. 12° Congreso Geológico Argentino y 2° Congreso de Exploración de Hidrocarburos 1:541–550

Bühn B, Pimentel MM, Matteini M, Dantas EL (2009) High spatial resolution analysis of Pb and U isotopes for geochronology by laser ablation multi-collector inductively coupled plasma mass spectrometry (LA-MC-IC-MS). Anais da Academia Brasileira de Ciências 81(1):1–16

Caminos R (1993) El Basamento Metamórfico Proterozoico-Paleozoico inferior. In: Ramos VA (ed) Geología y Recursos Naturales de Mendoza. XII Congreso Geológico Argentino and II Congreso de Exploración de Hidrocarburos, Buenos Aires, pp 11–19

Casquet C, Pankhurst RJ, Fanning CM, Baldo E, Galindo C, Rapela CW, González-Casado JM, Dahlquist JA (2006) U-Pb SHRIMP zircon dating of Grenvillian metamorphism in Western Sierras Pampeanas (Argentina): correlation with the Arequipa-Antofalla craton and constraints on the extent of the Precordillera terrane. Gondwana Res 9:524–529

Cingolani CA, Varela R (1999) The San Rafael Block, Mendoza (Argentina): Rb-Sr isotopic age of basement rocks. II South American Symposium on Isotope Geology, Anales 34 SEGEMAR, Actas 23–26. Carlos Paz, Córdoba

Cingolani CA, Llambías EJ, Ortiz LR (2000) Magmatismo básico pre-Carbónico de El Nihuil, Bloque de San Rafael, Provincia de Mendoza, Argentina. 9° Congreso Geológico Chileno 2:717–721, Puerto Varas

Cingolani CA, Llambías EJ, Basei MAS, Varela R, Chemale Jr F, Abre P (2005) Grenvillian and Famatinian-age igneous events in the San Rafael Block, Mendoza Province, Argentina: geochemical and isotopic constraints. Gondwana 12 conference, Abstracts 102, Mendoza

Cingolani CA, Basei MAS, Uriz NJ, Manassero MJ (2012a) Las metasedimentitas del Paleozoico Medio en el subsuelo de Mendoza (Corral de Lorca): su importancia en la evolución tectónica del Terreno Cuyania. XV Reunión de Tectónica. San Juan, Argentina. Octubre 2012, pp 38–39

Cingolani C, Uriz N, Marques J, Pimentel M (2012b) The Mesoproterozoic U-Pb (LA-ICP-MS) age of the Loma Alta Gneissic rocks: basement remnant of the San Rafael Block, Cuyania Terrane, Argentina. In: Proceedings of 8° South American Symposium on Isotope Geology. CD-ROM version. Medellín, Colombia. Abstract, p 140

Cingolani CA, Uriz NJ, Manassero M, Basei MAS (2014) La Formación Cerro Las Pacas al sur del Cerro Nevado, Mendoza: ¿Basamento Precámbrico o parte de la cuenca devónica de San Rafael? XIX Congreso Geológico Argentino. Córdoba. Argentina. Acta CD-ROM. Resumen: Tectónica Preandina, S21-11

Cingolani CA, Llambías EJ, Chemale Jr F, Abre P, Uriz NJ (this volume) Lower Paleozoic 'El Nihuil Dolerites': Geochemical and isotopic constraints of mafic magmatism in an extensional setting of the San Rafael Block, Mendoza, Argentina. In: Cingolani C (ed) Pre-Carboniferous evolution of the San Rafael Block, Argentina. Implications in the SW Gondwana margin. Springer, Berlin

Criado Roqué P (1972) Bloque de San Rafael. In: Leanza AF (ed) Geología Regional Argentina. Academia Nacional de Ciencias, Córdoba, pp 283–295

Criado Roqué P (1979) Subcuenca de Alvear. Segundo Simposio de Geología Regional Argentina v1. Academia Nacional de Ciencias, Córdoba, pp 811–836

Criado Roqué P, Ibáñez G (1979) Provincia geológica Sanrafaelino-Pampeana. Geología Regional Argentina, Academia Nacional de Ciencias de Córdoba 1:837–869

Dalla Salda L, Cingolani C, Varela R (1992) Early Paleozoic orogenic belt of the Andes in southwestern South America: result of Laurentia-Gondwana collision? Geology 20:617–620

Davicino RE, Sabalúa JC (1990) El Cuerpo Básico de El Nihuil, Dto. San Rafael, Pcia. de Mendoza, Rep. Argentina. 11° Congreso Geológico Argentino, Actas I, 43–47. San Juan

DePaolo DJ (1981) Neodymium isotopes in the Colorado Front Range and crust-mantle evolution in the Proterozoic. Nature 291:193–196

Dessanti RN (1956) Descripción geológica de la Hoja 27c-cerro Diamante (Provincia de Mendoza). Dirección Nacional de Geología y Minería. Boletín 85, 79 p. Buenos Aires

Drummond MS, Defant MJ (1990) A model for trondhjemite-tonalite-dacite genesis and crustal growth via slab melting: Archean to modern comparisons. Journal of Geophysical Research 95 (B13):21503–21521

Faure G (1986) Principles of isotope geology, 2nd edn. Wiley, New York

Hamilton PJ, O'Nions RK, Bridgwater D, Nutman A (1983) Sm-Nd studies of Archaean metasediments and metavolcanics from West Greenland and their implications for the Earth's early history. Earth Planet Sci Lett 62:263–272

Heredia S (1996) El Ordovícico del Arroyo Ponón Trehué, sur de la provincia de Mendoza. XIII Congreso Geológico Argentino y III Congreso de Exploración de Hidrocarburos, Actas I, 601–605

Heredia S (2006) Revisión estratigráfica de la Formación Ponón Trehué (Ordovícico), Bloque de San Rafael, Mendoza. INSUGEO, Serie Correlación Geológica 21:59–74

Heredia S, Mestre A (this volume) Ordovician conodont biostratigraphy of the Ponón Trehué Formation, San Rafael Block, Mendoza, Argentina. In: Cingolani C (ed) Pre-Carboniferous evolution of the San Rafael Block, Argentina. Implications in the SW Gondwana margin. Springer, Berlin

Holmberg E (1973) Descripción Geológica de la Hoja 29d, Cerro Nevado. Servicio Geológico Minero Argentino (SEGEMAR), Boletín 144. Buenos Aires, p 71

Keller M (1999). Argentine Precordillera. Sedimentary and plate tectonic history of a Laurentian crustal fragment in South America. The Geological Society of America, Special Paper 341, pp 1–131

Krogh TE (1973) A low-contamination method for hydrothermal decomposition of Zircon an extraction of U and Pb for isotopic age determinations. Geochemica et Cosmochimica Acta 37 (3):485–494

Kumar A, Heaman LH, Manikyamba C (2007) Mesoproterozoic kimberlites in south India: a possible link to ~1.1 Ga global magmatism. Precambrian Res 154:192–204

Ludwig KR (1999) Using Isoplot/Ex, version 2. A geochronological toolkit for Microsoft excel. Berkeley Geochronological Center, Special Publication 1a, 47 p

Ludwig KR (2001) Squid 1.02: a user manual. Berkeley Geochronology Center, Special Publication, 2, 19 p

Ludwig KR (2003) Isoplot/EX version 3.0, A geochronological toolkit for Microsoft Excel: Berkeley Geochronology Center Special Publication

Ludwig KR (2012) A geochronological toolkit for Microsoft Excel, version 3.76. Berkeley Geochronology Center, Special Publication No 5, Berkeley, 75 p

Morata D, Castro de Machuca B, Arancibia G, Pontoriero S, Fanning CM (2010) Peraluminous Grenvillian TTG in the Sierra de Pie de Palo, Western Sierras Pampeanas, Argentina: petrology, geochronology, geochemistry and petrogenetic implications. Precambr Res 177 (2010):208–322

Nuñez E (1962) Sobre la presencia del Paleozoico inferior fosilífero en el Bloque de San Rafael. Primeras Jornadas Geológicas Argentinas, II:185–189. Buenos Aires

Nuñez E (1976) Descripción geológica de la Hoja 28-C "Nihuil", Provincia de Mendoza. Servicio Geológico Nacional. Buenos Aires (unpublished report)

Nuñez E (1979) Descripción geológica de la Hoja 28d, Estación Soitué, Provincia de Mendoza. Servicio Geológico Nacional, Boletín 166:1–67

Padula E (1949) Descripción geológica de la Hoja 28-C "El Nihuil", Provincia de Mendoza. YPF (unpublished report)

Padula E (1951) Contribución al conocimiento geológico del ambiente de la Cordillera Frontal, Sierra Pintada, San Rafael (Mendoza). Revista de la Asociación Geológica Argentina 6(1):5–13. Buenos Aires

Polanski J (1949) El bloque de San Rafael. Dirección de Minas de Mendoza (unpublished report), Mendoza

Ramos VA (1995) Sudamérica: Un mosaico de continentes y océanos. Ciencia Hoy 6(32):24–29. Buenos Aires

Ramos VA, Dallmeyer RD, Vujovich G (1999) Time constraints on the Early Palaeozoic docking of the Precordillera, central Argentina. In Pankhurst RJ, Rapela CW (eds) The Proto-Andean Margin of Gondwana. Geological Society, Special Publications, 142, London, pp 143–158

Ramos VA (2004) Cuyania, an exotic block to Gondwana: review of a historical success and the present problems. Gondwana Res 7:1009–1026

Ramos VA, Jordan T, Allmendinger R, Mpodozis C, Kay S, Cortés J, Palma M (1986) Paleozoic terranes of the central Argentine-Chilean Andes. Tectonics 5:855–888

Ramos VA, Dallmeyer R, Vujovich GI (1998) Time constrains on the Early Paleozoic docking of the Precordillera central Argentina. In: Pankhurst RJ, Rapela CW (eds) The Proto Andean margin of Gondwana. Geological Society of London, Special Publication, 142, pp 143–158

Rapela CW, Pankhurst RJ, Casquet C, Baldo E, Galindo C, Fanning CM, Dahlquist JM (2010) The Western Sierras Pampeanas: protracted Grenville-age history (1330–1030 Ma) of intra-oceanic arcs, subduction-accretion at continental edge and AMCG intraplate magmatism. J South Am Earth Sci 29:105–127

Rapp RP, Shimizu N, Norman MD, Applegate GS (1999) Reaction between slab-derived melts and peridotite in the mantle wedge: experimental constraints at 3.8 GPa. Chem Geol 160:335–356

Rolleri EO, Criado Roqué P (1970) Geología de la Provincia de Mendoza. IV Jornadas Geológicas Argentinas (Mendoza, 1969), II:1–60. Buenos Aires

Rolleri EO, Fernández Garrasino CA (1979) Comarca septentrional de Mendoza. Segundo Simposio de Geología Regional Argentina, vol 1. Academia Nacional de Ciencias, Córdoba, pp 771–800

Sato AM, Tickyj H, Llambías EJ, Sato K (2000) The Las Matras tonalitic-trondhjemitic pluton, central Argentina: Grenvillian-age constraints, geochemical characteristics, and regional implications. J S Am Earth Sci 13:587–610

Sato AM, Tickyj H, Llambías EJ, Basei MAS, González PD (2004) Las Matras Block, Central Argentina (37°S-67°W): the Southernmost Cuyania Terrane and its Relationship with the Famatinian Orogeny. Gondwana Res 7(4):1077–1087

Sato K, Kawashita K (2002) Espectrometria de massas em Geologia Isotopica. Revista do Instituto de Geociencias-USP. Geol. USP Serie Científica, Sao Paulo 2:57–77

Sato K, Siga Jr O, Silva JA, McReath I, Liu D, Iizuka T, Rino S, Hirata T, Sproesser WM, Basei MAS (2009) In situ isotopic analyses of U and Pb in Zircon by remotely operated SHRIMP II, and Hf by LA-ICP-MS: an example of dating and genetic evolution of Zircon by $^{176}Hf/^{177}Hf$ from the Ita Quarry in the Atuba Complex, SE Brazil. Geol USP, Série Científica São Paulo 9:61–69

Stappenbeck R (1913) Apuntes hidrogeológicos sobre el sudeste de la Provincia de Mendoza. Bol. 6 (Serie B). Dirección General de Minas, Buenos Aires

Stacey JS, Kramers JD (1975) Approximation of terrestrial lead isotope evolution via two-stage model. Earth Planet Sci Lett 26:207–221

Steiger RH, Jager E (1977) Subcommission on geochronology: convention on the use of decay constants in geochronology and cosmochronology. Contrib Geol Time Scale, AAPG Stud Geol 6:67–71

Tanimizu M, Ishikawa T (2006) Development of rapid and precise Pb isotope analytical techniques using MC-ICP-MS and new results for GSJ rock reference samples. Geochem J 40:121–133

Taylor SR, McLennan SM (1985) The continental crust: its composition and evolution. Blackwell, Oxford, 312 p

Thomas WA, Tucker RD, Astini RA, Denison RE (2012) Ages of pre-rift basement and synrift rocks along the conjugate rift and transform margins of the Argentine Precordillera and Laurentia. Geosphere 8(6):1366–1383

Varela R, Basei MAS, González PD, Sato AM, Naipauer M, Campos Neto M, Cingolani CA, Meira VT (2011) Accretion of Grenvillian terranes to the southwestern border of the Río de la Plata craton, western Argentina. Int J Earth Sci 100:243–272

Vujovich GI, Van Staal CR, Davis W (2004) Age constraints on the tectonic evolution and provenance of the Pie de Palo complex, Cuyania composite terrane, and the Famatinian Orogeny in the Sierra de Pie de Palo, San Juan, Argentina. Gondwana Res 7(4):1041–1056

Wichmann R (1928) Reconocimiento geológico de la región de El Nihuil, especialmente relacionado con el proyectado dique de embalse del Río Atuel. Dirección Nacional de Minería (unpublished report). Buenos Aires

Williamson JH (1968) Least-squares fitting of a straight line. Can J Phys 46:1845–1847

Sedimentary Provenance Analysis of the Ordovician Ponón Trehué Formation, San Rafael Block, Mendoza-Argentina

Paulina Abre, Carlos A. Cingolani, Norberto Javier Uriz and Aron Siccardi

Abstract The present chapter deals with provenance analysis of a carbonate-siliciclastic Ordovician sedimentary unit of the San Rafael block, named the Ponón Trehué Formation (Darriwilian to Sandbian). This is the only sequence which exhibits a direct contact with the Mesoproterozoic basement through an unconformity, not only within the San Rafael block, but rather for the entire Cuyania terrane. When combining different provenance proxies, such as petrography, whole-rock geochemistry, Sm–Nd data, Pb–Pb analyses, and detrital zircon dating, it can be deduced that the source rocks are characterized by: (i) an upper continental crust composition, (ii) a subordinated influence of a more depleted composition, (iii) a dominantly Mesoproterozoic age, (iv) sedimentary recycling did not conspicuously affected the detrital source, and (v) weathering was relatively strong. All these characteristics point to the Mesoproterozoic Cerro La Ventana Formation basement as a main source of detritus to a restricted basin infilled during the Ordovician.

Keywords Geochemistry · Isotope geochemistry · Detrital zircon dating · Provenance · Ponón Trehué Formation · Cuyania terrane

---

P. Abre (✉)
Centro Universitario de la Región Este, Universidad de la República,
Ruta 8 Km 282, Treinta y Tres, Uruguay
e-mail: paulinabre@yahoo.com.ar

C.A. Cingolani · A. Siccardi
Centro de Investigaciones Geológicas and División Geología del Museo de La Plata,
CONICET-UNLP, Diag. 113 n. 275, CP1904 La Plata, Argentina
e-mail: carloscingolani@yahoo.com

A. Siccardi
e-mail: aron8112@gmail.com

C.A. Cingolani · N.J. Uriz
División Geología, Museo de La Plata, UNLP, Paseo del Bosque s/n,
B1900FWA La Plata, Argentina
e-mail: norjuz@gmail.com

© Springer International Publishing AG 2017
C.A. Cingolani (ed.), *Pre-Carboniferous Evolution of the San Rafael Block, Argentina*, Springer Earth System Sciences,
DOI 10.1007/978-3-319-50153-6_3

1 Introduction

The Darriwilian to Sandbian Ponón Trehué Formation crops out at the southern edge of the San Rafael Block (Cuyania terrane), Mendoza province, Argentina (Fig. 1a, b). It is an olistostromic carbonate–siliciclastic sequence unconformably overlying (Fig. 2a, b), the Mesoproterozoic basement known as the Cerro La Ventana Formation (Nuñez 1979; Criado Roqué and Ibañez 1979; Heredia 1996; Cingolani and Varela 1999; Beresi and Heredia 2003; Cingolani et al. 2005; Heredia 2006). The unit comprises outcrops of the previously known Lindero Formation (Nuñez 1979 and see discussion in Heredia 1996, 2006 and Abre et al. 2011).

The continental Carboniferous Pájaro Bobo Formation (correlated with El Imperial Formation towards the Northwest of San Rafael Block) overlies the Ponón Trehué sequence through either an unconformity or a fault contact.

As an Ordovician fossil-rich unit (see Fig. 3), it contains trilobites, brachiopods, ostracods, fragmentary crinoids, corals and conodonts (Nuñez 1962; Baldis and Blasco 1973; Rossi de García et al. 1974; Levy and Nullo 1975; Heredia 2006). The

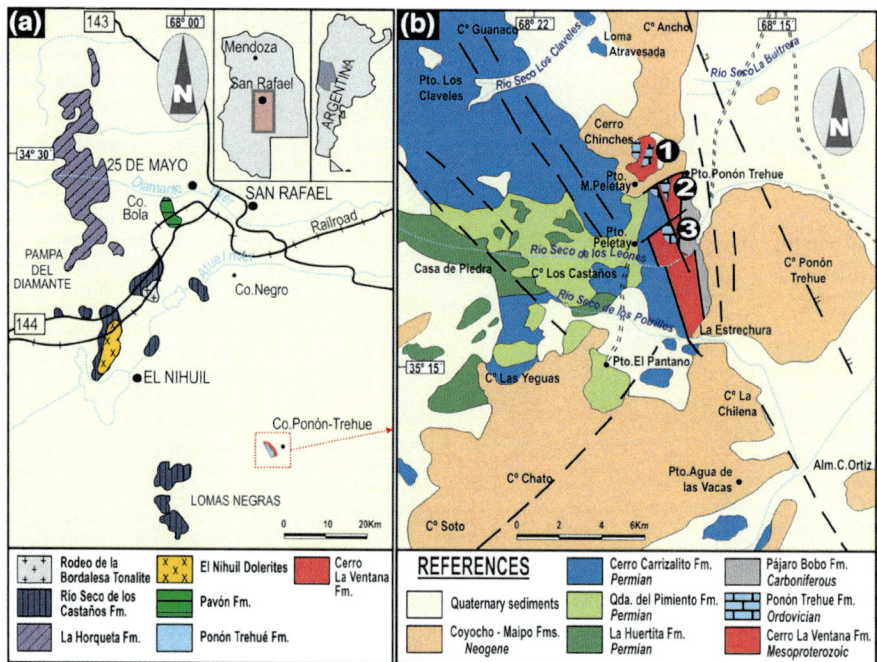

Fig. 1 **a** Geological sketch map of the pre-Carboniferous units within the San Rafael block (from González Díaz 1982). Outcrops of the Ponón Trehué Formation are located southwards and unconformably overlying the Cerro La Ventana basement. **b** Detailed geology of the Ponón Trehué area, with outcrops of the Ponón Trehué Formation in the central part (areas named 1, 2 and 3). Modified from Nuñez (1979), Cingolani and Varela (1999), and Cingolani et al. (2005)

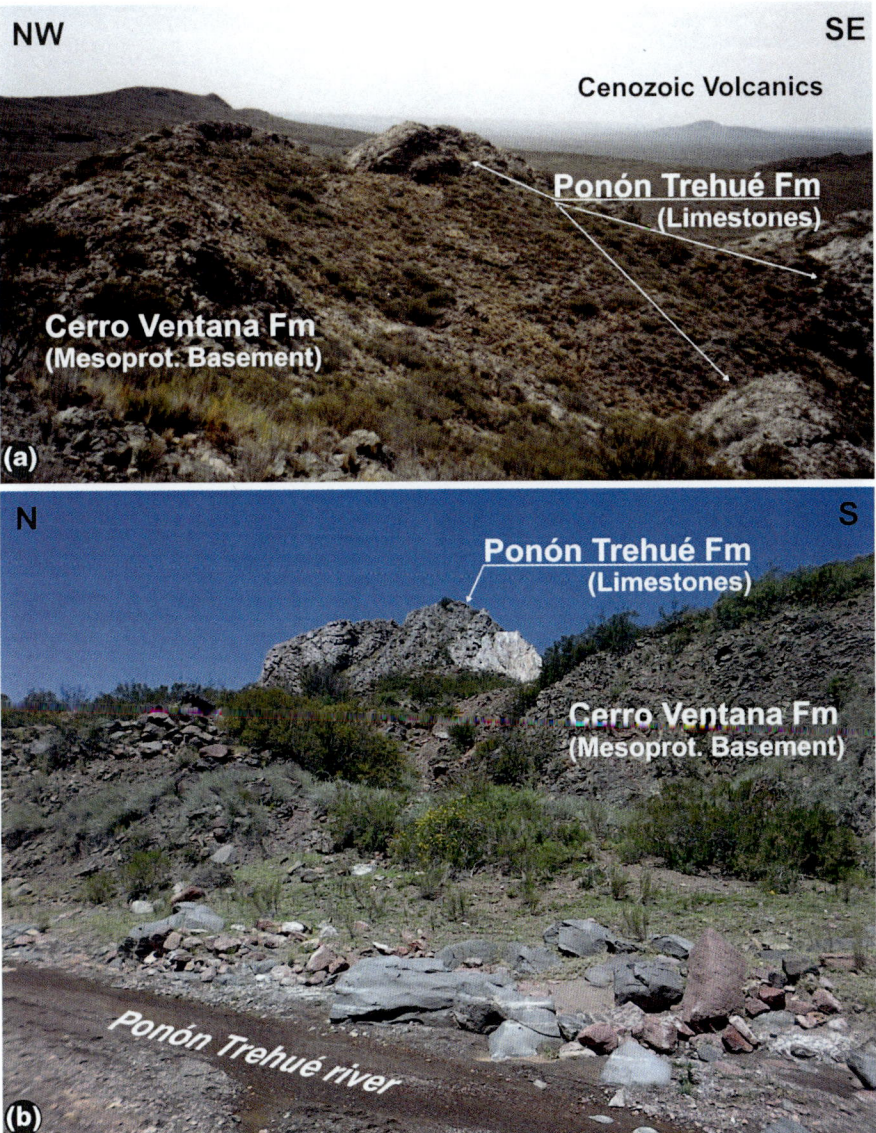

Fig. 2 a General view towards the North of the Ordovician Ponón Trehué limestones in contact with the Mesoproterozoic basement. **b** Outcrops of the Ponón Trehué fossiliferous limestones near the homonymous creek showing the remnants of the carbonate platform that slumped down the slope margin

first fossiliferous record was made by Nuñez (1962). This material was preliminary classified by Armando F. Leanza as the brachiopods *Obolus* and *Taffia*, and the trilobite *Lonchodomas* cf *salagastensis* (Rusconi). These records have been used to

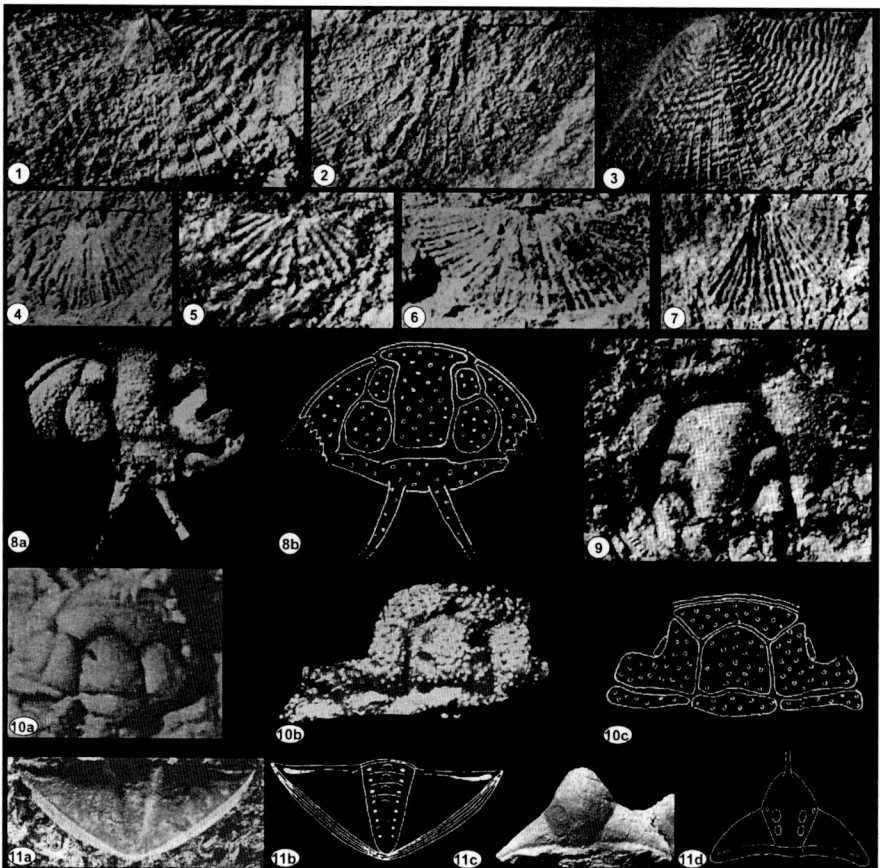

Fig. 3 Ordovician fossils known in the Ponón Trehué Formation (after Nuñez 1962; Baldis and Blasco 1973; Levy and Nullo 1975). BRACHIOPODS: *Ptychoglyptus mendocina* Levy and Nullo, *1075*: (*1*) ventral valve; (*2*) dorsal valve; (*3*) valve showing costulation. *Nugnecella rafaelensis* Levy and Nullo, 1975: (*4*) ventral valve; (*5–6*) dorsal valve; (*7*) ventral valve. TRILOBITES: "*Elbaspis*" (*Miraspis?*) *pintadensis* Baldis and Blasco, 1973: (*8a*) dorsal view of cranidium, (*8b*) reconstruction of the cranidium; (*9*) "*Flexycalymene*" (*Reacalymene*) *frontalis* Baldis and Blasco, 1973: cranidium; *Toernquistia chinchensis* Baldis and Blasco, 1973: (*10a–b*) dorsal view of cranidium (*10c*) reconstruction of the cranidium in dorsal view; *Ampyx nunezi* Baldis and Blasco, 1973: (*11a*) pygidium, internal mold, (*11b*) reconstruction of the pygidium, (*11c*) cranidium in dorsal view; (*11d*) reconstruction of the cranidium. Reproduced with permission of Asociación Paleontológica Argentina, Buenos Aires

correlate the Ponón Trehué unit with the Middle Ordovician San Juan Limestones, cropping out in the Precordillera region, as was first mentioned by Wichmann (1928) that considered the carbonates similar to those of Cerro de la Cal and Salagasta (near the city of Mendoza). Baldis and Blasco (1973) revised in detail the trilobite material and described the new genus *Elbaspis* (*Odontopleuridae*,

Selenopeltinae) (? = *Miraspis*; Ramsköld 1991; Jell and Adrain 2003) and the new species *Elbaspis pintadensis, Toernquistia chinchensis* (*Dimeropygidae*) (reassigned to *Paratoernquistia* by Chatterton et al. 1998), *Ampyx nunezi* (*Raphiophoridae*), and *Flexicalymene frontalis* (*Calymenidae*). Undeterminable species of *Monorakidae, Trinucleidae,* and *Illaeninae* are also present in the assemblage (see Fig. 3).

At the northern sector (in the way to the Chinches Hill) in small outcrops of Ponón Trehué Formation, some stromatolite structures were recognized in limestone rocks preserved as olistoliths.

The Ponón Trehué Formation is subdivided into two members: the lowermost (Peletay Member) is composed of conglomerates and conglomeratic arkoses, limestones, quartz arenites, and black shales, whereas the uppermost (Los Leones Member) is composed of mudstones, siltstones, arenites, and conglomeratic arenites.

The extension undergone by the Cuyania terrane after its accretion to Gondwana produced the brecciation of parts of the carbonate platform (Fig. 2a, b) that slumped down the slope, forming this breccia-type deposit (Astini 2002) within a fine clastic matrix. The absence of blocks from the Cambrian carbonate platform indicates that for the time of deposition of the Ponón Trehué Formation the basement was exposed (Heredia 2006).

Determining the paleogeographic and paleoclimatic conditions of deposition within a certain basin may be achieved by means of sedimentary provenance analysis, since sedimentary siliciclastic rocks record characteristics of their source rocks and areas. The final composition of a sedimentary rock does not depend solely on the original composition of the parent rock due to the effects that other processes such as weathering, sorting during transport, sedimentary recycling, and diagenesis would imprint. Therefore, several provenance techniques should be applied together, if possible, since each dataset would reveal different aspects of the parent material and/or the changes in composition that the sediments have suffered through their history. The provenance of the Ponón Trehué Formation was determined using petrography, whole-rock geochemistry and isotope geochemistry (including detrital zircon dating).

2 Petrography

The Ponón Trehué Formation at the La Tortuga section (35° 10′ 53″S–68° 18′ 13″W) comprises claystones, siltstones and fine-grained sublith- and subfeldspathic arenites (Dott 1964). The arenites are moderately sorted with scarce matrix. The framework minerals include: subrounded to subangular monocrystalline (with low sphericity) and polycrystalline (less abundant) quartz as well as subrounded K-feldspar commonly totally replaced by chlorite or clay minerals; detrital muscovite lamellae are very scarce. Sedimentary lithoclasts derived from siltstones, carbonates, mudstones, and cherts were also described. When present, the cement is composed

Table 1 Illite crystallinity index measured to determine the grade of diagenesis and/or very low-grade metamorphism of the Ponón Trehué Formation

Air-dried samples			
Sample	FWHM	CIS	Diagenesis/Metamorphisms
CT3	0.328	0.378	Low anchizone
CT6	0.318	0.370	Low anchizone
CT7	0.274	0.339	Low anchizone
CT8	0.334	0.382	Low anchizone
Ethylene-glycol attacked samples			
CT3	0.322	0.373	Low anchizone
CT6	0.313	0.367	Low anchizone
CT7	0.275	0.339	Low anchizone
CT8	0.329	0.379	Low anchizone

FWHM full-width-height-maximum. CIS crystallization index corrected according to standardized values. For details on methodology see Abre (2007). The term anchizone is equivalent to very low-grade metamorphism according to the recommendations of the Subcommission on the Systematics of Metamorphic Rocks from the IUGS (Árkai et al. 2003)

of calcite. The heavy minerals fraction comprises zircon, apatite, chromian spinel, tourmaline, rutile, and iron oxides such as hematite. X-ray diffraction analyses indicate that clay minerals within the three lithotypes are mainly chlorite, sericite, and illite (Abre 2007; Abre et al. 2011).

Sandstones of the Ponón Trehué Formation had shown relatively textural immaturity and mineralogical maturity, which altogether may imply that the detritus had suffered low transport but a certain degree of chemical weathering. The composition of the lithoclasts indicates sedimentary rocks as part of the source, while the presence of detrital chromian spinels clearly points to a mafic source. The bulk of the mineralogical composition indicates felsic sources.

The 'illite crystallinity index' (ICI; Kübler 1966 in Warr and Rice 1994) is defined as the width of the (001) XRD peak at half of its height, and it is used to determine either the grade of diagenesis or the very low-grade metamorphism that a clastic rock could have suffered. A well-crystallized illite, characteristic of a relatively high-temperature history, has sharp peaks, and therefore a low index, while low-temperature illite is more disordered, and has irregular peaks with large indexes. The ICI on four fine-grained clastic samples of the Ponón Trehué Formation shows that the unit was affected by very low-grade metamorphism (Table 1).

3 Whole-Rock Geochemistry

Geochemical analyses quantify bulk composition of a rock and give additional information to petrographic data regarding provenance and processes that might have modified the original composition. Major elements have been proved useful for weathering analysis, whereas certain trace elements (particularly REE) and their ratios are used to characterize the composition of the source (or the mix of sources) and would indicate tectonic setting. Furthermore, elements easily affected during weathering and diagenesis would permit to evaluate to which extent the bulk chemical composition of the detrital rock would have been affected by these processes.

Weathering: The degree of primary material transformation due to weathering can be estimated using the Chemical Index of Alteration (CIA; Nesbitt and Young 1982):

$$\text{CIA} = \{Al_2O_3/(Al_2O_3 + CaO^* + Na_2O + K_2O)\} \times 100,$$

where CaO* refers to the calcium associated with silicate minerals and the index is calculated using mole fractions. Index values vary between around 50 for unweathered crystalline rocks to 100 for kaolinitic residues. CIA values of the Ponón Trehué Formation (n = 8) ranges from 69 to 77 and samples are grouped toward the A-K boundary and close to the muscovite idealized composition, in accordance to XRD mineralogical data (Fig. 4a). Their distribution on the A-CN-K space does not correspond to a normal weathering path; furthermore, only two samples show a slight K_2O enrichment comparing to upper continental crust (UCC) values. Therefore, the very low-grade metamorphism indicated by ICI should explain the behavior of samples from the Ponón Trehué Formation, although effects of source mixing cannot be ruled out. Since the CIA values seem to have been modified, they can be recalculated using the expected weathering path of the average UCC, assuming such a composition for the source rocks (Fedo et al. 1995; Bock et al. 1998). In Fig. 3a, the dashed lines go from the K apex through the samples with the lowest and highest CIA and toward the normal weathering path for the UCC. Dotted lines end on the CIA scale indicating the corresponding CIA recalculated value for that interval, which would now range from 76 to 81, indicating strong weathering (Fedo et al. 1995), as it was deduced by petrographic analysis (Abre 2007; Abre et al. 2011).

The Th/U ratios have been used to estimate weathering effects in sedimentary rocks (McLennan et al. 1993), however, for the Ponón Trehué Formation results are not conclusive since some samples show Th/U ratios typical for unrecycled sediments (around 3.5–4; McLennan 1989), while others show U enrichment (therefore not indicative of weathering) and others do show values suggesting weathering (Abre et al. 2011).

Recycling: During reworking there is a tendency to increase stable heavy minerals content, particularly zircon (main Zr carrier), and therefore certain ratios such as the Zr/Sc would also increase (McLennan et al. 1993). The spread of Zr/Sc ratios of the Ponón Trehué Formation shows a cluster of data indicating that processes of recycling were not important, which is in accordance to petrographic characteristics such as relatively textural immaturity.

Source composition: Certain trace elements and their ratios are useful for provenance determination due to their preference either for felsic (Zr, Y, Th) or

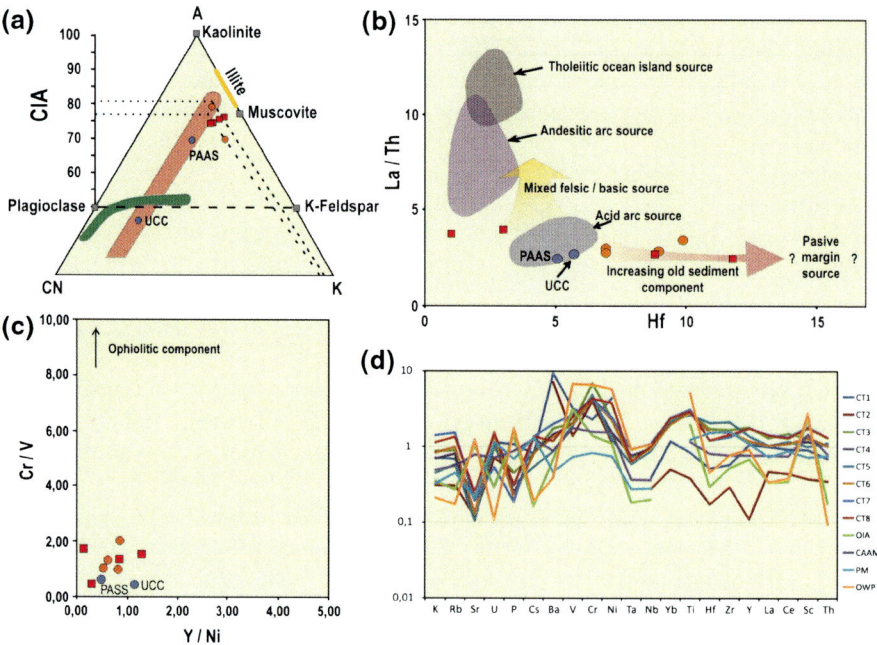

Fig. 4 Whole rock geochemical data obtained from Abre (2007) and Abre et al. (2011). **a** A–CN–K diagram; UCC, PAAS and idealized mineral compositions are according to Taylor and McLennan (1985). Field of *vertical lines* indicates the predicted weathering trend for the average UCC. Note that the *lower part* of the diagram with A < 40 is not shown. The *left side of the figure* shows the range of CIA values corrected according to Fedo et al. (1995). **b** La/Th versus Hf for samples of the Ponón Trehué Formation. La/Th ratios of 2–4 are common values for upper crust composition and indicate felsic compositions (McLennan et al. 1980). **c** The input of a mafic source could be discriminated using the Y/Ni and Cr/V ratios (McLennan et al. 1993). The Y/Ni ratio indicates the concentration of ferromagnesian trace elements (e.g. Ni) compared with a proxy for HREE (represented by Y). The Cr/V ratio indicates the enrichment of Cr over other ferromagnesian trace elements. Ophiolitic components would have Cr/V ratios higher than 10 (discussion in McLennan et al. 1993). UCC values according to McLennan et al. (2006), while PAAS is following Taylor and McLennan (1985). **d** Upper continental crust-normalized multielement patterns; elements are arranged from *left to right* in order of decreasing ocean residence time and comprise a relatively mobile group (from K to Ni) and a more stable group (from Ta to Th). Normalization values are from Taylor and McLennan (1985), except for P and Yb which are from McLennan et al. (2006). For a, b and c sandstones are represented by circles while mudstones are by squares

mafic (Sc, Cr, V) facies during melt crystallization (Taylor and McLennan 1985). Rare earth elements (REE) patterns in well-mixed sedimentary rocks also represent reliable provenance indicators (McLennan 1989). The Th/Sc ratios indicate that the source of detritus for the Ponón Trehué unit had dominantly a typical upper continental crustal composition (Abre et al. 2011). La/Th ratios between 2.58 and 4.05 (Fig. 4b) further support such a felsic dominant composition for the source rocks. However, contents of Sc (up to 20 ppm), Cr (up to 240 ppm) and V (up to 193 ppm), and Cr/V ratios higher than the UCC values, and Y/Ni ratios lower than the UCC indicate the influence of a source with a composition less evolved than the average UCC, although an ophiolitic source can be neglected (Fig. 4c). The mafic input is also documented by chromian spinels (Abre et al. 2011).

The chondrite normalized REE patterns for the Ponón Trehué Formation show certain enrichment of LREE (La_N/Yb_N of about 5.8 on average), as well as HREE (Tb_N/Yb_N of 1.17 on average) compared with the Post-Archaean Australian Shales (PAAS). The negative Eu-anomaly (Eu_N/Eu^* of about 0.52 on average) typical for detrital rocks derived from UCC is present.

Tectonic setting: Using Bhatia and Crook (1986) tectonic classification diagrams, a continental arc or an active continental margin was deduced for the Ponón Trehué Formation (Abre et al. 2011). Since the continental arc setting represent convergent plate margins and the active continental margin setting comprises the Andean-type and strike-slip continental margins (Bhatia and Crook 1986), such provenance proxie was not enough to determine the tectonic setting of the basin to aid on palaeogeographic reconstructions.

Upper continental crust-normalized multielement patterns have proven useful to describe the range of compositions of greywackes as a result of the tectonic environment (Floyd et al. 1991). The Ponón Trehué Formation shows (Fig. 4d): (1) negative Sr anomalies typical of passive margin settings, (2) positive V–Cr–Ni–Ti–Sc anomalies indicative of mafic inputs, (3) Nb similar to UCC with exception of samples CT1 and CT2, (4) positive Hf–Zr–Y anomalies that can be related to heavy mineral contents (particularly zircon) which typically increases in passive margin settings; noteworthy are the negative anomalies for samples CT1 and CT2, (5) the most stable element ratios are close to 1, feature generally indicative of a continental arc/active margin tectonic environment.

In conclusion, the multielement patterns confirm that the source comprised a mafic component typical of active margins, although anomalies related to passive margins are present. The behavior of samples CT1 and CT2 is clearly related to high silica contents (more than 79 %) linked to intense weathering in a continental active margin rather than in an oceanic setting.

4 Isotope Geochemistry

Sm–Nd: The Sm–Nd isotope system is a good provenance indicator since it aids to determine the grade of fractionation and the average crustal residence time of the detrital mix (McLennan et al. 1990), because the system is usually not resetted by erosion, sedimentation, and high-grade metamorphism (DePaolo 1981). According to Abre et al. (2011) the $\varepsilon_{Nd}(t)$ values, where t = 462 Ma (depositional age) ranges from −3.95 to −4.91 for the Ponón Trehué Formation (average −4.47 ± 0.39). $f_{Sm/Nd}$ ranges from −0.34 to −0.40 (average −0.37 ± 0.02), while T_{DM} ages ranges from 1.3 to 1.52 Ga (average 1.44 ± 0.078 Ga). These values are neither typical of an old upper crust nor of an arc component. $\varepsilon_{Nd}(t)$ values are similar to those from other Ordovician sedimentary rocks from the Cuyania terrane (Abre 2007; Gleason

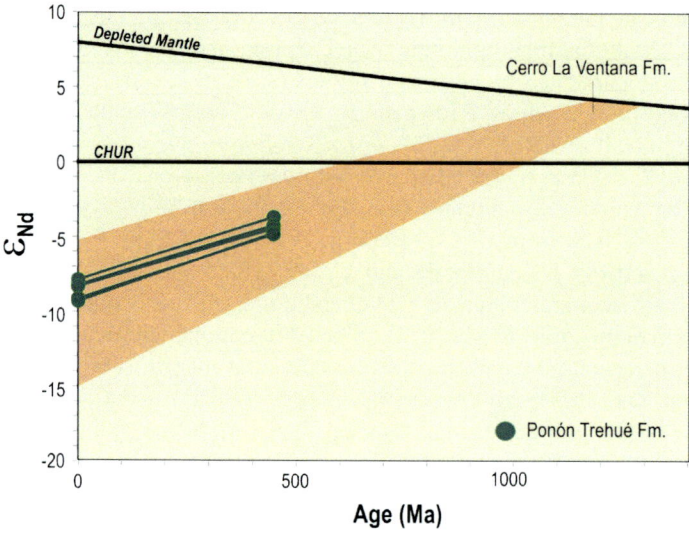

Fig. 5 ε_{Nd} versus age of the Ponón Trehué Formation. The range of the Mesoproterozoic Cerro La Ventana Formation Nd data is drawn for comparison (Cingolani et al. 2005). *CHUR* Chondritic Uniform Reservoir

et al. 2007; Abre et al., this volume), as well as from rocks of the Famatinian arc (Pankhurst et al. 1998) and are in the range of variation of data from the Cerro La Ventana Formation (Fig. 5) calculated to the time of deposition of the Ponón Trehué Formation.

Pb–Pb: $^{206}Pb/^{204}Pb$ ranges from 19.028 to 19.303 and $^{208}Pb/^{204}Pb$ ranges from 38.83 to 38.99 (Abre et al. 2011). The radiogenic lead ($^{207}Pb/^{204}Pb$) is interesting since most terrestrial ^{207}Pb was produced early in the Earth's history and therefore the high $^{207}Pb/^{204}Pb$ ratio of these samples (15.66–15.71) might suggest the input from an old Pb component to the sedimentary succession (Hemming and McLennan 2001). Analyzing both, uranogenic- and thorogenic-Pb it is evident that samples from the Ponón Trehué Formation plot slightly above the Stacey and Kramers (1975) second stage Pb evolution curve for average crust and they are more similar to those from the globally subducted sediment (GLOSS) average than the upper continental crust composition (Fig. 6) being more comparable with samples from trailing edges and continental collision zones (Hemming and McLennan 2001). The Pb system of the Ponón Trehué Formation differs consistently from the Mesoproterozoic Cerro La Ventana Formation (Fig. 6), although it is not possible to know whether this is due to isotopes remobilization during weathering and low-grade metamorphism or if it would point to a different source.

U–Pb detrital zircon: Age determination of individual detrital zircon grains has been proved a powerful tool for provenance discrimination, particularly when determining age of intermediate to felsic source rocks. Detrital zircon dates (n = 38) of the Ponón Trehué Formation cluster between 1065 and 1277 Ma with a main

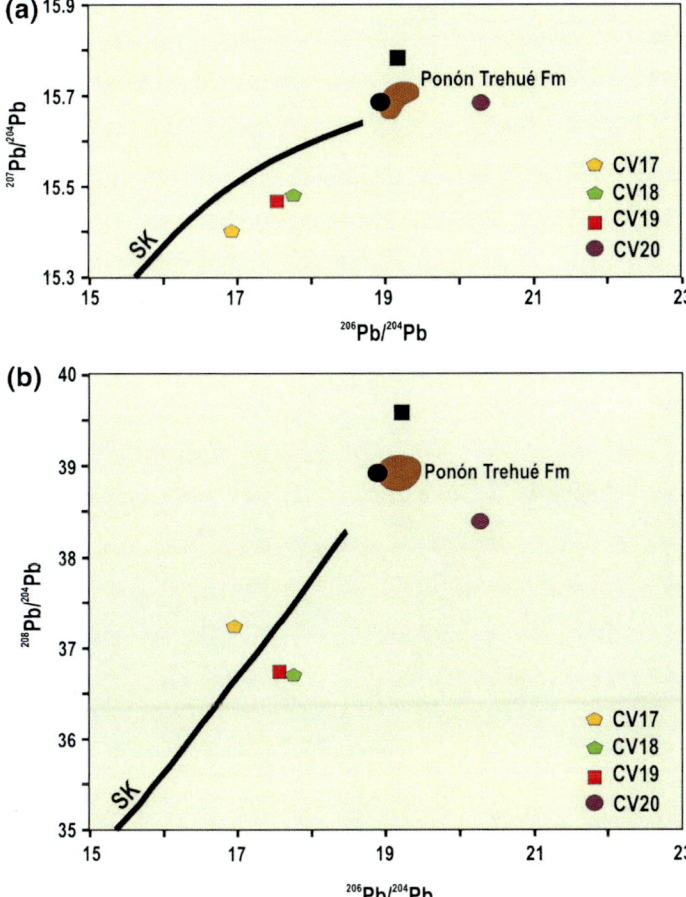

Fig. 6 **a** $^{207}Pb/^{204}Pb$ versus $^{206}Pb/^{204}Pb$ and **b** $^{208}Pb/^{204}Pb$ versus $^{206}Pb/^{204}Pb$ present-day ratios. *Brown areas* represent samples from the Ponón Trehué Formation; *solid square* is the average value for the upper crust and *solid circle* the average value for GLOSS, both from Hemming and McLennan (2001). *SK* Stacey and Kramers reference line. Samples from the Cerro La Ventana Formation are shown for comparison (data from Cingolani et al., this volume)

peak at about 1200 Ma (Fig. 7). Only one discordant grain has a younger age of 834 Ma (Abre et al. 2011). The very narrow range of detrital zircon ages implies a local and restricted provenance, most likely from the underlying Cerro La Ventana Formation and is in agreement with the low recycling deduced from petrographic and geochemical analyses. Th/U ratios measured in zircons along with cathodoluminescence images indicate a dominance of grains originated by magmatic processes rather than metamorphic. The Cerro La Ventana Formation, with ages between 1.1 and 1.2 Ga (Cingolani and Varela 1999; Cingolani et al. 2005), matches the detrital zircon ages and was a source of detritus. Other

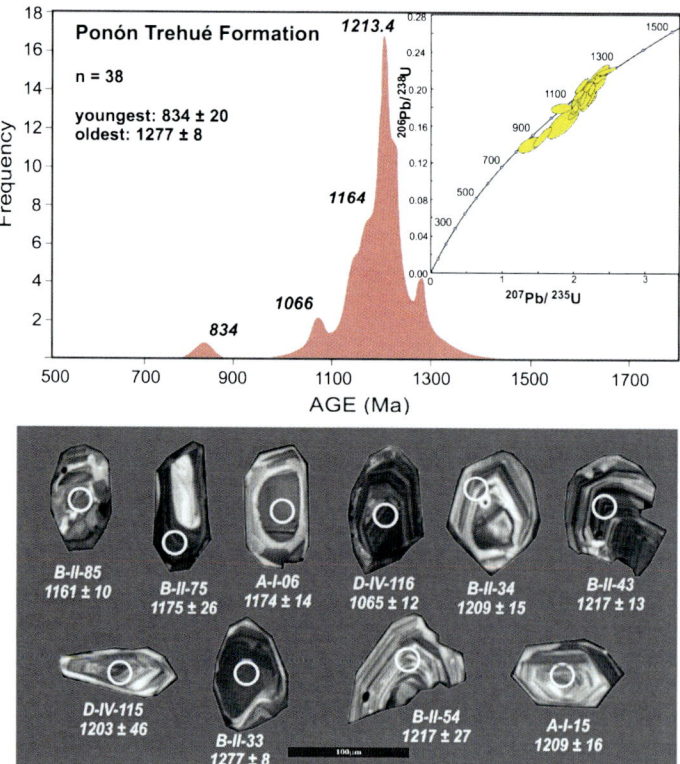

Fig. 7 U–Pb distribution of analyzed detrital zircons with probability curves for the Ponón Trehué Formation. Representative cathodoluminescense microphotographs of selected zircon grains used for detrital dating show the predominance of magmatic internal textures. Bar length is 100 μm and ages are in Ma (Abre et al. 2011)

Mesoproterozoic rocks within the basement of the Cuyania terrane are found at the Pie de Palo Range (1.0–1.2 Ga; McDonough et al. 1993) and the Umango, Maz and Espinal Ranges (1.0–1.2 Ga; Varela and Dalla Salda 1992; Varela et al. 1996; Casquet et al. 2006; Rapela et al. 2010; Varela et al. 2011).

5 Discussion

The petrographic analyses of the Ordovician Ponón Trehué Formation showed the dominance of monocrystalline quartz and K-feldspar and the presence of sedimentary lithoclasts (siltstones, carbonates, mudstones and chert) and spinels, which point to an upper continental crustal component (including the recycling of sedimentary rocks) and the influence of a source less evolved than the UCC.

Geochemical analyses of the Ponón Trehué Formation indicate strong weathering and other secondary processes, since high CIA values are present and potassium enrichments were detected. The dominance of an upper continental crustal component is clearly reflected, but relatively high abundances of compatible elements along with low Th/Sc ratios suggest a mafic input, although an ophiolitic source can be neglected based on Y/Ni and Cr/V ratios. Recycling was not important, according to low Zr/Sc ratios, pointing to source rocks closely related to the depositional basin.

Sm–Nd isotopes indicate a narrow range of variation with ε_{Nd} value of about −4.5, a $f_{Sm/Nd}$ of about −0.37 and T_{DM} of 1.44 Ga. Pb isotopes for the Ponón Trehué Formation indicate that at least a part of the components has a "Gondwanan Pb-signature," characterized by high $^{207}Pb/^{204}Pb$ ratios. Zircon dating constrains the age of the main sources to the Mesoproterozoic, with a main peak at 1.2 Ga.

Sedimentological characteristics indicate a dominant provenance from the underlying Mesoproterozoic Cerro La Ventana Formation (Heredia 2006). The basement consists of mafic to intermediate gneisses, foliated quartz diorites, and tonalites that partially graded to amphibolites and migmatites, as well as acidic to intermediate granitoids and pegmatitic and aplitic veins (Cingolani et al. 2005; Cingolani et al., this volume). Nd data presented by Cingolani et al. (2005) of the Cerro La Ventana Formation show ε_{Nd} values in the range of variation of data calculated to the time of deposition of the Ponón Trehué Formation. Sm–Nd, Rb–Sr, and U–Pb on zircons indicate Mesoproterozoic ages (1.1–1.2 Ga; Cingolani et al. 2012), which fit detrital zircon data from the Ponón Trehué Formation.

The Ponón Trehué Formation is an olistostromic sequence deposited within a restricted basin and the sediment supply was from a local source (the Cerro La Ventana Formation). Considering that the Cerro La Ventana Formation is the most probable source area, it would have been uplifted since at least the Darriwilian in order to provide detritus to the Ponón Trehué basin.

6 Conclusions

Petrographical and geochemical analyses of the Ponón Trehué Formation indicate a main provenance component with an upper continental crust composition but with a subordinate input from a less evolved source. The unit is strongly weathered. Th/Sc ratios indicate no important recycling.

The Nd system is similar to that from the Mesoproterozoic basement of the San Rafael block (Cerro La Ventana Formation). Pb isotopes of the Ponón Trehué Formation clearly account for an important influence of a source with high Pb ratios, particularly $^{207}Pb/^{204}Pb$. Detrital zircon dating further constrains the ages of the sources as almost exclusively Mesoproterozoic (peak at 1.2 Ga), in coincidence with a dominant provenance from the Cerro La Ventana Formation.

The Ponón Trehué Formation is an olistostromic sequence deposited within a restricted basin and the sediment supply was from a local source (the Cerro La Ventana Formation).

Acknowledgements The manuscript was improved due to sedimentary provenance comments done by Gonzalo Blanco (Universidad de la República, Uruguay) and paleontological aspects done by Franco Tortello and Héctor Leanza (University of La Plata, Argentina) of earlier drafts. Field and laboratory works for this study were funded by CONICET 647, 199 grants. Mario Campaña (Museo de La Plata, Argentina) helped with figures and photographs.

References

Abre P (2007) Provenance of Ordovician to Silurian clastic rocks of the Argentinean Precordillera and its geotectonic implications. PhD thesis, University of Johannesburg, South Africa (UJ web access)

Abre P, Cingolani CA, Zimmermann U, Cairncross B, Chemale Jr F (2011) Provenance of Ordovician clastic sequences of the San Rafael Block (Central Argentina), with emphasis on the Ponón Trehué Formation. Gondwana Res 19(1), 275–290

Abre P, Cingolani CA, Manassero MJ (this volume) The Pavón Formation as the upper Ordovician unit developed in a turbidite sand-rich ramp, San Rafael Block, Mendoza, Argentina. In: Cingolani C (ed) The pre-carboniferous evolution of the San Rafael Block, Argentina. Implications in the SW Gondwana margin. Springer, Berlin

Árkai P, Sassi FP, Desmons J (2003) Very low- to low-grade metamorphic rocks. Recommendations by the IUGS Subcommission on the Systematics of Metamorphic Rocks. Web version of 01-02-07. www.bgs.ac.uk/scmr/home.html

Astini RA (2002) Los conglomerados basales del Ordovícico de Ponón Trehué (Mendoza) y su significado en la historia sedimentaria del terreno exótico de Precordillera. Revista de la Asociación Geológica Argentina 57:19–34

Baldis BA, Blasco G (1973) Trilobites ordovícicos de Ponón Trehué, Sierra Pintada de San Rafael, provincia de Mendoza. Ameghiniana 10(1):72–88

Beresi M, Heredia S (2003) Ordovician calcified cianobacteria from the Ponón Trehué Formation, Mendoza Province, Argentina. 9 International Symposium on the Ordovician System. Serie Correlación Geológica, vol 17. Instituto Superior de Correlación Geológica (INSUGEO), pp 257–262

Bhatia MR, Crook KAW (1986) Trace element characteristics of graywackes and tectonic setting discrimination of sedimentary basins. Contrib Miner Petrol 92:181–193

Bock B, McLennan SM, Hanson GN (1998) Geochemistry and provenance of the Middle Ordovician Austin Glen Member (Normanskill Formation) and the Taconian Orogeny in New England. Sedimentology 45:635–655

Casquet C, Pankhurst RJ, Fanning CM, Baldo E, Galindo C, Rapela CW, González-Casado JM, Dahlquist JA (2006) U-Pb SHRIMP zircon dating of Grenvillian metamorphism in Western Sierras Pampeanas (Argentina): correlation with the Arequipa-Antofalla craton and constraints on the extent of the Precordillera terrane. Gondwana Res 9:524–529

Chatterton BDE, Edgecombe GD, Waisfeld BG, Vaccari NE (1998) Ontogeny and systematics of Toernquistiidae (Trilobita, Proetida) from the Ordovician of the Argentine Precordillera. J Paleontol 72:273–303

Cingolani CA, Varela R (1999) The San Rafael block, Mendoza (Argentina). Rb–Sr isotopic age of basement rocks. II South American Symposium on Isotope Geology, SEGEMAR Anales, vol 24. Córdoba, pp 23–26

Cingolani CA, Llambías EJ, Basei MAS, Varela R, Chemale Jr F, Abre P (2005) Grenvillian and Famatinian-age igneous events in the San Rafael Block, Mendoza Province, Argentina: geochemical and isotopic constraints. Gondwana 12 Conference, Abstracts 102, Mendoza

Cingolani C, Uriz N, Marques J, Pimentel M (2012) The Mesoproterozoic U-Pb (LA-ICP-MS) age of the Loma Alta Gneissic rocks: basement remnant of the San Rafael Block, Cuyania Terrane, Argentina. In: Proceedings of 8° South American Symposium on Isotope Geology, CD-ROM version, Medellín, Colombia, Abstract, p 140

Cingolani CA, Varela R, Basei MAS (this volume) The Mesoproterozoic basement at the San Rafael Block, Mendoza Province (Argentina): geochemical and isotopic age constraints. In: Cingolani C (ed) The Pre-Carboniferous evolution of the San Rafael Block, Argentina. Implications in the SW Gondwana margin. Springer, Berlin

Criado Roqué P, Ibañez G (1979) Provincia geológica Sanrafaelino-pampeana. Segundo Simposio de Geología Regional Argentina, vol 1. Academia Nacional de Ciencias, Córdoba, pp 837–869

DePaolo DJ (1981) Neodymium isotopes in the Colorado Front Range and crust-mantle evolution in the Proterozoic. Nature 291:193–196

Dott RH (1964) Wacke, graywacke and matrix-what approach to immature sandstone classification. J Sediment Petrol 34:625–632

Fedo CM, Nesbitt HW, Young GM (1995) Unraveling the effects of potassium metasomatism in sedimentary rocks and paleosols, with implications for paleoweathering conditions and provenance. Geology 23(10):921–924

Floyd PA, Shail R, Leveridge BE, Franke W (1991) Geochemistry and provenance of Rheno-hercynian synorogenic sandstones: implications for tectonic environment discrimination. Geol Soc Lond (Spec Publ) 57:173–188

Gleason JD, Finney SC, Peralta SH, Gehrels GE, Marsaglia KM (2007) Zircon and whole-rock Nd–Pb isotopic provenance of Middle and Upper Ordovician siliciclastic rocks, Argentine Precordillera. Sedimentology 54:107–136

González TED, Prieto JAF (1982) Aportaciones al conocimiento del género "Saxifraga" L., sección "Dactyloides" Tausch, de la Cordillera Cantábrica. In: Anales del Jardín Botánico de Madrid, vol 39, No. 2. Real Jardín Botánico, pp. 247–272

Hemming SR, McLennan SM (2001) Pb isotope compositions of modern deep sea turbidites. Earth Planet Sci Lett 184:489–503

Heredia S (1996). El Ordovícico del Arroyo Ponón Trehué, sur de la provincia de Mendoza. XIII Congreso Geológico Argentino y III Congreso de Exploración de Hidrocarburos, Actas I, 601–605

Heredia S (2006) Revisión estratigráfica de la Formación Ponón Trehué (Ordovícico), Bloque de San Rafael, Mendoza. INSUGEO, Serie Correlación Geológica 21:59–74

Jell PA, Adrain JM (2003) Available generic names for trilobites. Memoirs Queensland Museum 48(2):331–553

Levy R, Nullo F (1975) Braquiópodos ordovícicos de Ponón Trehué, Bloque de San Rafael (Provincia de Mendoza), Argentina. Primer Congreso Argentino de Paleontología y Bioestratigrafía, Actas, vol 1. San Miguel de Tucumán, pp 23–32

McDonough MR, Ramos VA, Isachsen CE, Bowring SA, Vujovich GI (1993) Edades preliminares de circones del basamento de la Sierra de Pie de Palo, Sierras Pampeanas Occidentales de San Juan: sus implicancias para el supercontinente proterozoico de Rodinia. XII Congreso Geológico Argentino y II Congreso de Exploración de Hidrocarburos, Actas III, pp 340–342

McLennan SM (1989) Rare earth elements in sedimentary rocks: influence of provenance and sedimentary processes. Miner Soc Am Rev Mineral 21:169–200

McLennan SM, Nance WB, Taylor SR (1980) Rare earth element-thorium correlations in sedimentary rocks, and the composition of the continental crust. Geochimica et Cosmochimica Acta 44:1833–1839

McLennan SM, Taylor SR, McCulloch MT, Maynard JB (1990) Geochemical and Nd–Sr isotopic composition of deep-sea turbidites: crustal evolution and plate tectonic associations. Geochim Cosmochim Acta 54:2015–2050

McLennan SM, Hemming S, McDaniel DK, Hanson GN (1993) Geochemical approaches to sedimentation, provenance, and tectonics. In: Johnsson MJ, Basu A (eds) Processes controlling the composition of clastic sediments: Geological Society of America, Special Paper, 284, pp 21–40

McLennan SM, Taylor SR, Hemming SR (2006) Composition, differentiation, and evolution of continental crust: constraints from sedimentary rocks and heat flow. In: Brown M, Rushmer T (eds) Evolution and differentiation of the continental crust, Cambridge, 377 pp

Nesbitt HW, Young GM (1982) Early Proterozoic climates and plate motions inferred from major element chemistry of lutites. Nature 199:715–717

Nuñez E (1962) Sobre la presencia del Paleozoico inferior fosilífero en el Bloque de San Rafael. Primeras Jornadas Geológicas Argentinas, II. Buenos Aires, pp 185–189

Nuñez E (1979) Descripción geológica de la Hoja 28d, Estación Soitué, Provincia de Mendoza. Servicio Geológico Nacional, Boletín 166:1–67

Pankhurst RJ, Rapela CW, Saavedra J, Baldo E, Dahlquist J, Pascua I (1998) The Famatinian magmatic arc in the central Sierras Pampeanas: an Early to Mid-Ordovician continental arc on the Gondwana margin. In: Pankhurst RJ, Rapela CW (eds) The Proto-Andean margin of Gondwana: Geological Society of London Special Publications, 142, pp 343–368

Ramsköld L (1991) Pattern and process in the evolution of the Odontopleuridae (Trilobita). The Selenopeltinae and Ceratocephalinae. Trans R Soc Edinburgh: Earth Sci 82:143–181

Rapela CW, Pankhurst RJ, Casquet C, Baldo EG, Galindo C, Fanning CM, Dahlquist J (2010) The Western Sierras Pampeanas: protracted Grenville-age history (1330–1030 Ma) of intra-oceanic arcs, subduction-accretion at continental edge and AMCG intraplate magmatism. J S Am Earth Sci 29:105–127

Rossi de García E, Proserpio C, Nuñez E (1974) Ostrácodos ordovícicos de Ponon Trehue, Provincia de Mendoza. Ameghiniana 11:400–411. Buenos Aires

Stacey JS, Kramers JD (1975) Approximation of terrestrial lead isotope evolution via two-stage model. Earth Planet Sci Lett 26:207–221

Taylor SR, McLennan SM (1985) The continental crust. Its composition and evolution. Blackwell, London, 312 pp

Varela R, Dalla Salda L (1992) Geocronología Rb-Sr de metamorfitas y granitoides del extremo sur de la Sierra de Pie de Palo, San Juan. Revista de la Asociación Geológica Argentina 47:271–275

Varela R, López de Luchi M, Cingolani CA, Dalla Salda L (1996) Geocronología de gneises y granitoides de la Sierra de Umango, La Rioja. Implicancias tectónicas. XIII Congreso Geológico Argentino y III Congreso de Exploración de Hidrocarburos: Actas, 3. Buenos Aires, pp 519–527

Varela R, Basei MAS, González PD, Sato AM, Naipauer M, Campos Neto M, Cingolani CA, Meira VT (2011) Accretion of Grenvillian terranes to the southwestern border of the Río de la Plata craton, western Argentina. Int J Earth Sci 100:243–272

Warr LN, Rice AHN (1994) Interlaboratory standardization and calibration of clay mineral crystallinity and crystallite size data. J Metamorph Geol 12:141–152

Wichmann R (1928) Reconocimiento geológico de la región de El Nihuil, especialmente relacionado con el proyectado dique de embalse del Río Atuel. Dirección Nacional de Minería (unpublished report). Buenos Aires

Ordovician Conodont Biostratigraphy of the Ponón Trehué Formation, San Rafael Block, Mendoza, Argentina

Susana Heredia and Ana Mestre

Abstract A review of the Middle Ordovician conodont fauna of the southern sector of the Ponón Trehué Creek, Mendoza Province, Argentina, is presented in this contribution. Different genera and species of conodonts were recovered from clastic–carbonate beds from the Ponón Trehué Formation. Two conodont zones and several subzones have been proposed for these outcrops, the *Pygodus serra* Zone with the *E. robustus* and *E. lindstroemi* Subzones, and the *Pygodus anserinus* Zone. An evaluation of conodont assemblages present in the last mentioned biozone allowed us recognizing two informal subzones.

Keywords Ordovician · San Rafael Block · Mendoza · Ponón Trehué Formation · Conodonts

1 Introduction

The outcrops in the Ponón Trehué area (35° 09′S–35° 14′S and 68° 18′W–68° 20′W, extending in a N–S belt 7.5 km long and up to 2 km wide) comprise three different units (Fig. 1). One is composed entirely of the Mesoproterozoic basement. The other two are sedimentary units that include varying amounts of reworked basement rocks. The existence of these sedimentary units confuses the recognition of proper basement. A well-established Ordovician succession of the Ponón Trehué Formation (Upper Darriwilian to Lower Sandbian) is composed of granite conglomerate, sandstone, and thin-bedded fossiliferous limestone. A distinct unconformity can be traced between the Ordovician clastic sequence and the underlying basement, exposed to the east. The well-bedded carbonate–clastic sequence, in turn, is overlain to the west by an olistostromic

S. Heredia (✉) · A. Mestre
CONICET—Instituto de Investigaciones Mineras, Universidad Nacional de San Juan,
Av. Libertador Gral. San Martin 1109, 5400 San Juan, Argentina
e-mail: sheredia@unsj.edu.ar

A. Mestre
e-mail: amestre@unsj.edu.ar

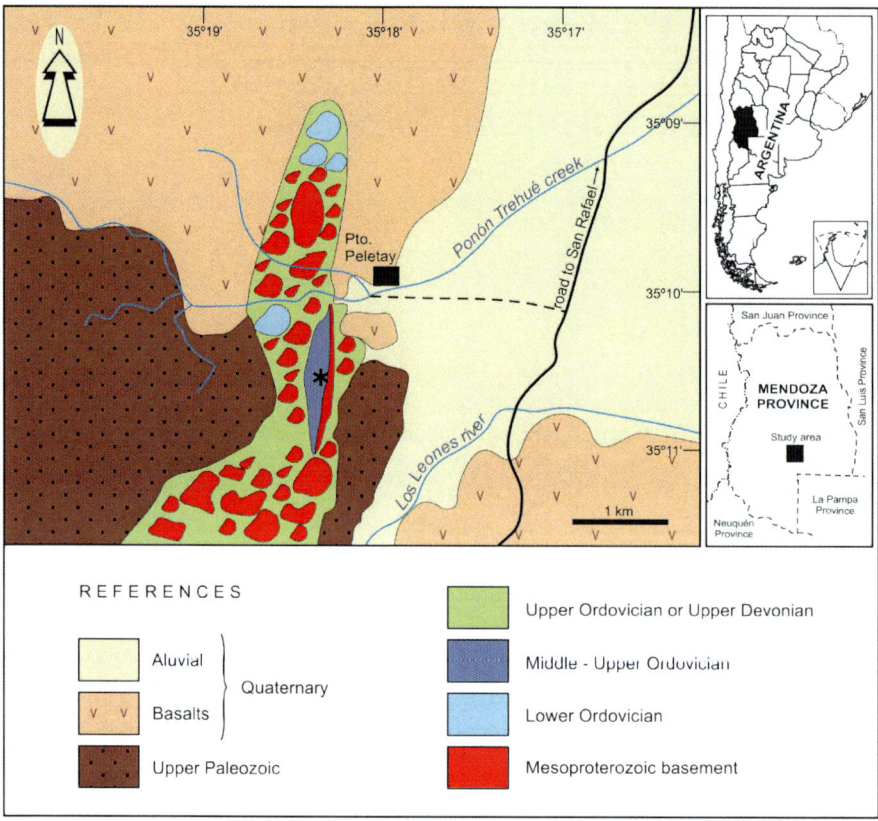

Fig. 1 Geologic sketch map of the Ponón Trehué Ordovician outcrops. The *asterisk points* the location of the La Tortuga type section

succession, which is composed of large olistoliths of basement rocks of various compositions (gneiss, amphibolite, and potassium feldspar-rich granitoid) and white limestone (Figs. 2 and 3b).

The purpose of this contribution is to provide a review of the biostratigraphic conodont information from the Ordovician rocks cropping out to the south of the Ponón Trehué creek, in the San Rafael Block, Mendoza Province, as a part of the Cuyania terrane.

2 Geological Setting

In the Ponón Trehué area, a Grenvillian granitic basement (Cingolani and Varela 1999) (Fig. 1), carboniferous sedimentary rocks, tertiary basalt rocks, and isolated Ordovician strata (Criado Roqué and Ibáñez 1979; Núñez 1979; Bordonaro et al.

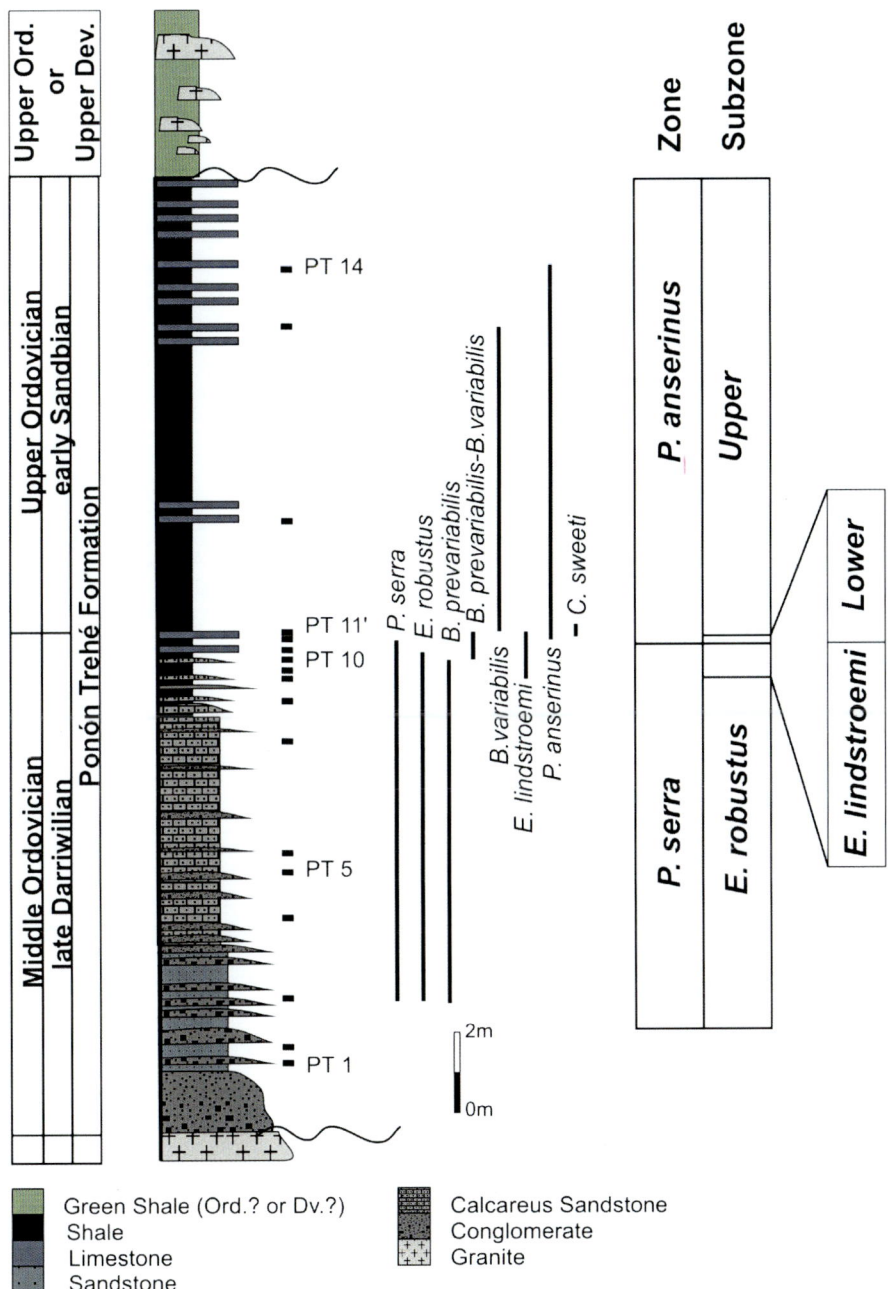

Fig. 2 Stratigraphic column in La Tortuga section with the ranges of the selected conodont species and conodont biostratigraphical scheme for the Ponón Trehué Formation

Fig. 3 Field aspects of the outcrops in the Ponón Trehué region. **a** Mesoproterozoic granite **A** and the contact with Lower Paleozoic *green shale* **B**. **b** Granite olistholiths in green shale, behind a big carbonate olistholith. **c** Stratigraphic relationship between a granite olistholith, *green shale*, and sandstone bed. **d** La Tortuga section, the basal part at left with the contact between granite and the sedimentary succession

1996) have been recognized. Ordovician strata are composed of limestone, conglomerate, sandstone, and shale deposits, and were named as Ponón Trehué Formation (Criado Roqué and Ibáñez 1979), or Lindero Formation (Núñez 1979). Bordonaro et al. (1996) and Lehnert et al. (1998) redefined these units, restricting the Ponón Trehué Formation to Tremadocian and Floian carbonate deposits cropping out northward the Ponón Trehué creek and the Lindero Formation to middle

Darriwilian clastic–carbonate beds present in a continuous strip southward the Ponón Trehué creek. Heredia (2006) proposed to retain the previous nomenclature (sensu Criado Roqué and Ibáñez 1979) for all Ordovician outcrops located both northward and southward the Ponón Trehué creek (Fig. 1). Sedimentary provenance data were presented by Abre et al. (2011). The basal contact of Ordovician deposits with the Mesoproterozoic granite is erosive in certain outcrops (Astini 2002) (Figs. 2 and 4a). These rocks are mainly siliciclastic and two members were defined (Heredia 2006). The lower one comprises a siliciclastic/carbonate succession from the upper Middle Ordovician (upper Darriwilian) to the lower Upper Ordovician (Sandbian) (Fig. 3d), while the upper member is an olistostromic deposit with Mesoproterozoic granites and Lower Ordovician carbonate olistoliths that exhibits breccia or sandstone beds in the basal part (Figs. 3b, c and 4c) (Astini 2002; Heredia and Beresi 2005). This olistostromic deposit probably correlates with the Upper Ordovician or the Upper Devonian. The former age is proposed here and it could be related to a major depositional olisthostromic event that has been identified in several sections in the Eastern and Western Precordillera. Recently, several contributions interpreted that Cambrian, Ordovician, Silurian, and lower Devonian olistholiths of different rock composition characterize the Los Sombreros Formation in Precordillera (Peralta and Heredia 2005; Peralta 2007; Peralta and Martínez 2014). Peralta (2007) proposed that this process involved a deeper erosion of the older platform margin demonstrated by the occurrence of basement and Cambrian blocks in deep-basin fine deposits. However, further studies in this sense must be carried out and relationships between these Mesoproterozoic and Ordovician outcrops in the San Rafael Block remain conjectural.

The Ponón Trehué Formation is characterized by an increasing amount of carbonate material, implying a rapid transition from predominantly shallow coarse clastic deposits to predominantly deepwater carbonate regime (Figs. 2 and 4a, b).

A fossil-rich assemblage was recovered from the Ponón Trehué Formation including trilobites, *Toernquinstia* cf. *T. chinchensis*, uncertain pygidia, and a possible pygidium of *Carolinites* sp. (Baldis and Blasco 1973), articulate and inarticulate brachiopods (Levy and Nullo 1975), sponge spicules (Beresi and Heredia 2000), crinoids, bryozoans, bivalves, ostracodes (Heredia 2001), cyanobacteria remains and *Epiphyton* sp. (Beresi and Heredia 2003), and conodonts.

The stratigraphic successions which outcrop in the Ponón Trehué area could be interpreted in terms of the following geologic history:

(1) Early to early Middle Ordovician: represented by carbonates of shallow marine platform which correlates to the San Juan Formation of the Precordillera.
(2) Early Middle Ordovician to late Middle Ordovician: this geological interval is not recorded in the succession, thus it is interpreted as time of exposure and erosion (until lower *P. serra* Zone) as a result of gentle uplift, which is thought to be regional, because there is no record of this time interval elsewhere or in the Precordillera.

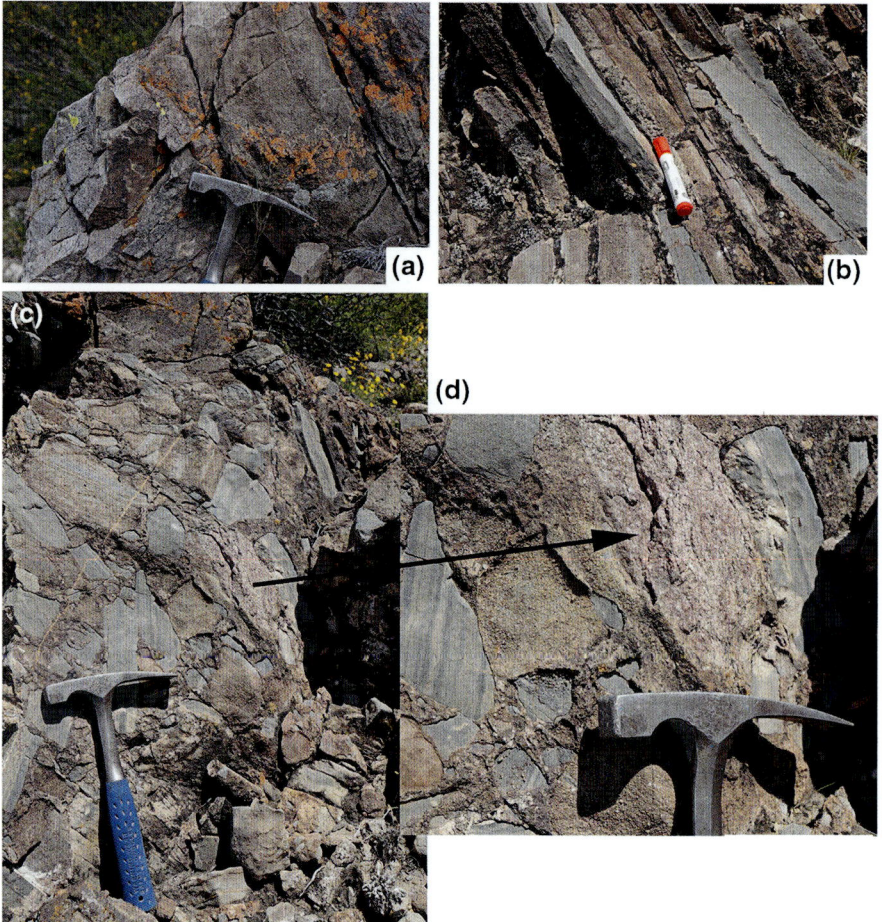

Fig. 4 Field aspects of the outcrops in the Ponón Trehué region. **a** Fifteen meters to north of La Tortuga section the outcrops clearly show the erosive contact between the Mesoproterozoic granite and the Ordovician coarse sandstone bed. **b** Typical Middle Ordovician carbonate beds of the Ponón Trehué Formation. **c, d** To the top of the Ponón Trehué Formation lies a breccia of 2 m thick in erosive contact. Note that the clasts are composed mainly of black carbonate, sandstone, and granite

(3) Middle to early Late Ordovician: gentle subsidence and extensive flooding surface, and siliciclastic deposition in gradually deepening marine environment which involved progressively more carbonates in the succession.
(4) Late Ordovician or Late Devonian: rapid increase of deepening rate (basin subsidence) to create a steep basin margin followed by collapse of carbonate and granitoid blocks into the basin. In order to create accommodation space for the thick olistostromic succession including large olistoliths and to origin a steep basin margin necessary to generate gravity slide processes, basin

subsidence must have been rapid and tectonically driven. This interpretation also provides a new answer for all these Ordovician deposits suggesting that they probably should be considered as olistholiths as well as partially the Mesoproterozoic basement exposed in the area (Heredia and Beresi 2005).

3 Methods and Materials

The La Tortuga Section was proposed as a reference section for the Ponón Trehué area (Heredia and Rosales 2006) (Fig. 1). It occurs in an intermediate sector of the outcrop strip, where conglomerates, sandstone, and limestone beds appear. Conodont sampling was restricted to carbonate layers (Fig. 2) which were subjected to the standard procedure (Stone 1987). The insoluble residue was recovered with sieve 100 and 200 (IRAM) and conodont elements were separated using a binocular microscope. The preservation is good and conodont elements frequently occur complete, except for those of large size or with long processes. Reworked simple-cone elements are usually broken.

The conodont collection is housed at the Museum of Paleontology (CORD-MP) of the National University of Córdoba, Argentina.

4 Conodont Biostratigraphy

The biostratigraphy of these Ordovician outcrops has been based on conodont assemblages only. In the northern sector Bordonaro et al. (1996) identified conodonts from the *P. striatus/C.quadraplicatus, P. proteus/A. deltatus, Oepikodus comunis/Prioniodus elegans,* and *Oepikodus evae* Zones, indicating Tremadocian and Floian ages for those sampled levels. Reviewing of this collection is not developed for this study since those conodonts were not illustrated, thus observations and comparisons cannot be done.

On the southern sector of the Ordovician outcrops Heredia (1982) identified the *Pygodus anserinus* Zone and Bordonaro et al. (1996) and Heredia (1996) recognized the *Pygodus serra* Zone. Leslie and Lehnert (1999) and Lehnert et al. (1999) pointed out the presence of *Cahabagnathus sweeti* (Bergström) and Heredia (2006) attributed a lowest Sandbian age for the top of the Ponón Trehué Formation. Several contributions dealing with the conodont fauna, biostratigraphy, and stratigraphy were carried out by Heredia (2001, 2003, 2006) and Heredia and Beresi (2000, 2005). The conodont species of the Ponón Trehué Formation (Fig. 2) are well known and widely described in the literature and allow linking this stratigraphical interval to the Baltic or North Atlantic scheme (Bergström 1971, 1990).

Pygodus serra Zone (Fig. 2, PT 3 to PT 10 sampled beds).

Eoplacognathus robustus Subzone. The PT 3 to PT 8 beds interval (Fig. 2) exhibits the species *Eoplacognathus robustus* in different evolutive stages. These beds contain *Pygodus serra* (Hadding) (Fig. 5r), *E. robustus* Bergström (Fig. 5a–e), *Baltoniodus prevariabilis* (Fåhræus) (Fig. 5p), *Periodon aculeatus* Hadding, *Ansella sinuosa* Stouge, *Ansella biserrata* Lehnert et Bergström, *Pseudooneotodus mitratus* (Moskalenko), *Spinodus spinatus* (Hadding), *Phragmodus polonicus* Dzik, *Strachanognathus parvus* Rhodes, *D. reclinatus* (Lindström), *Drepanoistodus* cf. *suberectus*, *Erismodus* sp, *Erraticodon* sp., *Panderodus* cf. *sulcatus*, *Protopanderodus rectus* (Lindström), and *Costiconus ethingtoni* (Fåhræus).

Eoplacognathus lindstroemi Subzone. Early forms of *E. lindstroemi* prevail in the conodont collection in PT 9 a PT 10' beds (Fig. 2). *Pygodus serra*, *E. robustus–E. lindstroemi* transition (Fig. 5k–m), *E. lindstroemi*, *Baltoniodus prevariabilis-variabilis* (sensu Dzik 1994), *A. sinuosa*, *A. biserrata*, *C. ethingtoni*, *S. parvus*, *P. aculeatus*, *Erraticodon* sp., *D. reclinatus*, *Phragmodus*? sp. and *Panderodus* sp. compose the conodont association.

Pygodus anserinus Zone (Fig. 2, PT 11–PT 15).

Lower Subzone. The occurrence of *P. anserinus* Lamont and Lindström (Fig. 5s) indicates the beginning of the homonymous biozone (PT 11), coexisting with few elements of *P. serra*. This bed (PT 11) contains *Pygodus anserinus*, *Baltoniodus prevariabilis-variabilis*, *Periodon aculeatus* Hadding, *Strachanognathus parvus* Rhodes, and *Pygodus serra*. This conodont association suggests the lower part of the *P. anserinus* Zone.

Upper Subzone. The occurence of *P. anserinus* along with late forms of *E. lindstroemi* (Fig. 5f–j), *Cahabagnathus sweeti* (Bergström) (Fig. 5n–o), and *Baltoniodus variabilis* (Bergström) (Fig. 5q) allows referring to the upper part of the *P. anserinus* Zone (bed PT 11') to the uppermost part of the section, following the Baltic biostratigraphic scheme (Bergström 1983; Lehnert et al. 1999).

Bergström (1971) proposed two informal subzones, named as lower and upper for *P. anserinus* Zone. The lower Subzone is indicated by the appearance of *P. anserinus* and *B. prevariabilis* and the upper Subzone by *P. anserinus*, *B. variabilis,* and *C. sweeti*. Later, Dzik (1976) suggested two "key" conodonts for defining these subzones, "*Amorphognathus*" *kielcensis* (Dzik) and "*A.*" *inaequalis* (Rhodes), criterion followed by Bergström (1983) and Bergström et al. (1987). However, Dzik (1994) reevaluated the generic assignation of these species and interpreted them as *Sagittodontina kielcencis* with long vertical distribution and *Rhodesognathus inaequalis* which appears exclusively in the *Amorphoganthus tvaerensis* Zone. This interpretation allowed rethinking these both conodont species as key conodonts for these biostratigraphic records.

Lehnert et al. (1999) described the conodont assemblages that represent the *Eoplacognathus reclinathus* and *E. lindstroemi* Subzones of the *P. serra* Zone and the *A. inequalis* Subzone of the *P. anserinus* Zone. These authors also proposed new conodont species from this outcrops, named *Amorphognathus sanrafaelensis* Lehnert and Bergström. A critical review of this conodont apparatus (in Lehnert et al. 1999; Pl.1, Figs. 1–3, 5–6, 8–9) allowed us identifying P elements of *Sagittodontina* sp. (in Lehnert et al. Pl.1, Figs. 1, 3, 6) and P elements of

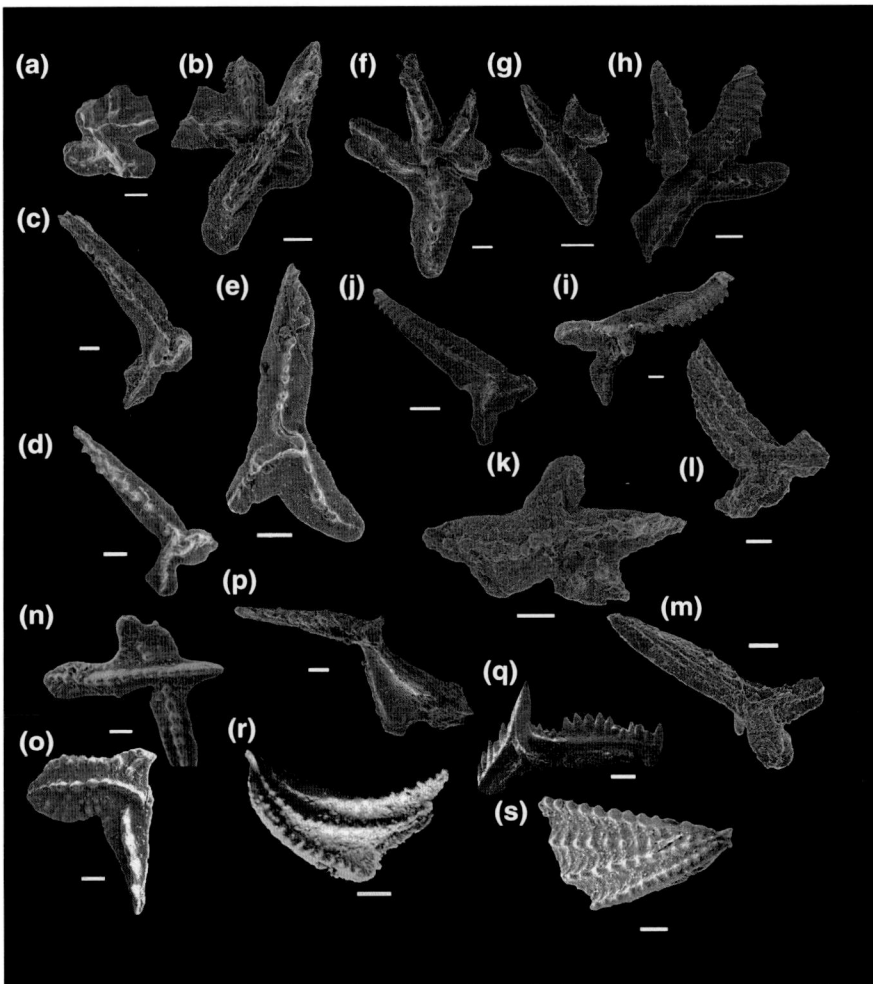

Fig. 5 Index conodonts from La Tortuga section, Upper Darriwilian, Ponón Trehué Formation (San Rafael Block, Mendoza). Scanning electron microscope photomicrographs, scale bar 0.1 mm. **a–e** *Eoplacognathus robustus* Bergström (*E. robustus* Subzone). **a** Dextral Pa element, CORD MP 2228(1), PT 8. **b** Sinistral Pa element, CORD MP 2225 (1), PT 9. **c** Dextral Pb element, CORD MP 2229(4) PT 9. **d** Dextral Pb element, CORD MP 2222(2) PT 8. **e** Sinistral Pb element, CORD MP 2223(5), PT 9. **f–j** *Eoplacognathus lindstroemi* Bergström (*E. lindstroemi* Subzone) all the elements from PT 11' sampled bed. **f–g** Dextral Pa elements, CORD MP 2364(1–2). **h** Sinistral Pa element, CORD MP 2364(3). **i** Sinistral Pb element CORD MP 2361(2). **j** dextral Pb element, CORD MP 2362(2). **k–m** *E. robustus–E. lindstroemi* transition, all the elements from PT 10 sampled bed. **k** Dextral Pa, CORD MP 2269(2). **l** Dextral Pb element, CORD MP 2270(1). **m** Sinistral Pb element CORD MP 2270(2). **n–o** *Cahabagnathus sweeti* (Bergström), dextral Pa and Pb elements, CORD MP 2363(3-2), PT 11'. **p** *Baltoniodus prevariabilis* Bergström, Pa element, CORD MP 2357(1), PT 9. **q** *Baltoniodus variabilis* Bergström, Pa element, CORD MP 2356(43), PT 11. **r** *Pygodus serra* (Hadding) (*P. serra* Zone), CORD MP 2236(1) PT 8. **s** *Pygodus anserinus* Lamont et Lindström (*P. anserinus* Zone), CORD MP 2359(4), PT 11

Complexodus sp. (in Lehnert et al., Pl.1, Figs. 2, 5, 8, 9). Those conodonts described as Pa elements of *E. reclinatus* (in Lehnert et al. Pl.1, Figs. 10, 14) are here interpreted as *E. robustus* and those Pb elements illustrated in the Pl.1 (Figs. 13 and 17, in Lehnert et al.) are considered in this study as reworked Pb elements of *Eoplacognathus suecicus* Bergström. The *E. lindstroemi* illustrated in the Pl.2 (Figs. 1–4, in Lehnert et al.) are reviewed as Pa and Pb elements of *E. robustus*.

Heredia (2003) described reworked conodonts in the La Tortuga section and noted that this association is abundant in the lower levels of the Ponón Trehué Formation (PT 4 to PT 5, Fig. 2). These reworked conodonts have been identified (Heredia 2001) as *Eoplacognathus pseudoplanus* (Viira), *Eoplacognathus suecicus* Bergström, *Eoplacognathus foliaceus* (Fåhræus), *Eoplacognathus reclinatus* (Fåhræus), *Histiodella* sp., among others.

5 Remarks

The Ponón Trehué Formation is of special biostratigraphic interest since it records the boundary between the Middle-Upper Ordovician Series. Despite the absence of graptolite record in this locality, we propose the presence of *Pygodus anserinus* along with late forms of *E. lindstroemi*, *Cahabagnathus sweeti* (Bergström), and *Baltoniodus variabilis* (Bergström) as the key conodonts from the upper part of *P. anserinus* Zone, allowing drawing the base of the Upper Ordovician Series.

Acknowledgements We are indebted to CONICET for supporting the study in the Ponón Trehué area (PIP308). Particular thanks are due to Dr. Matilde Beresi for years of good discussions about the geology of this locality. The MEByM, CCT-Mendoza are thanked for the microphotographs.

References

Abre P, Cingolani CA, Zimmermann U, Cairncross B, Chemale Jr F (2011) Provenance of ordovician clastic sequences of the San Rafael block (central Argentina), with emphasis on the Ponón Trehué Formation. Gondwana Res 19(1):275–290

Astini R (2002) Los conglomerados basales del Ordovícico de Ponón Trehué (Mendoza) y su significado en la historia sedimentaria del terreno exótico de Precordillera. Rev Asoc Geol Argentina 57(1):19–34. Buenos Aires

Baldis B, Blasco G (1973) Trilobites ordovícicos de Ponón Trehué, Sierra Pintada de San Rafael, provincia de Mendoza. Asoc Paleontol Argentina Ameghiniana 10(1):72–88

Bergström S (1971) Conodont biostratigraphy of the middle and upper ordovician of Europe and Eastern North America. In: Sweet WC, Bergström S (eds) Symposium on conodont biostratigraphy. Geological society of America memoir, vol 127, pp 83–161

Bergström S (1983) Biogeography, evolutionary relationships and biostratigraphic significance of Ordovician platform conodonts. Fossils Strata 15:35–58. Oslo

Bergström S (1990) Relations between conodont provincialism and changing palaeogeography during the early Palaeozoic. In: McKerrow WS, Scotese CR (eds) Palaeozoic palaeogeography and biogeography. Geological society of London memoir, vol 12, pp 105–121

Bergström SM, Rhodes FHT, Lindström M (1987) Conodont biostratigraphy of the Llanvirn-Llandeilo and Llandeilo-Caradoc series boundaries in the Ordovician system of Wales and the Welsh Borderland. In: Austin RL (ed). Conodonts: investigative techniques and applications, vol 18. Ellis Horwood Limited, Chichester, pp 294–315

Beresi M, Heredia S (2000) Sponge spicule assemblages from the middle Ordovician of Ponón Trehué, Southern Mendoza, Argentina. Rev Esp Paleontol 15(1):37–48. Oviedo

Beresi M, Heredia S (2003) Ordovician calcified cyanobacteria from the Ponón Trehué Formation, Mendoza province, Argentina. Ser Correl Geol 17:257–262

Bordonaro O, Keller M, Lehnert O (1996) El Ordovícico de Ponón Trehué en la Provincia de Mendoza (Argentina): Redefiniciones estratigráficas. In: XIII Congreso Geológico Argentino y III Congreso de Exploración de Hidrocarburos, Actas I, pp 541–550

Cingolani CA, Varela R (1999) The San Rafael Block, Mendoza (Argentina). Rb-Sr isotopic age of basement rocks. II SSAGI, Anales 34 SEGEMAR, Actas, pp 23–26. Carlos Paz, Córdoba

Criado Roqué P, Ibáñez G (1979) Provincia geológica Sanrafaelino-pampeana. In: Academia Nacional de Ciencias. Segundo Simposio de Geología Regional Argentina, vol 1, pp 837–869

Dzik J (1976) Remarks on the evolution of Ordovician conodonts. Acta Palaeontol Pol 21:395–455

Dzik J (1994) Conodonts from the Mójcza Limestone. In: Dzik J, Olempska E, Pisera A (eds) Ordovician carbonate platform ecosystem of the holy cross mountains. Palaeontol Pol 53:43–128

Heredia S (1982) Pygodus anserinus Lamont et Lindström (Conodonto) en el Llandeillano de la formación Ponón Trehué. Ameghiniana 19(3–4):101–104

Heredia S (1996) El Ordovícico del Arroyo Ponón Trehué, sur de la provincia de Mendoza. In: XIII Congreso Geológico Argentino y III Congreso de Exploración de Hidrocarburos. Actas I, pp 601–605

Heredia S (2001) Late Llanvirn conodonts from Ponón Treuhué formation, Mendoza, Argentina. GAIA Rev Geocienc 16:101–117

Heredia S (2003) Upper Llanvirn—Lower Caradoc conodont biostratigraphy, Southern Mendoza, Argentina. In: Aceñolaza FG (ed) Aspects of the ordovician system in Argentina. Serie Correlación Geológica, vol 16, pp 167–176. Instituto Superior de Correlación Geológica (INSUGEO)

Heredia S (2006) Revisión estratigráfica de la Formación Ponón Trehué (Ordovícico), Bloque de San Rafael, Mendoza. In: Temas de la Geología Argentina I. Serie Correlación Geológica, vol 21, pp 59–74

Heredia S, Beresi M (2000) Conodont biostratigraphy and paleoenvironment of the Late Llanvirn—Lower Caradoc (Ordovician), south of the province of Mendoza, Argentina: 17th Geoscientific Latinamerica Colloquium. Institut für Geologie und Paläontologie. Universität Stuttgart. Profil, Band 18, p 48

Heredia S, Beresi M (2005) New insights on the Cerro La Ventana Formation (Mesoproterozoic), the crustal basement at the San Rafael Block, Mendoza province (Argentina). 12 Gondwana Conference, Abstracts: 198. Mendoza, Argentina

Heredia S, Rosales C (2006) Biofacies de Conodontes de la Formación Ponón Trehué y la importancia bioestratigráfica como sección tipo para el límite del Ordovícico Medio-Ordovícico Superior de Cuyania (Argentina). In: Temas de la Geología Argentina I. Serie Correlación Geológica, vol 21, pp 7–16

Lehnert O, Bergström S, Keller M, Bordonaro O (1999) Middle Ordovician (Darriwilian-Caradocian) conodonts from the San Rafael region, west-central Argentina: Biostratigraphic, paleoecologic and paleogeographic implications. Boll Soc Paleontol Ital 37 (2–3):199–214. Modena

Lehnert O, Keller M, Bordonaro O (1998) Early Ordovician conodonts from the southern Cuyania Terrane (Mendoza Province, Argentina). In: Szaniawski H (ed) Proceedings of the sixth European conodonts symposium, (ECOS VI). Palaeontologia Polonica, vol 58, pp 47–65

Leslie S, Lehnert O (1999) New insight into the phylogeny and paleogeography of Cahabagnathus (Conodonta). Acta Univ Carol Geol 1999, 43:443–446

Levy R, Nullo F (1975) Braquiópodos Ordovícicos de Ponón Trehué. Bloque de San Rafael (Provincia de Mendoza, Argentina). 1° Congreso Argentino de Paleontología y Bioestratigrafía. Actas, vol 1, pp 23–32. Tucumán

Núñez E (1979) Descripción Geológica de la Hoja 28d, Estación Soitué. Provincia de Mendoza. Ser Geol Nac Bol 166:1–67

Peralta SH (2007) Extensional history of the Devonian Basin of Precordillera: its tecto sedimentary significance in the evolution of the Cuyania Terrane. Devonian Land-Sea Interaction: Evolution of ecocsytems and Climates (DEVEC): 102–105. Meeting of the IGCP 409 Unesco

Peralta S, Heredia S (2005) Depósitos de olistostroma del Devónico (inferior?-medio?), Formación Los Sombreros, en la Quebrada de San Isidro, Precordillera de Mendoza, Argentina. XVI Congreso Geológico Argentino. IV, pp 621–626. La Plata

Peralta SH, Martínez M (2014) Silúrico y Devónico de la Quebrada de Las Vegas, Sierra de La Tranca, Precordillera Central de San Juan. In: Martino R, Lira R, Guereschi A, Baldo E, Franzese J, Krohling D, Manassero M, Ortega G, Pinotti L (eds) Actas del XIX Congreso Geológico Argentino T1.39. Córdoba, Argentina

Stone J (1987) Review of investigative techniques used in the study of conodonts. In: Austin R (ed) Conodonts: investigative techniques and applications. Ellis Horwood Limited, Chichester, pp 17–34

The Pavón Formation as the Upper Ordovician Unit Developed in a Turbidite Sand-Rich Ramp. San Rafael Block, Mendoza, Argentina

Paulina Abre, Carlos A. Cingolani and Marcelo J. Manassero

Abstract The Pavón Formation crops out in the central-east region of the San Rafael block; it is composed of massive green-reddish-grey sandstones, wackes, quartz sandstones, siltstones and shales. The sand-dominated facies show tabular bedding with sharp contacts and scarce syndepositional deformational structures, and were deposited in a turbidite sand-rich ramp. The graptolite fauna, in particular the presence of *Climacograptus bicornis* Biozone indicates a Sandbian age (Upper Ordovician). Sandstone petrography records a stable craton or a faulted continental basement as probable source areas. Illite crystallinity index suggest anchimetamorphic conditions underwent by the sequence. Relatively high CIA values and K/Cs ratios indicate intermediate to advanced weathering. Geochemical provenance proxies display detrital compositions derived from an average upper continental crust, but relatively high abundances of compatible elements (Cr, Ni, V, Ti and Sc) along with low Th/Sc ratios suggest mixing with a less fractionated component. The Zr/Sc ratios indicate that recycling was not important. An ophiolitic source can be neglected based on Y/Ni and Cr/V ratios. Chemical analyses of detrital chromian spinels indicate that they were formed within mid-ocean ridge basalts and conti-

P. Abre (✉)
Centro Universitario de la Región Este, Universidad de la República (UdelaR),
Ruta 8 Km 282, Treinta y Tres, Uruguay
e-mail: paulinabre@yahoo.com.ar

C.A. Cingolani · M.J. Manassero
Centro de Investigaciones Geológicas, CONICET-UNLP,
Diag. 113 n. 275, CP1904 La Plata, Argentina
e-mail: carloscingolani@yahoo.com

M.J. Manassero
e-mail: mj.manassero@gmail.com

C.A. Cingolani
Centro de Investigaciones Geológicas (CONICET-UNLP) and División Geología,
Museo de La Plata, UNLP, Paseo del Bosque s/n, B1900FWA La Plata, Argentina

nental flood basalts. Nd model ages of Pavón sandstones scatter around 1.4 Ga, $\varepsilon_{Nd}(t)$ (t = 455 Ma) values range from −0.4 to −4.1 and $f_{Sm/Nd}$ is of −0.40 ± 0.06 on average, indicating an affinity to Grenvillian-age crust. Detrital zircon grains dated confirmed a main Mesoproterozoic source (with peaks at 1.1 and 1.4 Ga), with subordinated inputs from Neoproterozoic and Paleoproterozoic crystalline rocks. The complete provenance dataset suggest the basement of the San Rafael block (Cerro La Ventana Formation) as the main source of detritus, but derivation from the Western Pampeanas Ranges was also probable. The siliciclastic sequence was deposited in a foreland basin at latitude of around 26°S, and linked to the accretion of the Cuyania terrane towards west of Gondwana; this accretion caused uplift by thrusting of the Mesoproterozoic crust to the east at *ca.* 460 Ma.

Keywords Geochemistry · Isotope geochemistry · Detrital zircon dating · Provenance · Paleomagnetism · Ordovician · Cuyania terrane

1 Introduction and Geological Setting

The Pavón Formation (Holmberg 1948 *emend.* Cuerda and Cingolani 1998) crops out at the eastern slope of the Cerro Bola hill, in the central region of the San Rafael Block, (Mendoza province, central Argentina; Fig. 1a–c). It is a sandy marine turbidite 700 m thick siliciclastic unit trending NW–SE for 3.5 km and with a maximum width of 1.2 km. It is intruded by rhyolites of Permian–Triassic age and partially covered by Permian volcaniclastic rocks.

The sequence is composed of an alternate green, reddish-grey massive arenites (either wackes or quartz-feldspathic arenites) and siltstones and minor black shales. Within black shales and siltstones a rich graptolite fauna was found and date the unit as Sandbian (Early Caradoc; Cuerda and Cingolani 1998). Sedimentary tractive structures are absent (Manassero et al. 1999). The sequence is gently folded forming a large anticline with an axe plunging 15° towards north; the eastern homoclinal flank has a strike of N170°/30–50°E. The central part of the homoclinal is folded and faulted (Cingolani et al. 1999). The western flank is intruded by a rhyolitic laccolith known as Cerro Bola hill (Fig. 2).

The illite crystallinity values measured by X-Ray diffraction in shales using international standards suggest anchimetamorphic grade conditions, confirmed by cleavage development, deformation and siliceous recrystallization (Cuerda and Cingolani 1998). Dykes and bedded intrusives of rhyolitic composition (2–3 m thick) as well as thin hydrothermal quartz veins are common (Manassero et al. 1999). It is not in contact with the Ponón Trehué Formation (Darriwilian to Sandbian; Abre et al. this volume), and the base of the unit is not exposed.

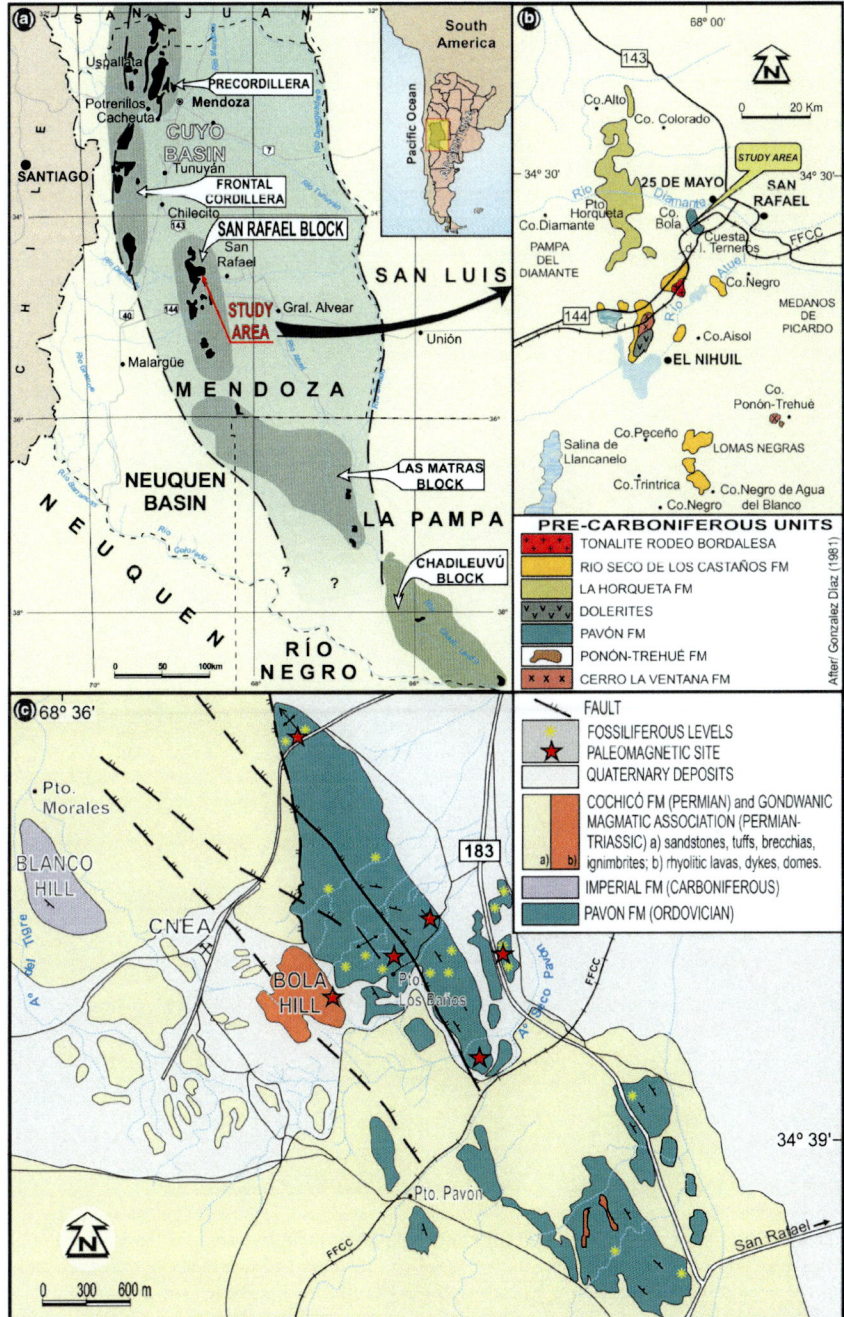

Fig. 1 a Location of the San Rafael Block in central-western Mendoza Province, western Argentina. b Sketch map showing outcrops of pre-Carboniferous rocks within the San Rafael block after González Díaz (1981). c The Pavón Formation develops on the eastern slope of the Cerro Bola; the studied geological sections as well as the paleomagnetic sampling sites are denoted. Modified from Cingolani et al. (2003)

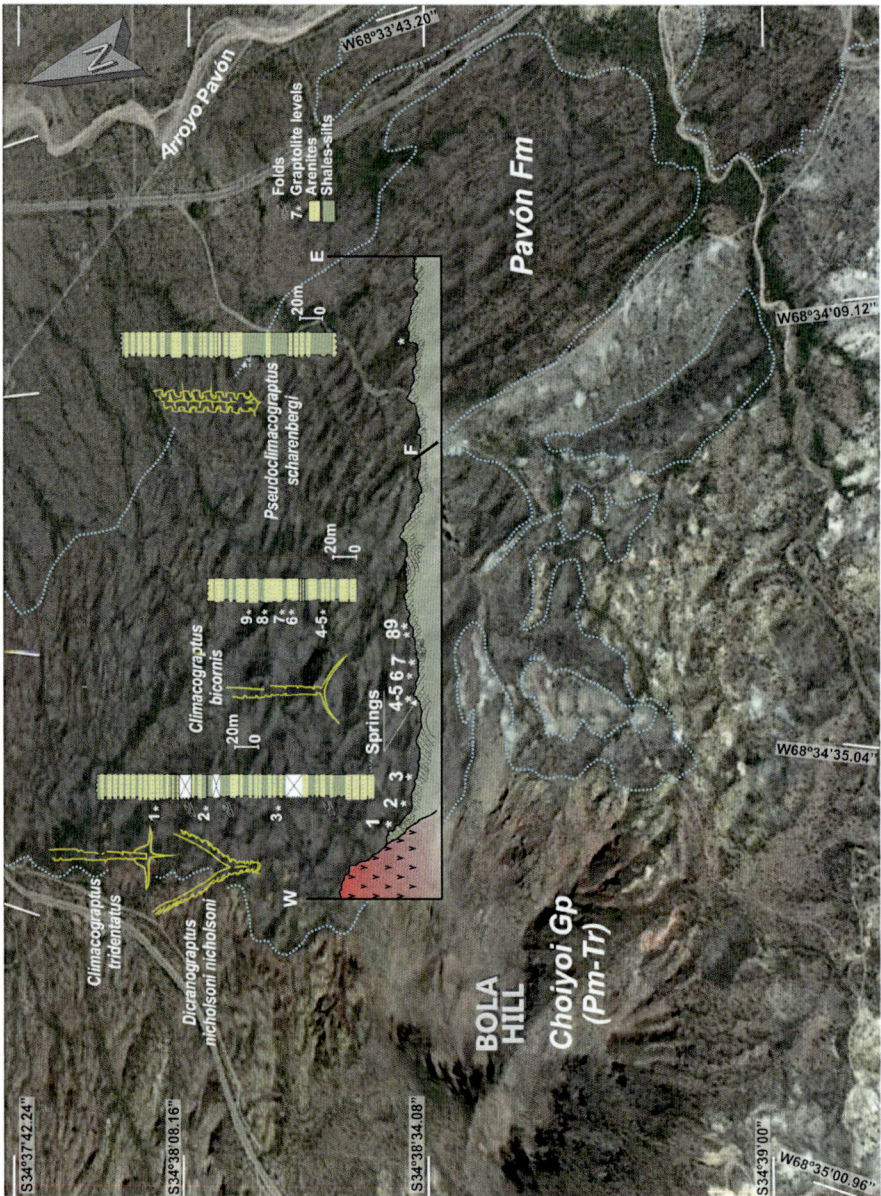

Fig. 2 Satellite image of the Bola Hill region showing the main outcrops of the Pavón Fm. The W–E section along the Baños (Springs) creek showing the Gondwanian tectonic vergence and the stratigraphic columns are based on Cuerda and Cingolani (1998). Some relevant graptolites are represented

Detailed stratigraphical, petrographical, biostratigraphical, paleomagnetic and provenance studies (including geochemistry, heavy minerals, Sm–Nd and U–Pb detrital zircon dating) were carried out on this unit, to gain insights on the tectonic evolution of the proto-Andean Gondwana margin during the Ordovician. A review of all the information available is here presented.

2 Stratigraphy and Paleontological Record

The unit is characterized by alternate arenites and pelites in tabular strata, laterally continuous, with sharp contacts (Fig. 3a–d). The arenites are mainly moderately sorted wackes although low matrix arenites are also present, indicating association of turbidite and granular flows; substratal sedimentary structures such as flow casts, load casts and ripples are common, as well as lamination and current-ripples. All these characteristics point to turbidite deposits within a sand-rich ramp, with predominance of sandy-facies proximal regarding system feeding source (Manassero et al. 1999). The deposition of the unit occurred within a progradational system, showing rather vertical than lateral facial changes indicating sedimentary transport through a linear trough. The coarser grain size recorded is very coarse arenite. In a broad sense, it is a coarsening and thickening upwards sequence; the arenite levels

Fig. 3 a General view of outcrops on the northern section of the Pavón Fm. Coarsening and thickening upwards trends with beds reaching up to 1 m towards the top can be observed. **b** and **c** Show the folding of the unit with east vergence as a consequence of the 'Chanic tectonic phase'. **d** Substratal structures are a common feature

are 0.2–2 m thick, being the commonest the less than 0.5 m thick levels, but strata showing thickness of up to 12 m were also reported (Manassero et al. 1999).

Paleocurrents indicate towards west (N240°–310°) depositional direction. According to Manassero et al. (1999) five lithofacies can be identified: black shales with graptolites, finely stratified arenites and pelites, green siltstones, medium stratified arenites and coarse stratified arenites.

Graptolites found within black shales were mentioned by Marquat and Menéndez (1985), and then described by Cuerda and Cingolani (1998) and Cuerda et al. (1998). They are scarce and poorly preserved and rhabdosomes show evidence of deformation. The 25 different taxa belong to the families of Glossograptidae, Nemagraptidae, Dicranograptidae, Diplograptidae, Orthograptidae, Lasiograptidae and Retiolitidae. Particularly important regarding age determination is the presence of *Climacograptus bicornis*, *Climacograptus tridentatus*, since they point to Sandbian age (Fig. 2). From base to top, the graptolites are arranged in three assemblages based on which the unit was correlated to Empozada, Portezuelo del Tontal, Sierra de la Invernada, Las Plantas and La Cantera Formations of the Precordillera *s.s.* (Cuyania terrane), as well as to the Lagunitas Formation in the eastern flank of the Frontal Cordillera (Tickyj et al. 2009).

3 Petrography

A total of 43 samples of sandstones and siltstones were analyzed. Under the microscope arenites show grain sizes ranging from very fine to medium, are moderately to poorly selected and show subangular clasts with low sphericity (Manassero et al. 1999; Cingolani et al. 2003). Main mineralogical constituents are: monocrystalline (dominant) and polycrystalline quartz. K-feldspar (microcline and orthose) are commonly altered to sericite, chlorite, vermicular kaolinite and other clay-minerals. Plagioclases are scarce. Lithoclasts recorded were derived from low grade metamorphic rocks, as well as from sedimentary (chert, chalcedony and pelites) and igneous rocks (Manassero et al. 1999; Cingolani et al. 2003). Following Dott (1964), samples are classified as feldspathic and quartz wackes, with matrix content ranging from 15 to 30%. Their characteristics as well as the determination of tectonic setting using discrimination ternary diagrams based on petrographical data allowed Manassero et al. (1999) to determine a provenance from crustal plutonic and metamorphic rocks, more probably from a stable craton or a faulted continental basement. Illite crystallinity index indicate anchimetamorphic conditions for the finest rocks of the sequence, which are in turn composed mainly of illite and chlorite, with subordinated smectite and kaolinite (Manassero et al. 1999).

4 Heavy Minerals

Analysis of detrital heavy minerals is a reliable tool to constrain the composition of the sources, particularly if petrography, geochemistry and Nd isotopes had shown source mixing (Mange and Maurer 1992), as it is the case for the Pavón Formation (Cingolani et al. 2003). The detrital heavy mineral assemblages found in five wacke samples are predominantly composed of zircon, spinel and rutile in order of abundance; less common are apatite and sphene (Abre et al. 2003, 2005; Abre 2007). Spinel grains contained in sedimentary rocks can provide important information regarding source areas closely related to the depositional site, via geochemical analysis of individual grains (Dick and Bullen 1984; Barnes and Roeder 2001; Kamenetsky et al. 2001).

The spinels occur in fine grained, poor to moderately sorted wackes of the Pavón Formation. The Cr concentration of the unit varies between 65 and 292 ppm and has an average of 139 ppm (Cingolani et al. 2003), which represents an enrichment compared with the average Cr concentration of wackes (Cr = 67 ppm) and shales (Cr = 88 ppm) according to Taylor and McLennan (1985).

Based on chemical composition of the separated chromian spinels two groups of grains were identified (Fig. 4). These groups are morphologically indistinguishable, showing sizes ranging from 40 to 110 µm with an average of 60 µm for both, and the proportion of subhedral, anhedral and euhedral grains are also equal in both groups. They do not show any visible zonation, intergrowth or inclusions and are black in colour. No compositional variations were determined, as there is no

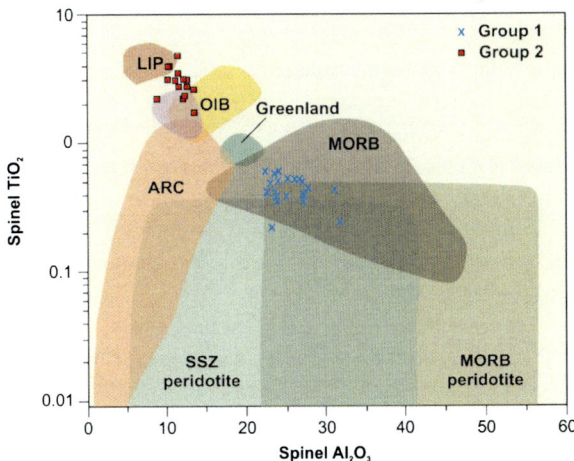

Fig. 4 Relationship between spinel contents of TiO_2 versus Al_2O_3 showing Pavón Formation samples plotting within the MORB field (Group 1) or clustering between the LIP and OIB fields (Group 2). The *purple area* represents data from the Don Braulio Formation (from Abre 2007). *SSZ* supra-subduction zone; *OIB* ocean island basalts; *LIP* large igneous provinces; *MORB* mid-ocean ridge basalt. Modified from Abre et al. (2009)

difference between core and rim, neither intergrowths nor exsolution of other mineral species (Abre 2007; Abre et al. 2009).

Group 1 (60% of total grains) has a range in Cr# = Cr/(Cr + Al) between 0.47 and 0.58, a Mg# = Mg/(Mg + Fe^{2+}) atomic ratio ranging from 0.50 to 0.80, TiO_2 concentrations of less than 0.60% and Fe^{3+}# ratio (Fe^{3+}# = Fe^{3+}/(Cr + Al + Fe^{3+})) ranging from 0.04 to 0.11, while MnO, V_2O_5, NiO and ZnO concentrations are very low. Group 2 (40% of the spinel grains) has a range in Cr# between 0.69 and 0.73, a Mg# ranging from 0.21 to 0.50, TiO_2 concentrations are very high (between 1.7 and 4.7%) and Fe^{3+}# ratios are between 0.09 and 0.19; MnO, V_2O_5, NiO and ZnO concentrations are also very low (Abre 2007; Abre et al. 2009).

Using several tectonic setting discriminatory diagrams based on Fe^{3+}#, Fe^{2+}#, TiO_2, Cr, Al_2O_3 and their ratios (Barnes and Roeder 2001; Kamenetsky et al. 2001) deduced that Group 1 is related to ocean floor basalts from a mid-ocean ridge, while Group 2 was derived from flood basalts in an oceanic or continental intraplate setting (Fig. 4; Abre 2007; Abre et al. 2009). Unfortunately, the source rocks of these spinels were yet not identified, but mafic rocks from the Western Precordillera belt, particularly its extension within the San Rafael Block (namely the Mesoproterozoic section of the El Nihuil Mafic Unit; Cingolani et al. this volume) remain as the more probable sources. Chromian spinels from the Don Braulio Formation (Precordillera *s.s.*; Cuyania terrane) had an origin in ocean island basalts, being therefore different to those comprised in the Pavón Formation (Fig. 4; Abre et al. 2006; Abre 2007).

5 Whole-Rock Geochemistry

Differential concentration of elements rather in mafic (Sc, Cr, Co) than felsic (La, Th) rocks, REE (rare earth elements) patterns and the character of the Eu-anomaly have been used for provenance and tectonic determinations (Taylor and McLennan 1985). The use of geochemical analyses for provenance determination could be particularly useful in the case of wackes. However, since the signature of the source rock may be modified by weathering, hydraulic sorting and diagenesis, it is then necessary to determine the effects that these factors had on the geochemical composition of sedimentary rocks (Nesbitt and Young 1982; Cox et al. 1995). Geochemical analyses of the Pavón Formation were presented by Cingolani et al. (2003).

Samples from the Pavón Formation have SiO_2 concentrations ranging from 61.1 to 91.7%, Al_2O_3 is between 4.2 and 12.7%, Fe_2O_3 ranges from 0.4 to 9%, CaO and Na_2O are present in low concentrations (0.42 and 0.74% on average, respectively), whereas K_2O is between 0.84 and 3.72%.

Weathering: CIA (Chemical Index of Alteration) values range from 67 to 85 indicating intermediate to advanced weathering conditions and samples plotted in an ACNK diagram follow a general weathering trend parallel to the A–CN boundary; Potassium enrichments are not shown (Fig. 5a; Abre et al. 2011). Since Cs tends to be fixed in weathering profiles whereas K tends to be lost in solution,

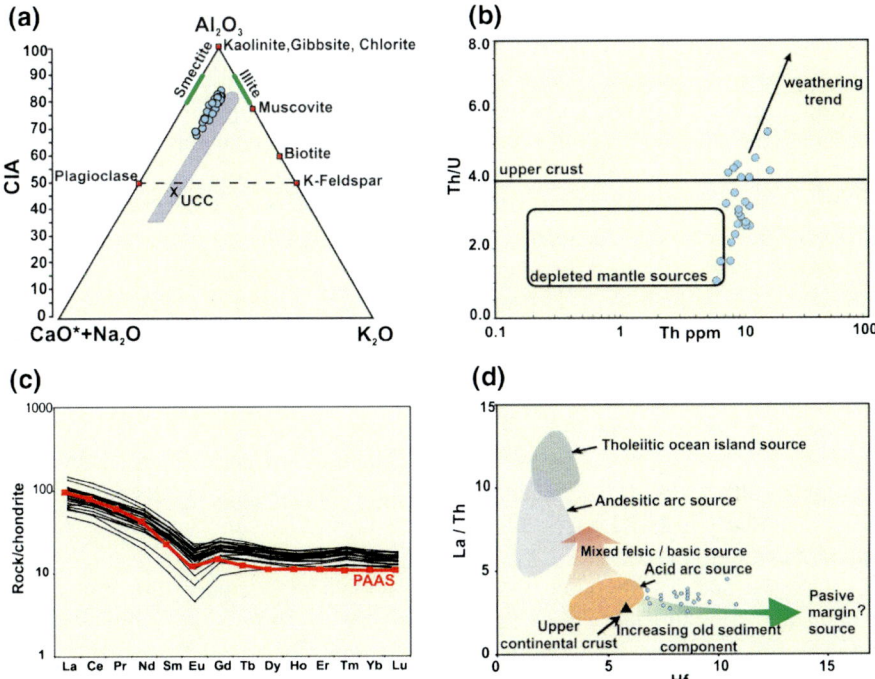

Fig. 5 a The Chemical Index of Alteration is calculated using molecular proportions as $(Al_2O_3/(Al_2O_3 + CaO^* + Na_2O + K_2O)) \times 100$, where CaO^* refers to the calcium associated with silicate minerals; samples plot along a vertical array parallel to the expected weathering trend (field of *vertical lines*) for average upper crustal rocks. UCC values according to Taylor and McLennan (1985). b Th/U versus Th based on McLennan et al. (1993) showing distribution of samples from the Pavón Formation. c Chondrite normalized REE patterns; PAAS = post-Archaean Australian shales pattern (Nance and Taylor 1976) is draw for comparison. Eu-anomaly calculated as $Eu_N/Eu^* = Eu_N/(0.67Sm_N + 0.33Tb_N)$, where "N" denotes values normalized to chondrite. d La/Th versus Hf after McLennan et al. (1980); common values of the La/Th ratios for upper crust components are between 2 and 4

the K/Cs ratio is an indicator of weathering (McLennan et al. 1990); the K/Cs ratio for the Pavón Formation ranges from 1322 to 6410, with an average value of 4621 indicating strong weathering conditions (calculated from data presented in Cingolani et al. 2003).

During weathering, there is a tendency for an increase in the ratio between Th and U to greater than upper crustal igneous values of 3.5–4.0, due to the oxidation of U^{4+} to the more soluble U^{6+}. A low Th/U ratio can be a consequence of U enrichment (McLennan 1989). The Pavón Formation have Thorium concentrations below the average for the upper continental crust (9.3 ppm, with Th concentrations as low as 5.77 ppm), but a few samples are enriched (Cingolani et al. 2003). Although most of the samples are enriched in U compared with the upper continental crust (UCC), some samples are depleted (average U concentrations is of 3.2 ppm). The Th/U ratios are either below 3.5 or between 3.5 and 4, although a very few samples have high Th/U ratios of up to 5.2, pointing to weathering conditions (Fig. 5b; Abre et al. 2011).

Provenance: Average values of trace elements recorded for the Pavón Formation indicate a dominant provenance from a UCC composition, although the unit is enriched in Sc, V and Cr (14, 112 and 139 ppm on average respectively), which typically would indicate a depleted source; Cingolani et al. 2003).

Further discriminations of such mafic component could be achieved using the Y/Ni and Cr/V ratios (McLennan et al. 1993). For the Pavón Formation, the Y/Ni ratio ranges from 0.07 to 1.26 whereas the Cr/V ratio is between 0.83 and 2.5, confirming the influence of a mafic source, being represented by the detrital chromian spinels, but ruled out an ophiolitic precursor (Cingolani et al. 2003).

The chondrite normalized REE patterns for samples of the Pavón Formation (Fig. 5c) show a moderately enriched Light-REE pattern (La_N/Yb_N of about 5.7 on average), a negative Eu-anomaly (Eu_N/Eu^* ranges from 0.47 to 0.77) and a flat Heavy-REE (Tb_N/Yb_N of 1.2 on average). The samples are enriched in Heavy-REE compared with the PAAS (Cingolani et al. 2003). The La/Th ratio falls between 2.6 and 4.6, similarly to UCC rocks, and the relationship to Hf concentrations indicate limited recycling (Fig. 5d).

The Zr/Sc and Th/Sc ratios are powerful tools to decipher different source components of sedimentary rocks (McLennan et al. 1993). The samples of the Pavón Formation have Zr/Sc ratios ranging from 11.7 to 40.1 confirming that recycling played a subordinated role, whereas the Th/Sc ratios (0.43–0.89) include low values, which clearly point to a depleted source, besides the main derivation from average UCC composition (Cingolani et al. 2003).

According to the tectonic classification from Bhatia and Crook (1986), samples of the Pavón Formation plot within the continental island arc and the active continental margin fields (Cingolani et al. 2003).

6 Isotope Geochemistry

Sm–Nd: The grade of fractionation and the average crustal residence time of the detrital mix can be determined using the Sm–Nd isotope system as applied to sedimentary rocks (Nelson and DePaolo 1988; McLennan et al. 1990). Five samples of the Pavón Formation were analyzed and analytical data can be found in Cingolani et al. (2003). ε_{Nd} (t) values, where t = 455 Ma (depositional age) range from −0.4 and −4.1(Fig. 6), $f_{Sm/Nd}$ is between −0.34 and −0.52 (average −0.40 ± 0.06) and T_{DM} ages range from 1.1 and 1.51 Ga.

ε_{Nd} (t) values obtained are neither typical of UCC nor of a juvenile input despite one sample showing a less negative ε_{Nd} (t) of −0.4 and a low Th/Sc ratio (0.43) which could be indicating the input from a juvenile source. The $f_{Sm/Nd}$ values could be assigned to either an old upper crust or to an arc component and generally indicate that the Sm–Nd system is not fractionated, except in the case of one sample with a $f_{Sm/Nd}$ of −0.52 (Abre 2007; Abre et al. 2011).

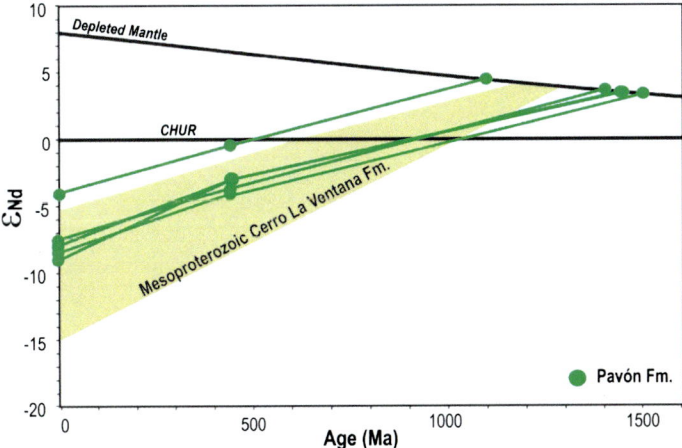

Fig. 6 ε_{Nd} versus age of the Pavón Formation and in yellow the Mesoproterozoic Cerro La Ventana area evaluated as probable source. *CHUR* Chondritic Uniform Reservoir. ε_{Nd} (*t*) = {[(^{143}Nd/^{144}Nd) sample $_{(t)}$/(^{143}Nd/^{144}Nd)CHUR $_{(t)}$] − 1} × 10,000. ^{143}Nd/^{144}Nd$_{CHUR}$ = 0.512638. Data from Cingolani et al. (2003)

Results presented from the five samples analyzed are consistent with a Grenvillian basement as the main source of detritus (Fig. 7; Cingolani et al. 2003), but it does not explain the −0.4 value displayed by one sample of the Pavón Formation, implying either that not all the Mesoproterozoic basement rocks were detailed studied (e.g. mafic compositions within the Cerro La Ventana Formation and gabbroic and mafic cumulates recorded within the El Nihuil Mafic Unit), or inputs from other sources. In this regard, the data available from the Western Pampeanas Ranges (e.g. Umango Range; Porcher et al. 2004; Vujovich et al. 2005; Varela et al. 2011) display positive ε_{Nd} values and point to such rocks as probable detrital sources. Furthermore, T_{DM} ages of Pavón Formation and of basement Mesoproterozoic rocks from the Cuyania terrane are comparable (Kay et al. 1996; Cingolani et al. 2003), as well as they are similar to T_{DM} ages of supracrustal Ordovician to Silurian rocks (Gleason et al. 2007; Abre 2007; Abre et al. 2012).

U–Pb detrital zircon: Another provenance approach is to determine the ages of detrital zircon grains in order to constrain the possible source rocks for the Pavón Formation basin, in particular regarding felsic to intermediate crystalline rocks. Zircons were obtained from a subfeldspathic–arenites and data and analytical techniques were presented by Abre et al. (2011).

Discrimination between igneous and metamorphic zircon grains may be achieved by measuring the Th/U ratio of single grains, since this ratio is of about 0.1 or lower for metamorphic zircons, whereas it is >0.2 or >0.5 for igneous zircons (Vavra et al. 1999; Hoskin and Schaltegger 2003). The detrital zircon dating of the Pavón Formation shows that all the zircon grains analyzed except one have Th/U ratios indicative of a magmatic origin. Such a conclusion is supported by cathodoluminescence images showing that most of the grains are subhedral and

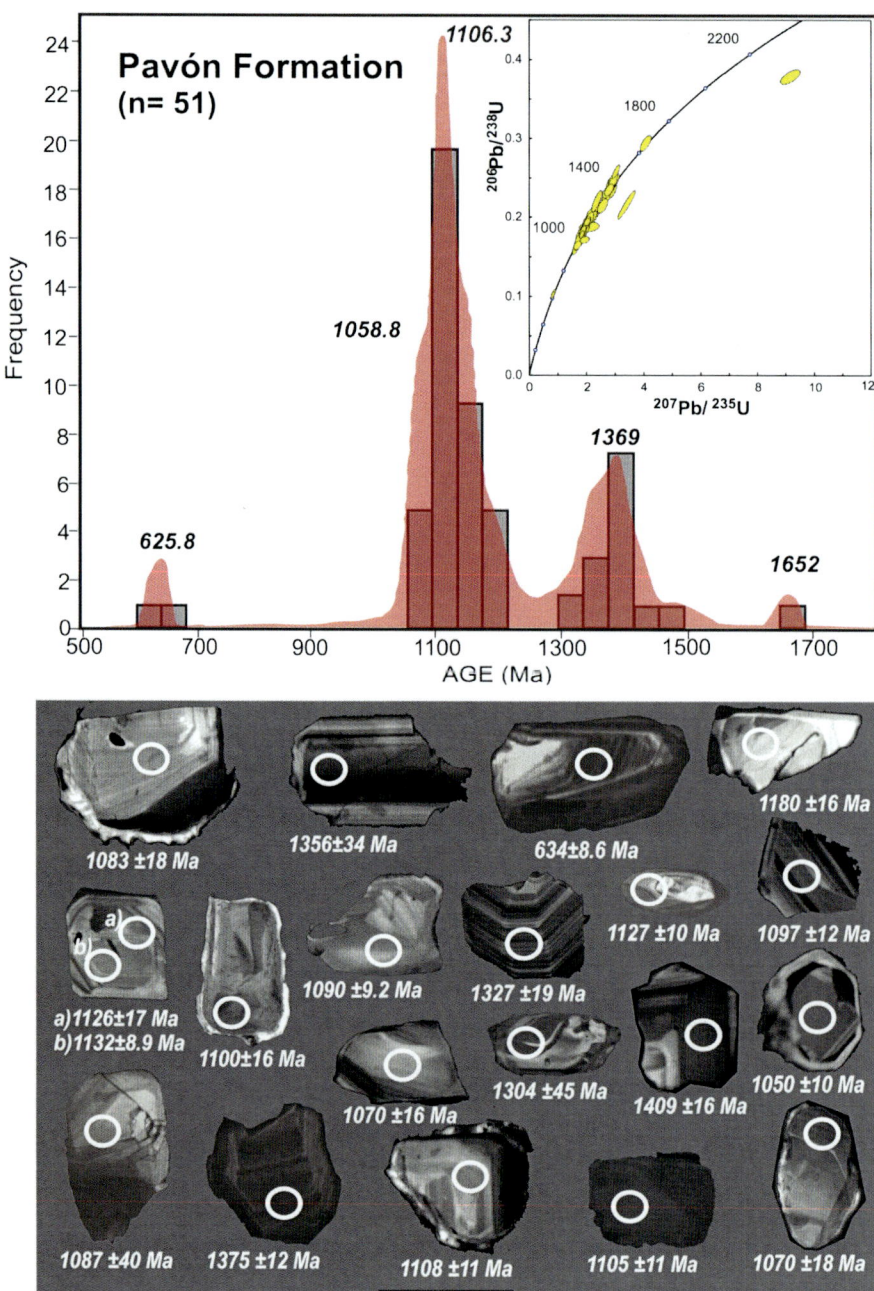

Fig. 7 U–Pb distribution of analyzed detrital zircons with probability curves for the Pavón Formation; representative cathodoluminescence microphotographs of selected zircon grains. Based on Abre et al. (2011)

display oscillatory zoning interpreted as magmatic in origin, whereas only a few have patchy metamorphic zoning (Fig. 7; Abre 2007; Abre et al. 2011).

The zircon dating of the Pavón Formation (n = 53) indicate a main population between 1.0 and 1.3 Ga comprising 35 grains (about 69% of the total measured grains), a population with ages between 1.3 and 1.6 Ga which comprise 13 grains (about 25%), whereas two grains are Neoproterozoic (634 and 615 Ma) and one grain is Paleoproterozoic with an age of 1652 Ma (Abre 2007; Abre et al. 2011).

Source rocks of Mesoproterozoic age that could have provided the bulk of detrital zircons are known from several neighbouring areas, such as the basement of the Cuyania terrane (Cerro La Ventana Formation; Cingolani and Varela 1999; Cingolani et al. this volume) and the Western Pampeanas Ranges (Varela and Dalla Salda 1992; Varela et al. 1996; Pankhurst et al. 1998; Casquet et al. 2006). These probable sources also comprise rocks of Paleoproterozoic age. The Neoproterozoic zircons could be linked to the Pampean/Brazilian Orogen.

7 Paleomagnetism

Eighty-three oriented samples collected at twelve sites from the Pavón Formation were used for paleomagnetic studies (Fig. 1c; Rapalini and Cingolani 2004). After standard demagnetization techniques, two components with geologic significance were determined. (1) Component A: isolated at most sites sampled of the Pavón Formation as well as at a Permo-Triassic rhyolitic dome intruding the unit. This component is a secondary magnetization acquired in the latest Permian-Early Triassic during the widespread Choiyoi magmatic event that affected this region, since the paleomagnetic pole position is consistent with the Late Permian reference pole for South America. (2) Component B: determined at four sites (22 samples) of the Pavón Formation. It is carried either by hematite (B1) or magnetite (B2) and the presented dual polarities and the positive fold test suggests a primary detrital or early diagenetic origin, therefore this component accurately records the Earth Magnetic Field during the Sandbian (c. 455 Ma) and indicates that the sediments of the Pavón Formation were deposited at latitude of 25.7° ± 2.9°S. This Upper Ordovician paleomagnetic data record the first Cuyania terrane paleopole (Rapalini and Cingolani 2004).

This paleolatitude can be reconciled with the Gondwana reference pole by assuming a 30° clockwise rotation of the sampling localities around a vertical axis located in the study area. However, if the San Rafael block, as the southern extension of the Cuyania terrane is located close to Southeast Laurentia, paleomagnetic data agree with the Late Ordovician reference pole of Laurentia, particularly if a 500 km stretching is considered between the Cuyania terrane and the Ouachita Embayment along the Texas and Alabama-Oklahoma transforms faults (Rapalini and Cingolani 2004).

8 Tectonic Setting

Sedimentologic characteristics indicate deposition of sandy turbidites within a foreland basin (Manassero et al. 1999; Cingolani et al. 2003), formed during the extensional regime that followed the accretion of Cuyania terrane to Gondwana in the Middle Ordovician (Astini 2002; Cingolani et al. 2003); eastern palaeocurrents invalidate western sources (Manassero et al. 1999; Cingolani et al. 2003). Further constraints are provided by petrographical, geochemical and isotopic analyses which indicate that the sources components were dominantly unrecycled UCC and subordinately a less fractionated one. The depleted component is at least partially represented by detrital spinels derived from MORB and flood basalts in oceanic or continental intraplate settings, although the source rocks of such detrital grains were not identified (Abre et al. 2009). The age of the main sources is Mesoproterozoic, with minor contributions from Paleoproterozoic and Neoproterozoic sources.

Evidence presented link the Mesoproterozoic Cerro La Ventana Formation as a provenance component to the Ponón Trehué Formation (Darriwilian to Sandbian) of the San Rafael Block (Abre et al. 2011; Abre et al. this volume). Such a provenance is also very likely regarding the Pavón Formation, although the current information available needs to invoke another eastern source area in order to fully explain provenance proxies of the Pavón Formation; such area is most probably the western side of the Pampia terrane (Fig. 8), since the Umango, Maz and Espinal ranges comprise rocks that could account for detrital zircon ages, Sm–Nd data and tentatively for the host rocks of detrital chromian spinels. However, a certain detrital derivation from the Mesoproterozoic rocks of the El Nihuil Mafic Unit may have also occurred; detailed studies of its gabbros and mafic cumulates are needed to further support this. The absence of an important recycling (with some exceptions) tend to ruled out sources located further afield with respect to the depositional

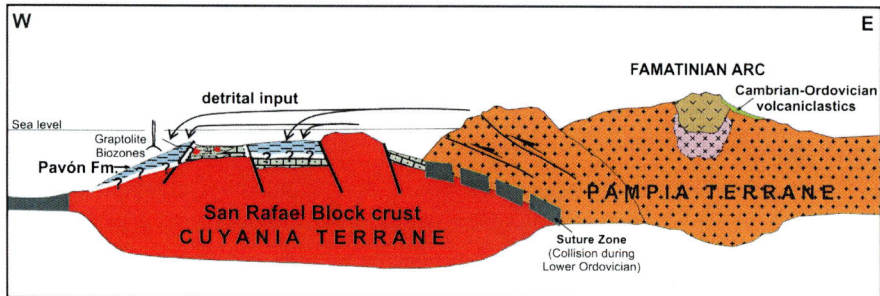

Fig. 8 Interpretative schematic cross section showing that the Ordovician Ponón Trehué Formation received an input restricted to the Cerro La Ventana Formation (see Abre et al. this volume). Progressive subsidence of the basin and deposition of the Pavón Formation during Sandbian age; the Western side of Pampia terrane was uplifted and acted as source of detrital material along with the Cuyanian basement. The Ponón Trehué and the Pavón Formations are not in contact. Modified from Abre et al. (2011)

basin. The sources identified would imply that the Cuyania terrane would have collided to Gondwana at least immediately before the beginning of the Ordovician clastic deposition (Abre et al. 2011).

9 Conclusions

Detailed bed-by-bed studies identified five lithofacies within the sand-dominated sequence of the Pavón Formation. Regular tabular bedding, sharp contacts and scarcity of flow and load casts suggest a relatively deep marine environment and deposition by gravity flows. Black shales and siltstones contain poorly preserved (deformed) graptolites which indicate a Sandbian age due to the development of the *Climacograptus bicornis* Biozone.

The petrographic analyses of the Pavón Formation showed the dominance of monocrystalline quartz and K-feldspar as well as sedimentary, metamorphic and igneous lithoclasts, which point to an UCC component and the influence of a depleted source. Such a derivation is confirmed by geochemical provenance proxies including REE distribution, Eu-anomaly, La/Th and Th/Sc ratios, as well as by heavy minerals analysis and isotope chemistry. The unit was subject to anchimetamorphism, according to Illite crystallinity data. CIA values, K/Cs and Th/U ratios indicate strong weathering. Recycling was not important, according to Zr/Sc ratios. Y/Ni and Cr/V ratios allow discarding an ophiolitic source.

Based on chemical characteristics, the detrital chromian spinel dataset is subdivided in two groups: Group 1 is characterized by intermediate Cr# values, low TiO_2, Fe^{2+}# and Fe^{3+}#, indicating a mid-ocean ridge emplacement of their initial host rocks; on the contrary, Group 2 show high TiO_2, Fe^{2+}# and Fe^{3+}#, and Cr# values of c. 0.7 suggesting an intraplate environment. El Nihuil Mafic Unit could be identified as the definite source.

The Sm–Nd isotopic characteristics of the Pavón Formation provide evidence of upper crust source materials and a depleted source; certain isotope fractionation was also detected. T_{DM} model ages and ε_{Nd} for the provenance protoliths are consistent with Mesoproterozoic source rocks such as the Cerro La Ventana Formation and the western Pampia terrane.

The detrital zircon dating indicates a main age population between 1.0 and 1.3 Ga, a second population with ages between 1.3 and 1.6 Ga, two Neoproterozoic grains and one Paleoproterozoic, confirming the source areas proposed.

Two magnetic components were determined and the one of primary origin indicate a paleomagnetic pole of around 26°S for the Sandbian record. The Pavón Formation was deposited in a foreland basin generated as a consequence of the accretion of the Cuyania terrane to the Gondwanan active margin.

Acknowledgements Field and laboratory works for this study were funded by CONICET and ANPCyT grants. The authors wish to thank Norberto Uriz for his help during the preparation of the manuscript and figures. Discussions with Pedro Farias and Joaquín García-Sansegundo (University of Oviedo, Spain) were helpful in the interpretation of this work.

References

Abre P (2007) Provenance of Ordovician to Silurian clastic rocks of the Argentinean Precordillera and its geotectonic implications. Ph.D. thesis, University of Johannesburg, South Africa. UJ web free access

Abre P, Zimmermann U, Cingolani CA (2003) Pavón Formation zircons under scanning electronic microscope: preliminary results on the Ordovician siliciclastics from the San Rafael block, Mendoza. In: Albanesi GL, Beresi MS, Peralta SH (eds) Ordovician from the Andes, INSUGEO, Tucumán, Serie Correlación Geológica, vol 17, pp 217–221

Abre P, Cingolani CA, Zimmermann U, Cairncross B (2005) Espinelas cromíferas detríticas en la Formación Pavón (Ordovícico): indicadoras de corteza oceánica en el terreno Precordillera, Argentina. In: XVI Congreso Geológico Argentino, Actas, vol 1, p 8

Abre P, Cingolani CA, Zimmermann U, Cairncross B (2006) Espinelas detríticas en la Formación Don Braulio (Ordovícico-Silúrico), Terrane Precordillera, Argentina. In: XVII Congreso Geológico Boliviano, Sucre, Bolivia. Actas CD Rom

Abre P, Cingolani CA, Zimmermann U, Cairncross B (2009) Detrital chromian spinels from Upper Ordovician deposits in the Precordillera terrane, Argentina: a mafic crust input. J S Am Earth Sci 28:407–418

Abre P, Cingolani CA, Zimmermann U, Cairncross B, Chemale Jr F (2011) Provenance of Ordovician clastic sequences of the San Rafael Block (Central Argentina), with emphasis on the Ponón Trehué Formation. Gondwana Res 19(1):275–290

Abre P, Cingolani C, Cairncross B, Chemale Jr F (2012) Siliciclastic Ordovician to Silurian units of the Argentine Precordillera: constraints on provenance and tectonic setting in the Proto-Andean margin of Gondwana. J S Am Earth Sci 40:1–22

Abre P, Cingolani CA, Uriz NJ (this volume) Sedimentary provenance analysis of the Ordovician Ponón Trehué Formation, San Rafael Block, Mendoza-Argentina. In: Cingolani C (ed) Pre-carboniferous evolution of the San Rafael Block, Argentina. Implications in the SW Gondwana margin. Springer, Berlin

Astini RA (2002) Los conglomerados basales del Ordovícico de Ponón Trehué (Mendoza) y su significado en la historia sedimentaria del terreno exótico de Precordillera. Rev Asoc Geol Argentina 57:19–34

Barnes SJ, Roeder PL (2001) The range of spinel compositions in terrestrial mafic and ultramafic rocks. J Petrol 42(12):2279–2302

Bhatia MR, Crook KAW (1986) Trace element characteristics of graywackes and tectonic setting discrimination of sedimentary basins. Contrib Miner Petrol 92:181–193

Casquet C, Pankhurst RJ, Fanning CM, Baldo E, Galindo C, Rapela CW, González-Casado JM, Dahlquist JA (2006) U-Pb SHRIMP zircon dating of Grenvillian metamorphism in Western Sierras Pampeanas (Argentina): correlation with the Arequipa-Antofalla craton and constraints on the extent of the Precordillera terrane. Gondwana Res 9:524–529

Cingolani CA, Varela R (1999) The San Rafael block, Mendoza (Argentina). Rb–Sr isotopic age of basement rocks. In: II South American Symposium on Isotope Geology, SEGEMAR anales, vol 24, pp 23–26. Córdoba

Cingolani CA, Cuerda A, Manassero M (1999) Litoestratigrafía de los depósitos siliclásticos ordovícicos del Cerro Bola, Bloque de San Rafael, Mendoza. In: XIV Congreso Geológico Argentino, Actas, vol 1, pp 409–413. Salta

Cingolani CA, Manassero M, Abre P (2003) Composition, provenance and tectonic setting of Ordovician siliciclastic rocks in the San Rafael Block: Southern extension of the Precordillera crustal fragment, Argentina. J S Am Earth Sci 16:91–106

Cingolani CA, Basei MAS, Varela R, Llambías EJ, Chemale Jr F, Abre P, Uriz NJ, Marques JCh (this volume) The Mesoproterozoic basement at the San Rafael Block, Mendoza Province (Argentina): Geochemical and isotopic age constraints. In: Cingolani C (ed) Pre-carboniferous evolution of the San Rafael Block, Argentina. Implications in the SW Gondwana margin, Springer, Berlin

Cox R, Lowe DR, Cullers RL (1995) The influence of sediment recycling and basement composition on evolution of mudrock chemistry in the southwestern US. Geochim Cosmochim Acta 59(14):2919–2940

Cuerda AJ, Cingolani CA (1998) El Ordovícico de la región del Cerro Bola en el Bloque de San Rafael, Mendoza: Sus faunas graptolíticas. Ameghiniana 35(4):427–448

Cuerda AJ, Cingolani CA, Manassero M (1998) Caradoc graptolite assemblages and facial relations from the Cerro Bola section, San Rafael block, Mendoza Province Argentina. In: Sixth International Graptolite Conference. IUGS Subcommission on Silurian Stratigraphy, vol 23, pp 170–175. Madrid

Dick HJB, Bullen T (1984) Chromian spinel as a petrogenetic indicator in abyssal and alpine-type peridotites and spatially associated lavas. Contrib Miner Petrol 86:54–76

Dott RH (1964) Wacke, Graywacke and Matrix; what Approach to Immature Sandstone Classification? J Sediment Petrol 34:625–632

Gleason JD, Finney SC, Peralta SH, Gehrels GE, Marsaglia KM (2007) Zircon and whole-rock Nd–Pb isotopic provenance of Middle and Upper Ordovician siliciclastic rocks, Argentine Precordillera. Sedimentology 54:107–136

González Díaz EF (1981) Nuevos argumentos a favor del desdoblamiento de la denominada "Serie de la Horqueta", del Bloque de San Rafael, provincia de Mendoza. In: Actas 8° Congreso Geológico Argentino, vol 3, pp 241–256. San Luis

Holmberg E (1948) Geología del Cerro Bola. Contribución al conocimiento de la tectónica de la Sierra Pintada. Secretaría de Industria y Comercio de la Nación. Dirección General de Industria y Minería. Boletín 69:313–361. Buenos Aires

Hoskin PWO, Schaltegger U (2003) The composition of zircon and igneous and metamorphic petrogenesis. In: Hanchar JM, Hoskin PWO (eds) Zircon: reviews in mineralogy and geochemistry, vol 53, pp 27–62

Kamenetsky VS, Crawford AJ, Meffre S (2001) Factors controlling chemistry of magmatic spinel: an empirical study of associated olivine, Cr-spinel and melt inclusions from primitive rocks. J Petrol 42(4):655–671

Kay SM, Orrell S, Abruzzi JM (1996) Zircon and whole rock Nd–Pb isotopic evidence for a Grenville age and a Laurentian origin for the basement of the Precordillera in Argentina. J Geol 104:637–648

Manassero M, Cingolani CA, Cuerda AJ, Abre P (1999) Sedimentología, Paleoambiente y Procedencia de la Formación Pavón (Ordovícico) del Bloque de San Rafael, Mendoza. Rev Asoc Argent Sedimentol 6(1–2):75–90

Mange MA, Maurer HFW (1992) Heavy minerals in colour. Chapman and Hall, London 147 pp

Marquat FJ, Menéndez AJ (1985). Graptofauna y edad de la Formación Lutitas del Cerro Bola, Sierra Pintada. Departamento San Rafael, Provincia de Mendoza, Argentina. Centro Cuyano de Documentación Científica, 11 pp, Mendoza

McLennan SM (1989) Rare earth elements in sedimentary rocks: influence of provenance and sedimentary processes. Mineral Soc Am Rev Mineral 21:169–200

McLennan SM, Nance WB, Taylor SR (1980) Rare earth element-thorium correlations in sedimentary rocks, and the composition of the continental crust. Geochim Cosmochim Acta 44:1833–1839

McLennan SM, Taylor SR, McCulloch MT, Maynard JB (1990) Geochemical and Nd–Sr isotopic composition of deep-sea turbidites: crustal evolution and plate tectonic associations. Geochim Cosmochim Acta 54:2015–2050

McLennan SM, Hemming S, McDaniel DK, Hanson GN (1993) Geochemical approaches to sedimentation, provenance, and tectonics. In: Johnsson MJ, Basu A (eds) Processes controlling the composition of clastic sediments: Geological Society of America, Special Paper, vol 284, pp 21–40

Nance WB, Taylor SR (1976) Rare earth element patterns and crustal evolution I: Australian post-Archean sedimentary rocks. Geochim Cosmochim Acta 40:1539–1551

Nelson BK, DePaolo DJ (1988) Application of Sm–Nd and Rb–Sr isotopes systematics to studies of provenance and basin analysis. J Sediment Petrol 58:348–357

Nesbitt HW, Young GM (1982) Early Proterozoic climates and plate motions inferred from major element chemistry of lutites. Nature 199:715–717

Pankhurst RJ, Rapela CW, Saavedra J, Baldo E, Dahlquist J, Pascua I (1998) The Famatinian magmatic arc in the central Sierras Pampeanas: an early to mid-Ordovician continental arc on the Gondwana margin. In: Pankhurst RJ, Rapela CW (eds) The proto-Andean margin of Gondwana, vol 142. Geological Society of London Special Publications, pp 343–368

Porcher C, Fernandes LAD, Vujovich G, Chernicoff CJ (2004) Thermobarometry, Sm/Nd ages and geophysical evidence for the location of the suture zone between Cuyania and the Western Proto-Andean Margin of Gondwana. Gondwana Res 7:1057–1076

Rapalini AE, Cingolani CA (2004) First late Ordovician paleomagnetic pole for the Cuyania (Precordillera) terrane of western Argentina: a microcontinent or a Laurentian plateau? Gondwana Res 7:1089–1104

Taylor SR, McLennan SM (1985) The continental crust. Its composition and evolution. Blackwell, London 312 pp

Tickyj H, Cingolani CA, Raising Rodríguez, Alfaro MB, Uriz NJ (2009) Graptolitos ordovícicos en el sur de la Cordillera Frontal de Mendoza. Rev Asoc Geol Argent 64(2):295–302

Varela R, Dalla Salda L (1992) Geocronología Rb–Sr de metamorfitas y granitoides del extremo sur de la Sierra de Pie de Palo, San Juan. Rev Asoc Geol Argent 47:271–275

Varela R, López de Luchi M, Cingolani CA, Dalla Salda L (1996) Geocronología de gneises y granitoides de la Sierra de Umango, La Rioja. Implicancias tectónicas. In: XIII Congreso Geológico Argentino y III Congreso de Exploración de Hidrocarburos: Actas, vol 3, pp 519–527. Buenos Aires

Varela R, Basei MAS, González PD, Sato AM, Naipauer M, Campos Neto M, Cingolani CA, Meira VT (2011) Accretion of Grenvillian terranes to the southwestern border of the Río de la Plata craton, western Argentina. Int J Earth Sci 100:243–272

Vavra G, Schmid R, Gebauer D (1999) Internal morphology, habit and U–Th–Pb microanalysis of amphibolite-to-granulite facies zircons: geochronology of the Ivrea Zone (Southern Alps). Contrib Miner Petrol 134:380–404

Vujovich GI, Porcher CC, Chernicoff CJ, Fernandes LAD, Pérez DJ (2005) Extremo norte del basamento del terreno Cuyania: nuevos aportes multidisciplinarios para su identificación. Asoc Geol Argent Serie D Publ Espec 8:15–38

Lower Paleozoic 'El Nihuil Dolerites': Geochemical and Isotopic Constraints of Mafic Magmatism in an Extensional Setting of the San Rafael Block, Mendoza, Argentina

Carlos A. Cingolani, Eduardo Jorge Llambías, Farid Chemale Jr., Paulina Abre and Norberto Javier Uriz

Abstract The 'El Nihuil Mafic Unit' is exposed at the Loma Alta region northwards of the El Nihuil dam. This igneous body consists mainly of mafic rocks assigned to the Precambrian and Lower Paleozoic according to different authors. The mafic unit shows an elongated shape with a NNE–SSW orientation on the western side of the San Rafael Block (SRB), developed for a length of 17.5 km and with a maximum width of 4.2 km and is composed of deformed gabbros, amphibolites, and tonalites that represent the Mesoproterozoic continental crust, and dykes and sills of undeformed Lower Paleozoic porphyritic dolerites. We present the petrology, geochemistry, isotope data, and determinations of emplacement conditions of the dolerites that could represent a sliver of Cuyania-Chilenia terranes suture. The dolerites show classical porphyritic texture, with elongated subhedral plagioclase (andesine) and clinopyroxene phenocrysts. Geochemical analyses of El Nihuil Dolerite samples indicate that

---

C.A. Cingolani (✉) · E.J. Llambías
Centro de Investigaciones Geológicas, CONICET-UNLP,
Diag. 113 n. 275, CP1904 La Plata, Argentina
e-mail: carloscingolani@yahoo.com

E.J. Llambías
e-mail: llambias@cig.museo.unlp.edu.ar

F. Chemale Jr.
Programa de Pós-Graduação Em Geologia, Universidade Do Vale Do Rio Dos Sinos,
São Leopoldo, RS, Brazil
e-mail: faridcj@unisinos.br; faridchemale@gmail.com

P. Abre
Centro Universitario de la Región Este, Universidad de la República,
Ruta 8 Km 282, Treinta y Tres, Uruguay
e-mail: paulinabre@yahoo.com.ar

N.J. Uriz
División Geología, Museo de La Plata, UNLP,
Paseo Del Bosque s/n, B1900FWA La Plata, Argentina
e-mail: norjuz@gmail.com

© Springer International Publishing AG 2017
C.A. Cingolani (ed.), *Pre-Carboniferous Evolution of the San Rafael Block, Argentina*, Springer Earth System Sciences,
DOI 10.1007/978-3-319-50153-6_6

the rocks are MORB-type basalts. In the P_2O_5 versus Zr diagram, the dolerites plot in the tholeiitic field similarly to western Cuyania basalts, and in the Th–Hf/3–Ta tectonic discrimination diagram the dolerite dykes plot mainly as E-MORB. Dolerite samples were dated by K–Ar (whole rock) systematic and the ages are 448.5 ± 10 and 434.2 ± 10 Ma (Upper Ordovician and close to the Lower Silurian boundary). The dolerites represent the unique Lower Paleozoic mafic rock outcrops within the SRB. $Nd_{(TDM)}$ ages are in between 0.51 and 0.80 Ga; $\varepsilon Nd_{(0)}$ record positive values ranging from +3.85 and +7.84; $\varepsilon Nd_{(t)}$ record +4.27 to +12.42. $^{87}Sr/^{86}Sr$ ratios are in between 0.7032 and 0.7050 in agreement with values for ocean ridge tholeiites. These mafic rocks are interpreted as a part of a dismembered 'Famatinian ophiolite belt' emplaced during the Lower Paleozoic extensional environment within a thinned Mesoproterozoic continental crust on western Cuyania terrane.

Keywords Lower Paleozoic · Extensional event · Mafic rocks · El Nihuil · Western Cuyania

1 Introduction

The pre-Andean region in west-central Mendoza province, Argentina, comprises the SSE-NNW Cenozoic structured San Rafael Block (SRB); towards East, it passes into the Pampean plains being covered by modern sedimentary and basaltic back-arc volcanic units. The northern and southern boundaries are limited by the Cuyo and Neuquén sedimentary basins, respectively. To the West, the Andean foothill bounds it (Fig. 1a). Paleontological and geological evidence allow interpreting the SRB as a southern extension of the Cuyania terrane (Ramos 2004). The Upper Paleozoic regional unconformity (Dessanti 1956) limits the so-called 'pre-Carboniferous units,' which includes diverse igneous-metamorphic and sedimentary units of Precambrian-Middle Paleozoic age. One of these igneous units, described and mapped as the intrusive 'gabbro' by Dessanti (1956) is exposed northwards of the El Nihuil dam (Fig. 1b). More recent studies (Cingolani et al. 2000, 2012) have demonstrated that the 'gabbro' is composed of mafic rocks of two different ages. The oldest rocks are Precambrian and consist of deformed gabbros, amphibolites, tonalites, whereas the youngest rocks are dolerites of Lower Paleozoic age. The dolerites record a NE–SW belt cropping out mainly in the Loma Alta sector (1434 m, 34°58'50"S and 68°45'00"W) and exposed at low hills within an overall smooth and rounded relief (Fig. 2).

This paper focuses in the Lower Paleozoic dolerites. Figure 1b shows the elongated shape of the mafic bodies on the western side of the SRB, developed for a length of 17.5 km and with a maximum width of 4.2 km. A southern extension is recorded on an isolated outcrop located to the south of the El Nihuil Lake (Fig. 1b).

The dolerites were studied using several techniques including geochemistry and geochronology in order to determine emplacement conditions; its importance relies in the fact that these rocks could represent a sliver of a terrane suture. Other lithological components of the mafic belt such as gabbros, amphibolites, and

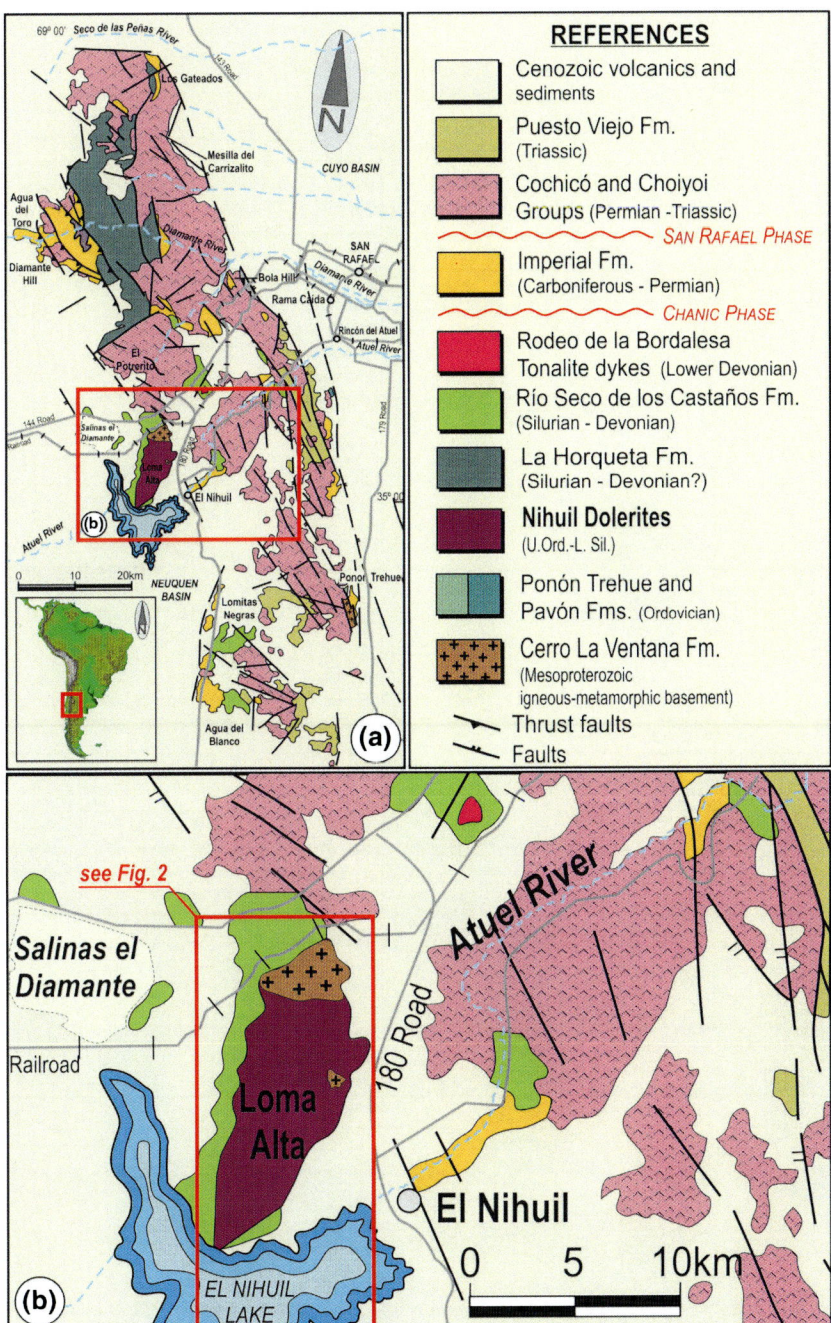

Fig. 1 a Geological sketch map of the San Rafael Block showing the location of the El Nihuil Dolerites and **b** details of the Lower Paleozoic dolerites exposed mainly in Loma Alta sector

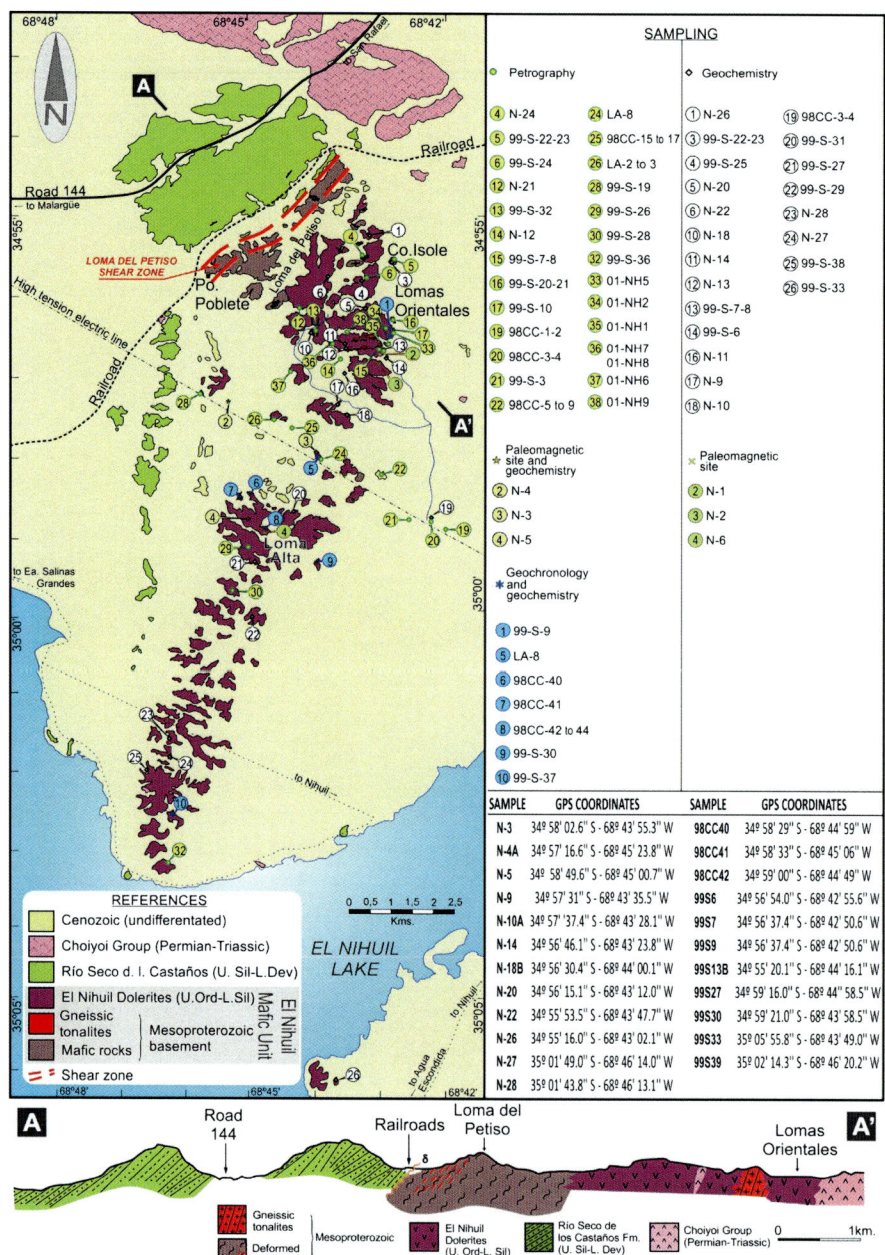

Fig. 2 Geological sketch map of the 'El Nihuil Mafic Unit' based on Cingolani et al. (2000) and location of the dolerite samples obtained for different studies (petrography, geochemistry, isotope analyses and paleomagnetism). A–A′: NW–SE cross-section showing the contact relationships between the Lower Paleozoic dolerites and other units

tonalites with ductile deformation are described in the chapter about the Mesoproterozoic basement (Cingolani et al. this volume).

The present information still at reconnaissance level could be important to encourage more detailed works on these interesting mafic rocks of the SRB, which are important to unravel some aspects of the nature and history of tectonic terranes. It is known that the dolerite (diabase or microgabbro) is a mafic, holocrystalline, subvolcanic rock equivalent to volcanic basalt or plutonic gabbro. Dolerite dykes and sills are typically shallow intrusive bodies and often exhibit fine grained to aphanitic chilled margins and dark-colored rocks. The term 'diabase' is often used as a synonym of dolerite by American geologists, however, in Europe the term is usually only applied to altered dolerites. In ophiolitic complexes the dolerites are related with gabbros (base) and pillow lavas (top).

2 Previous Works

Mafic rocks within the SRB were first described during the geotechnical Atuel river studies by Wichmann (1928) and for the oil industry by Padula (1949). Later on, it was mapped as 'intrusive gabbro in the La Horqueta Serie' by Dessanti (1956) and assigned tentatively to the Precambrian. When Nuñez (1976) completed the field-base mapping, he described the mafic rocks as 'gabbros and porphyritic dolerites' cut by andesite dykes from the Choiyoi Group (Permian-Triassic). González Díaz (1981) mentioned gabbros with saussuritization processes intruded by lamprophyres and record the first Ordovician K–Ar ages.

Other authors (Criado Roqué and Ibañez 1979) described dolerite dykes interpreted as a part of a 'phacolit-type' igneous body with chilled margins. The first geochemical prospection work over the entire 'El Nihuil Mafic Unit' was presented by Davicino and Sabalúa (1990) they recognized porphyritic gabbros and dolerites, cataclastic rocks, mylonites and spilites and affected by two Tertiary faulted systems (N50–140° and N100–190°). Cingolani et al. (2000) reported more geochemical data along the mafic unit, that allows separating two different magmatic events: one composed of deformed gabbros, amphibolites, and tonalites; and the other of undeformed dolerites. More information about the regional geology of this sector of the SRB was presented by Sepúlveda (1999) and Sepúlveda et al. (2007).

3 Geological Aspects

'El Nihuil Mafic Unit' comprises two units: (i) Mesoproterozoic gabbros, amphibolites, and tonalites (Cerro La Ventana Fm.) developed mainly at the 'Loma del Petiso shear zone' and Lomas Orientales, (ii) Upper Ordovician-Lower Silurian undeformed porphyritic dolerites (Fig. 2) developed along the central and southern sector of the body (Cingolani et al. 2000). Similar structural style and

Mesoproterozoic isotopic ages obtained from isolated deformed tonalite rocks (equivalents to deformed gabbros mentioned before) from Lomas Orientales as recorded by Cingolani et al. (2012) suggest that all deformed rocks with overprinted ductile deformation correspond to the Mesoproterozoic continental crust relicts.

Furthermore, deformed mafic rocks outcropping at the 'Loma del Petiso shear zone' in unconformity with the Río Seco de los Castaños Fm (Silurian-Devonian), display cross-cutting relationships supporting that the dolerite dykes were afterwards intruded (Fig. 3a, b). The sedimentary Upper Ordovician (Pavón Fm) that was the most probably country rocks of the dolerites were not recorded along the 'El Nihuil Mafic Unit.' The Pavón Fm records detrital chromian spinels derived from mid-ocean ridge and continental flood basalts. The El Nihuil Mafic Unit was considered as a probable source of these spinels (Abre et al. 2009), and although the dolerites were emplaced in the same tectonic setting as the source rocks of the spinels, such a derivation is not supported by isotopic ages, because dolerites are younger than the Pavón Fm. Detrital zircon ages from Pavón Fm confirmed a main Mesoproterozoic source (Abre et al. this volume), therefore a provenance of chromian spinels from the Mesoproterozoic section of the El Nihuil mafic body cannot be yet ruled out.

In summary, it is clear that the exposed country rocks of the Lower Paleozoic undeformed extrusive dolerites are only the deformed gabbro, amphibolites, and

Fig. 3 **a** A general view of the dolerite outcrops taken towards the North, near Loma Alta region in the 'El Nihuil Mafic Unit,' **b** details of the dolerite (D) that is intrusive in the deformed gabbros (G), **c** and **d** both pictures showing the typical texture of the dolerites with plagioclase and clynopiroxene phenocrysts

4 Petrology and Geochemistry

In hand specimen the color of undeformed porphyritic dolerites is dark, medium or dark-gray, or greenish (Fig. 3a). The porphyritic texture is clearly observed in several outcrops with fine and centimeter white/pale phenocrysts (Fig. 3c, d).

The dolerites show classical porphyritic texture (Fig. 4), with elongated subhedral plagioclase (andesine) and clinopyroxenes phenocrysts; the finer matrix has a subophitic texture and it is composed of anhedral plagioclase enclosing clinopyroxenes or amphiboles. Hornblende and tremolite-actinolite amphiboles are commonly replacing pyroxenes. Accessory minerals are apatite, opaque minerals and finely disseminated pyrite. Olivine crystals were also recognized in some samples.

Fifteen dolerite whole rock samples were selected to be analysed for major, minor, and trace elements (including REE) at ACTLABS, Canada; using a lithium metaborate/tetraborate fusion procedure with measurements done by ICP-MS; results are shown on Table 1. SiO_2 concentrations varying from 46.07 to 49.34% and $Na_2O + K_2O$ contents less than 4 wt% allow classifying the dolerites as basalts (Le Maitre 1989). MgO concentrations vary between 5.88 and 7.77%, while FeO ranges from 6.92 to 8.94%. Alkalis show low concentrations, ranging from 0.25 to 0.77% for K_2O and between 2.11 and 3.37% for Na_2O; wt% TiO_2 is relatively high (1.12–2.19%) and wt% MnO and P_2O_5 values are less than 0.29%; Na_2O concentration have an average of 2.58 wt%, that could indicate absence of albitization processes. Loss on ignition (LOI) is in between 1.09 and 4.5% indicating moderate to high alteration. $Cr = 207$, $Sc = 45$, and $Ni = 101$ ppm averages indicate moderate concentrations (Table 1).

To classify the dolerite rocks we employed diagrams based in immobile elements. According to the Zr/TiO_2 versus Nb/Y rock classification diagram (Winchester and Floyd 1977) the El Nihuil Dolerite samples plotted in the field of andesite/basalts (Fig. 5a), although according to their maximum SiO_2 content (anhydrous base) of 51.5% they are recorded as basalts. The distribution of major oxides indicates the effects of secondary processes (weathering), but despite this the samples have the chemical characteristics of the tholeiitic series, as shown in the AFM diagram (Irvine and Baragar 1971; Fig. 5b). Following Pearce and Cann (1973), the Ti/100 – Zr – Y/3 and Ti/100 – Zr – Sr/2 diagrams indicate that the dolerites correspond to MORB-type rocks (Fig. 6a, b). La/Yb normalized ratios between 0.93 and 1.44 (Table 2) fall within the range of values occurring in mid-ocean ridge volcanic rocks (Gale et al. 2013). The Th/U ratios between 3.53 and 4.56 are close to E-MORB and Zr/Y ratios ranging from 2.15 to 2.86 suggest continental crust contamination (Arévalo and McDonough 2010).

In the P_2O_5 versus Zr diagram (Fig. 7a) for basalts classification after Winchester and Floyd (1976), the El Nihuil Dolerite samples plot in the tholeiitic

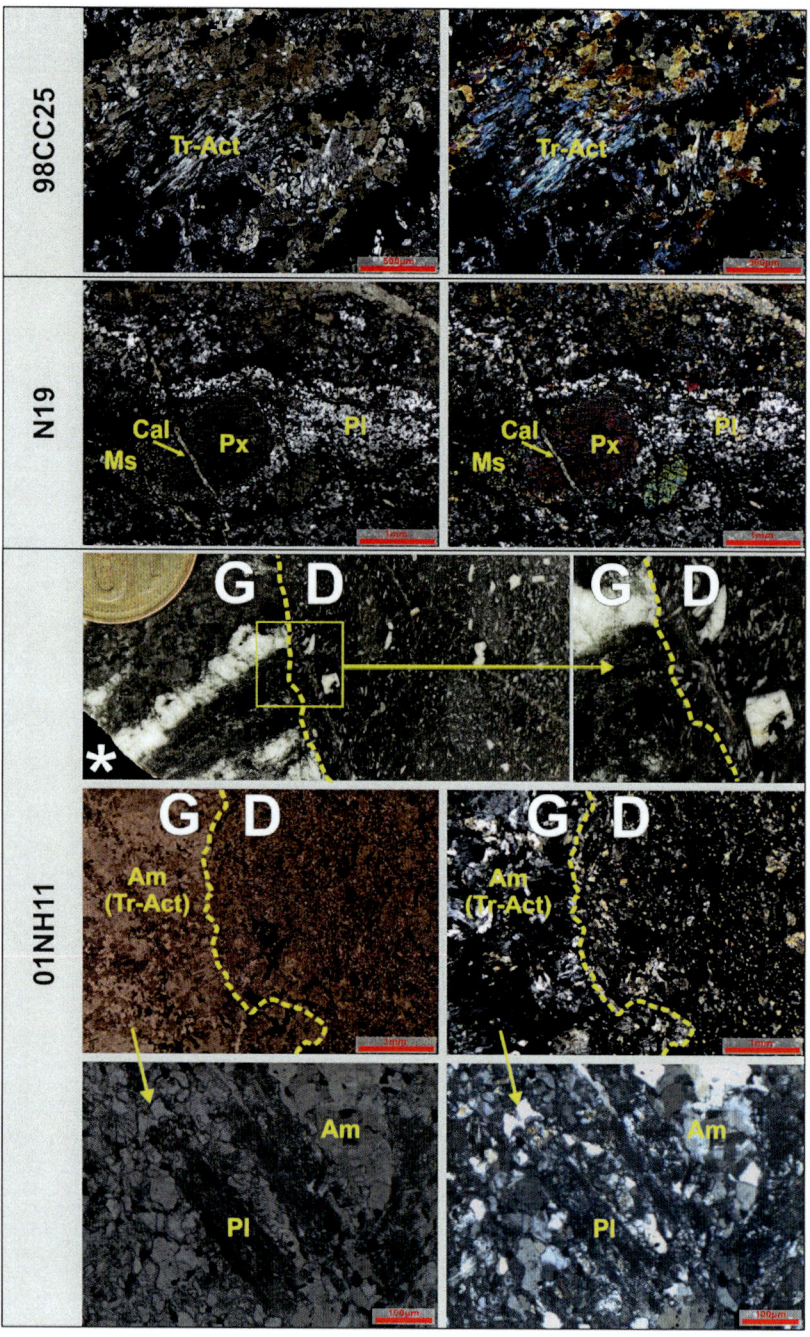

◀**Fig. 4** Photomicrographs of dolerite samples (98CC25, N19 and 01NH11) showing the texture and main mineralogical composition. On the *left*, //nicols; on the *right* X nicols. Sample 01NH11 records the intrusive contact between dolerite (D) and gabbro (G) like rocks. *Tr-Act* tremolite-actinolite; *Pl* plagioclase; *Am* amphibole; *Px* Pyroxene; *Ms* muscovite; *Cal* calcite. * shown gabbro-dolerite contact in the polished hand specimen. The coin as scale is 1.5 cm

field close to the western Cuyanian basalts (Boedo et al. 2013). As shown in Fig. 7b, the Th–Hf/3–Ta tectonic diagram (Wood et al. 1979), that can discriminate silicic magmas derived from E-type MORB or WPB and those associated with destructive plate margins or remelted continental crust) the dolerite dykes are plotted mainly in E-MORB field (intraplate oceanic basalts). Basaltic dykes, sills, and flows from Precordillera mafic-ultramafic belt after Boedo et al. (2013) are also plotted for comparison (green-shaded field).

In Fig. 8a the primordial mantle normalized trace elements show enrichment in Cs, Rb, Ba, and Nb compared to the average pattern of N-MORB. Chondrite-normalized REE diagrams (Fig. 8b) show El Nihuil dolerite patterns parallel to average N-MORB, but enrichments of the lighter elements (LREE) are evident, particularly for La and Ce. The La_n/Lu_n ratio is between 0.99 and 1.48. The normalized ratios data presented in Table 2 are similar to those from the Precordilleran basalts (Ramos et al. 2000; Boedo et al. 2013).

From the comparison to the N-MORB pattern, an enrichment of LIL (large ion lithophile) elements is evident, whereas the HFSE (high field strength elements) does not show such behavior. LIL elements could be enriched due to alteration of the rock or contamination with crustal material.

5 Isotopic Data

5.1 K–Ar

First K–Ar whole rock data was presented by González Díaz (1981), and indicate ages of 474 and 484 Ma for the "gabbro Loma Alta," however precision on sample locations was lacking (Linares et al. 1987). Two new fresh samples from El Nihuil Dolerites were dated by K–Ar whole rock systematic at the Centro de Pesquisas Geocronológicas, Universidade de São Paulo (Brazil). As we can see in the Table 3 the obtained ages are 448.5 ± 10 and 434.2 ± 10 Ma, dating the extrusive event as Upper Ordovician and close to the Lower Silurian boundary. Such age determination undoubtedly confirmed that the dolerites represent the unique Lower Paleozoic mafic belt within the SRB.

Table 1 Geochemical data (ACTLABS, Canada) of major, minor, and trace elements (including REE) of the studied dolerite rocks. Negative values mean below detection limits

SAMPLE	SiO$_2$ %	Al$_2$O$_3$ %	Fe$_2$O$_3$ ox %	FeO calc.	Fe$_2$O$_3$ calc.	MnO %	MgO %	CaO %	Na$_2$O %	K$_2$O %	TiO$_2$ %	P$_2$O$_5$ %	LOI %	TOTAL %
N-3	48.77	14.23	13.68	0.66	8.46	0.214	6.47	10.96	2.37	0.42	1.802	0.19	1.21	100.32
N-4A	46.07	13.65	14.13	0.66	8.63	0.196	7.25	7.47	3.37	0.33	1.433	0.15	4.53	98.58
N-5	49.00	14.38	11.97	0.67	7.42	0.205	6.92	11.37	2.39	0.28	1.553	0.17	1.13	99.36
N-9	47.90	15.38	11.44	0.66	7.04	0.181	7.22	10.72	2.35	0.77	1.416	0.16	2.15	99.67
N-10A	48.47	13.93	13.42	0.67	8.33	0.225	6.84	11.11	2.44	0.25	1.824	0.21	1.77	100.50
N-14	47.62	15.62	11.79	0.67	7.37	0.193	6.04	11.55	2.11	0.54	1.514	0.18	2.60	99.75
N-18	48.38	13.55	13.71	0.67	8.50	0.218	6.57	10.48	2.20	0.57	1.963	0.23	1.25	99.12
N-20	47.51	14.96	11.18	0.67	6.97	0.185	7.77	10.83	2.33	0.41	1.349	0.16	2.54	99.21
N-22	47.65	13.79	14.57	0.66	8.94	0.227	5.88	9.19	3.00	0.25	2.169	0.29	2.45	99.47
N-26	47.64	15.61	11.71	0.67	7.30	0.187	6.12	11.32	2.28	0.46	1.482	0.16	3.03	99.99
N-27	49.34	14.92	11.14	0.67	6.92	0.172	7.59	12.01	2.12	0.39	1.209	0.11	1.16	100.17
N-28	48.93	14.87	11.60	0.67	7.22	0.184	7.17	11.98	2.24	0.31	1.387	0.14	1.09	99.90
98CC42	48.7	13.92	12.28			0.19	6.71	11.34	2.63	0.35	1.62	0.17	1.21	99.1
98CC40	47.76	15.21	11.77			0.19	6.43	10.67	2.7	0.7	1.48	0.14	2.17	99.22
98CC41	48.15	16.11	10.8			0.17	6.9	11.95	2.44	0.28	1.12	0.11	1.88	99.92

SAMPLE	Hf ppm	Ti ppm	Y ppm	V ppm	Cr ppm	Co ppm	Ni ppm	Cu ppm	Zn ppm	Ga ppm	Tl ppm	Pb ppm	La ppm	Ce ppm	Pr ppm	Nd ppm	Sm ppm	Eu ppm	Gd ppm	Tb ppm	Dy ppm	Ho ppm	Er ppm	Tm ppm	Yb ppm	Lu ppm	U ppm	Th ppm	Ba ppm	Rb ppm	Cs ppm	K ppm	Sr ppm	Ta ppm	Nb ppm	Zr ppm
N-3	3.1	6.576	40.4	395	195	49	321	114	71	19	0.10	-5	5.92	15.1	2.33	12.5	3.99	1.50	5.24	1.05	6.74	1.40	4.03	0.603	3.94	0.595	0.11	0.42	127	12	0.9	0.35	149	0.6	8.7	114
N-4A	2.3	4.482	31.8	279	152	19	69	241	-30	11	-0.05	-5	4.58	11.8	1.82	9.85	3.23	1.00	4.41	0.85	5.46	1.19	3.33	0.497	3.12	0.474	0.07	0.28	268	4	0.5	0.27	129	0.3	4.6	79
N-5	2.9	6.822	37.6	346	216	41	109	89	46	18	-0.05	-5	5.59	14.5	2.20	11.9	3.92	1.49	5.23	1.02	6.60	1.36	3.92	0.610	3.76	0.585	0.10	0.34	95	8	1.0	0.23	135	0.5	5.5	97
N-9	2.5	6.432	30.7	302	255	45	118	98	61	17	0.42	5	5.37	13.5	2.01	10.8	3.38	1.32	4.53	0.85	5.46	1.14	3.14	0.490	3.06	0.469	0.09	0.33	266	20	0.6	0.64	182	0.4	4.9	81
N-10A	2.8	6.666	39.0	371	184	29	106	64	-30	16	-0.05	-5	6.92	17.0	2.49	13.1	4.09	1.57	5.43	1.04	6.67	1.39	3.95	0.580	3.81	0.565	0.13	0.45	6	135	0.3	0.21	179	0.6	5.8	91
N-14	2.5	6.93	32.7	315	184	18	97	51	-30	15	-0.05	-5	6.62	15.9	2.31	11.9	3.68	1.42	4.74	0.90	5.48	1.15	3.28	0.492	3.12	0.470	0.12	0.42	220	15	1.1	0.45	234	0.6	5.9	87
N-18	3.5	6.288	44.4	396	181	47	141	76	101	19	0.15	-5	7.38	18.5	2.77	14.7	4.45	1.68	5.99	1.15	7.36	1.56	4.45	0.680	4.13	0.656	0.13	0.48	234	15	2.0	0.47	149	0.7	6.7	116
N-20	2.2	6.498	29.6	290	285	43	138	83	36	16	0.06	-5	5.21	12.6	1.86	9.98	3.18	1.22	4.17	0.77	4.96	1.06	2.95	0.437	2.78	0.419	0.10	0.38	229	10	0.4	0.34	197	0.6	4.6	72
N-22	4.2	5.514	48.4	389	128	46	116	66	93	19	0.08	-5	9.62	23.8	3.44	18.2	5.39	1.82	7.09	1.30	8.37	1.76	4.98	0.752	4.69	0.716	0.18	0.69	107	6	0.7	0.21	144	0.7	9.2	138
N-26	2.5	6.792	33.2	334	203	45	123	95	85	18	0.09	-5	5.13	13.0	1.95	10.7	3.52	1.33	4.54	0.89	5.68	1.22	3.44	0.522	3.25	0.503	0.10	0.35	186	12	1.3	0.38	212	0.6	4.6	82
N-27	1.8	7.206	25.8	256	218	15	75	22	-30	11	-0.05	-5	3.62	9.0	1.41	7.60	2.53	1.01	3.44	0.67	4.28	0.94	2.66	0.401	2.62	0.384	0.08	0.28	140	10	0.5	0.32	126	0.5	2.7	55
N-28	2.1	7.188	28.8	280	164	17	70	12	-30	12	-0.05	-5	4.77	11.6	1.72	9.32	2.98	1.13	3.95	0.77	4.92	1.07	3.05	0.457	2.79	0.434	0.08	0.33	2	9.4	0.4	0.26	158	1.0	3.6	66
98CC42	2.8		35	345	403	49	63	49	57	18	-0.1		5.87	14.6	2.194	11.6	3.92	1.4	5.38	0.98	6.16	1.32	4.08	0.602	3.79	0.58	0.09	0.41	138	5.4	0.5		158	0.55	5.4	95
98CC40	2.5		29	316	377	47	51	99	72	19	0.1		4.89	12.6	1.877	10.1	3.35	1.241	4.8	0.82	5.18	1.11	3.44	0.511	3.15	0.496	0.1	0.45	159	17	1.5		179	0.57	4.7	83
98CC41	1.7		24	282	384	36	55	81	29	14	-0.1		3.78	9.4	1.392	7.22	2.53	1.01	3.64	0.64	4.03	0.89	2.71	0.398	2.49	0.398	0.06	0.26	84	6.1	1		153	0.56	3.2	52

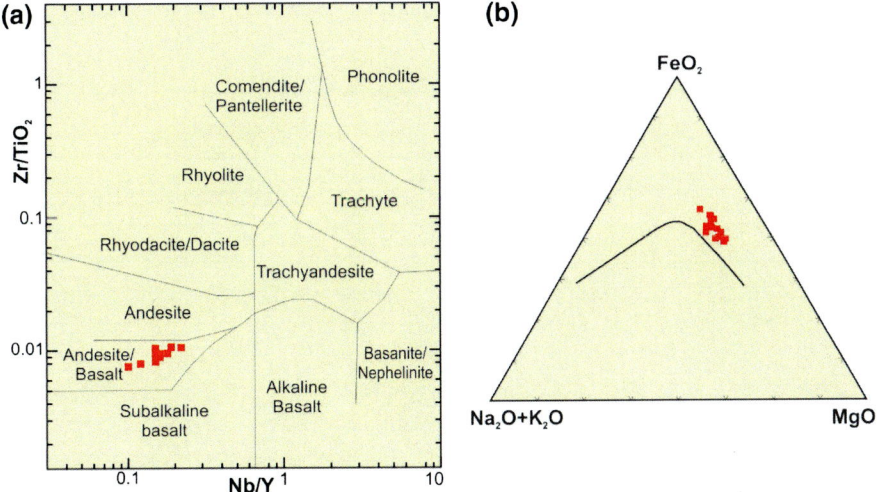

Fig. 5 **a** Classification diagram after Winchester and Floyd (1977) and **b** AFM diagram showing that samples belong to the tholeiitic series (Irvine and Baragar 1971)

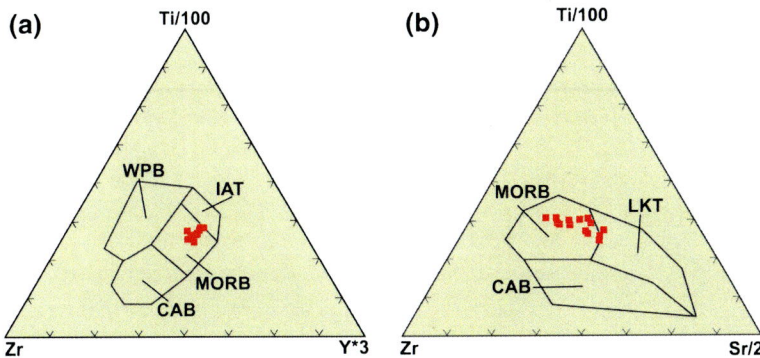

Fig. 6 In both tectonic discrimination diagrams **a** Ti/100 – Zr – (Y + 3) (after Pearce and Cann 1973) and **b** Ti/100 – Zr – Sr/2, dolerite rocks plot within the field of MORB (mid-ocean-ridge basalt). *WPB* within-plate basalt, *CAB* calc-alkaline basalt, *IAT* island arc tholeiites, *LKT* low-K tholeiites

5.2 Sm–Nd

Ten whole rock dolerite samples were used for Sm–Nd isotopic analyses (Table 4). The isotope dilution technique for Sm–Nd (using combining ^{149}Sm–^{150}Nd spike during August 2000) as well as the mass spectrometry for Sm and Nd, were carried out at the Laboratorio de Geología Isotópica, Universidade Federal do Río Grande do Sul, Porto Alegre (Brazil). The isotopic ratios were measured using the VG354

Table 2 Geochemical selected ratios of dolerite samples (n): normalized to chondrite after Taylor and McLennan (1985)

Samples	La/Yb (n)	La/Sm (n)	Sm/Yb (n)	La/Ta	Th/U	Zr/Y
N-3	1.02	0.94	1,09	9.60	3.90	2.82
N-4A	0.99	0.89	1.11	14.67	4.10	2.48
N-5	1.00	0.90	1 12	10.47	3.53	2.59
N-9	1.18	1.00	1.19	12.53	3.75	2.64
N-10A	1.23	1.06	1.15	11.76	3.54	2.32
N-14	1.44	1.13	1.27	11.52	3.55	2.66
N-1B	1.21	1.05	1.16	10.08	3.74	2.63
N-20	1.27	1.03	1.23	9.23	3.73	2.44
N-22	1.39	1.12	1.24	13.21	3.88	2.86
N-27	0.93	0.90	1.03	7.03	3.77	2.15
N-28	1.16	1.01	1.15	4.89	3.88	2.30
98CC42	1.05	0.94	1.11	10.67	4.56	2.71
98CC40	1.05	0.92	1,14	8.58	4.50	2.86
98CC41	1.03	0.94	1.09	6.75	4.33	2.17

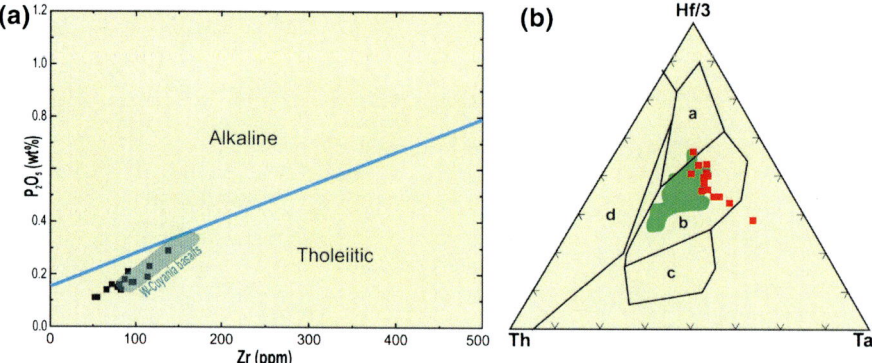

Fig. 7 a Classification diagram after Winchester and Floyd (1976). *Squares* in *black* are the El Nihuil Dolerites. For comparison the field of Western Cuyania basalts is shown (Boedo et al. 2013) and **b** tectonic discrimination diagram after (Wood et al. 1979). *Red squares* are the El Nihuil Dolerite samples. Fields *a* N-MORB; *b* E-MORB; *c* WPB; *d* VAB. Samples studied by Boedo et al. (2013) are shown for comparison as a *green-shaded area*

mass spectrometer with multiple collector system. The Nd model ages were calculated using single stage after DePaolo (1981) and second stage after Liew and Hofmann (1988) (Table 4). Epsilon Nd $_{(0)}$ values for the samples are in between +3.85 and +7.84 while $\varepsilon Nd_{(t\,=\,450\,Ma)}$ record positive values (Fig. 9) ranging from +4.27 to +12.42. Nd model ages (two stage depleted mantle) are in between 0.51 and 0.80 Ga. These values are very close due to fractionation of Sm–Nd from −0.01 to −0.03 (Table 4). Sample N4-A has a higher $\varepsilon Nd_{(t\,=\,450\,Ma)}$ value and a

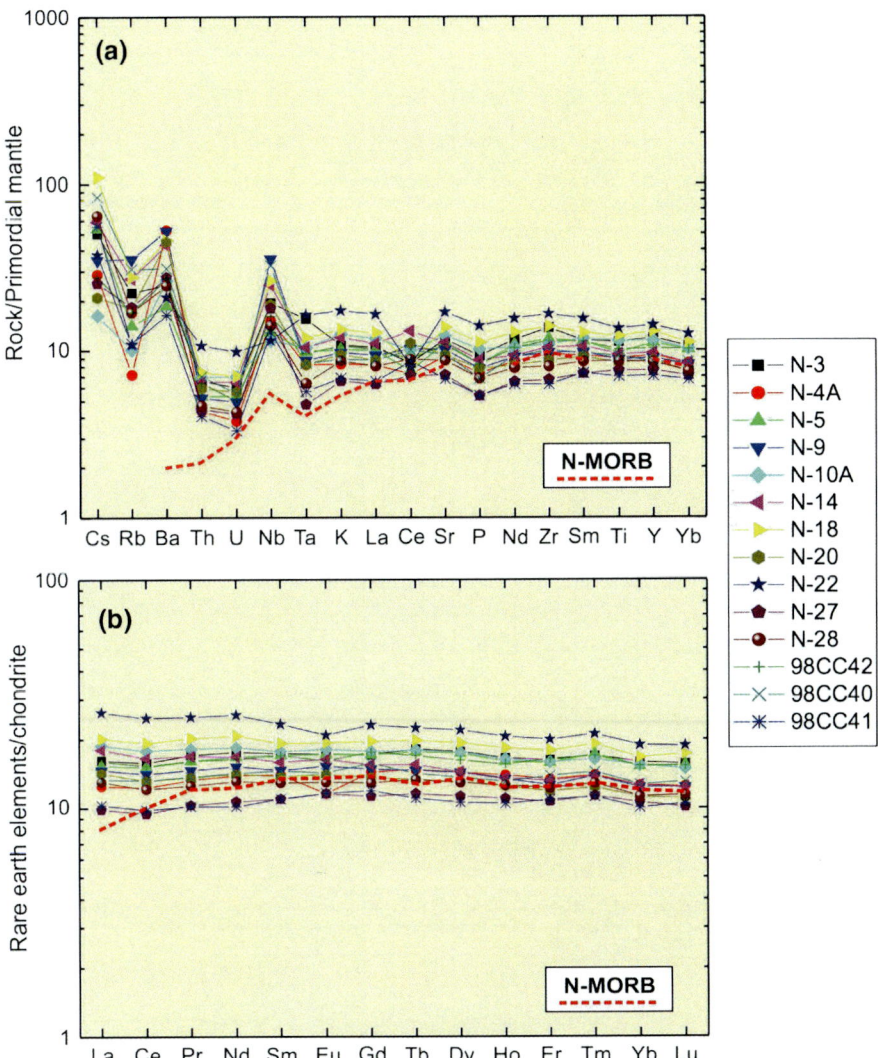

Fig. 8 a Primordial mantle normalized trace element diagram obtained from El Nihuil Dolerite samples and **b** chondrite-normalized REE diagram (Taylor and McLennan 1985). In both diagrams N-MORB average pattern is shown in *red*

Table 3 K–Ar laboratory (CPGeo, USP, Brazil) data for two dolerite samples

K–Ar EL Nihuil Dolerites									
SPK	Field No.	Material	Rock type	K%	Error	Ar40Rad ccSTP/g	Ar40 ATM	Age Ma	Error Ma
7729	99S9 A60	Whole rock	Dolerite	0.4017	0.6667	7.944	11.23	448.5	10
7730	99S30 A60	Whole rock	Dolerite	0.3095	0.5	5.901	15.23	434.2	10

Table 4 Sm–Nd whole rock data

Sample	Sm (ppm)	Nd (ppm)	$^{147}Sm/^{144}Nd$	$^{143}Nd/^{144}Nd$	±2s	$eNd(0)$*	$eNd(t)$**	$fSm-Nd$	T_{DM} (Ga)***	$T_{DM\ 2.Stage}$ (Ga)****
N4-A	1.7	10.1	0.099	0.512987	20	6.81	12.42	-0.49	0.22	
N5	3.8	11.8	0.193	0.513006	24	7.17	7.37	-0.02	0.86	0.55
N27	2.6	7.7	0.202	0.513001	24	7.07	6.77	0.03	1.35	0.60
N28	3.1	9.5	0.194	0.513040	19	7.84	8.02	-0.02	0.67	0.51
99S29	3.1	9.2	0.203	0.513012	59	7.29	6.96	0.03	1.29	0.59
99S33	2.8	8.9	0.189	0.512835	41	3.85	4.27	-0.04	1.62	0.80
99S6	2.6	8.4	0.189	0.512922	41	5.53	5.98	-0.04	1.17	0.67
99S27	2.1	6.3	0.200	0.512968	55	6.43	6.24	0.02	1.49	0.65
N9	2.7	8.5	0.190	0.512954	48	6.17	6.53	-0.03	1.05	0.62
99S30	4.2	12.7	0.198	0.512875	14	4.62	4.56	0.01	1.99	0.78

Mean of 100 isotope ratios with ionic intensity of 1.0V for ^{146}Nd and multicollector with ^{146}Nd axial collector

*Calculated assuming $^{143}Nd/^{144}Nd$ today = 0.512638 with data normalized to $^{146}Nd/^{144}Nd$ = 0.7219). Epsilon Nd(0) = (($^{143}Nd/^{144}Nd$[sample, now]/0.512638) − 1) × 10^4

**$eNd(t)$ = (($^{143}Nd/^{144}Nd$[sample, t]/143Nd/144Nd[CHUR, t]) − 1) × 10^4. t = crystallization age based on Ar–Ar WR age

***Calculated following model of DePaolo (1981)

****Two stage depleted mantle Nd model age calculated after Liew and Hofmann (1988)

Fig. 9 εNd diagram showing second stage T_{DM} ages and positive εNd values for $t = 450$ Ma (see text for explanation)

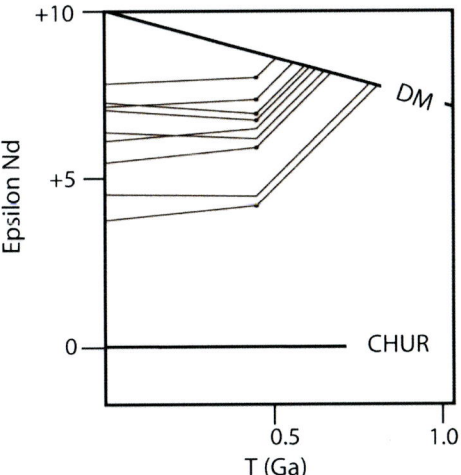

Fig. 10 Plot of samples in the $^{87}Sr/^{86}Sr$ versus $1/Sr$ diagram

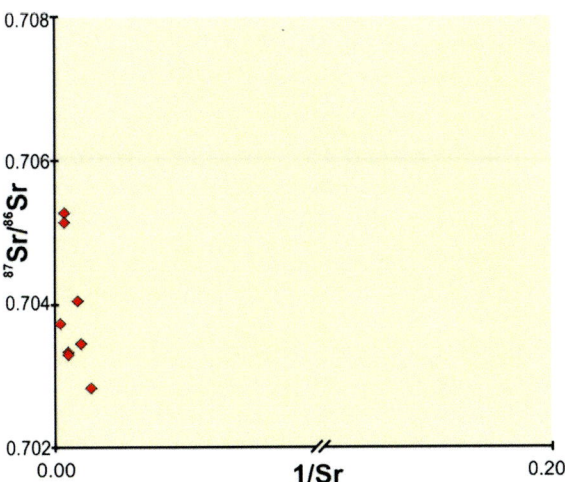

$Nd_{T_{DM}}$ age younger than the crystallization age, but the f_{Sm-Nd} is indicating that these aberrant values are consequence of REE remobilization, probably due to alteration as deduced by LILE enrichments.

5.3 $^{87}Sr/^{86}Sr$ Ratios

The $^{87}Sr/^{86}Sr$ isotopic ratios were applied using eight whole rock dolerite samples obtained along the central and southern region of the 'El Nihuil Mafic Unit'

Table 5 Rb–Sr laboratory data

LAB no.	Sample no.	Rb (ppm)	Sr (ppm)	1/Sr	$^{87}Sr/^{86}Sr$ (Y)	$^{87}Rb/^{86}Sr$ (X)	T (Ma) $\lambda = 1.42$	$(^{87}Sr/^{86}Sr)_0$ T (Ma)
CIG 1103	N9	23.4	200.9	0.0049776	0.70643	0.337	450	0.70427
CIG 1106	99S27	24.8	176.7	0.0056593	0.70633	0.406	450	0.70372
CIG 1195	99S7	7.6	190.5	0.0052493	0.70458	0.116	450	0.70384
CIG 1196	99S13B	6.8	342.6	0.0029189	0.70369	0.057	450	0.70332
CIG 1198	99S30	20.6	168.6	0.0059312	0.70566	0.354	450	0.7034
CIG 1199	99S39	8.7	129.3	0.007734	0.7042	0.195	450	0.70295
CIG 1200	N3	13.1	151.5	0.0066007	0.70433	0.25	450	0.70323
CIG 1202	N27	12.6	135.9	0.0073584	0.70509	0.268	450	0.70509

(Fig. 2). The sample preparation and extraction of natural Sr through cation exchange columns was performed at the Centro de Investigaciones Geológicas, Universidad Nacional de La Plata, Argentina and mass-spectrometer analyses were performed at the Centro de Pesquisas Geocronológicas, University of São Paulo, Brazil. As it is shown in Fig. 10 and Table 5 the $^{87}Sr/^{86}Sr$ ratios are in between 0.7032 and 0.7050 in agreement with ocean ridge tholeiites.

6 The Western Precordillera Mafic Belt: A Comparison

Haller and Ramos (1984), Ramos et al. (1999) and Ramos (2004) were the authors that linked the mafic-type rocks and their extension to the tectonic suture zone between Chilenia and Cuyania terranes in the proto-Andean margin of Gondwana (Fig. 11). However, the ophiolitic signature of the mafic and ultramafic belt developed along western Precordillera was established by Borrello (1963, 1969). This mafic belt that extends over ca. 1000 km from south to north was named by Haller and Ramos (1984, 1993) as the 'Famatinian ophiolites' emplaced as a disrupted ophiolite during the Early Paleozoic (Ramos et al. 1984, 1986) Mahlburg Kay et al. (1984) based on geochemical data proposed that the western Precordillera mafic rocks could have formed in a broad back-arc basin or at a mid-ocean ridge with an enriched source or as an early oceanic rift next to a continental margin. Davis et al. (1999) challenged this interpretation suggesting the occurrence of two different sections of an ophiolite assemblage in Western Precordillera a Proterozoic one and another of Early Paleozoic age. More recently Boedo et al. (2013) discussed the E-MORB (enriched-type MORB) like signature of the mafic dykes and sills studied along the western Precordillera mafic-ultramafic belt as part of a non-subduction related ophiolite. González Menéndez et al. (2013) argued that the studied mafic rocks related to subduction or either N-MORB or OIB environment derived from primordial garnet-spinel transition mantle sources. The model supports a thinned continental margin between Chilenia and Cuyania terranes

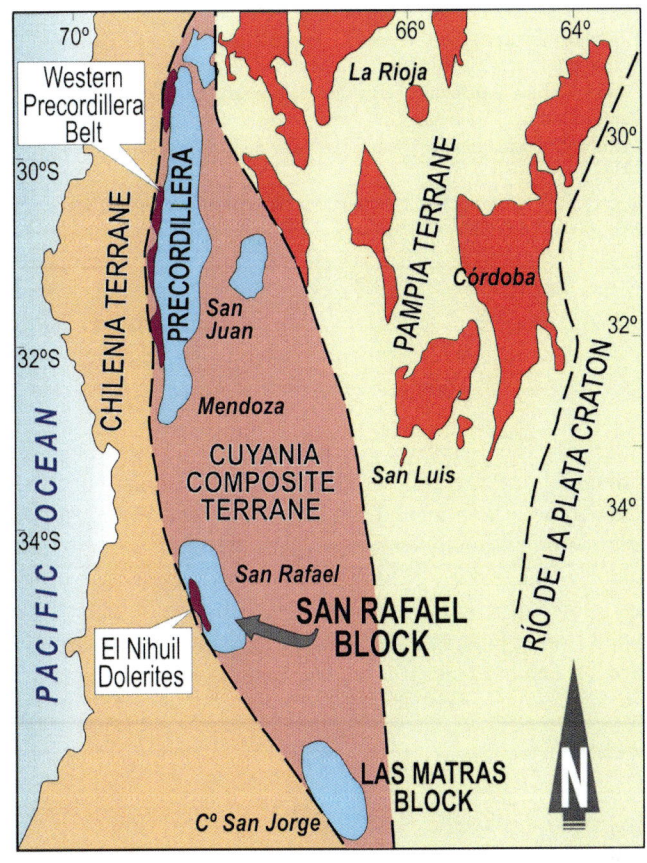
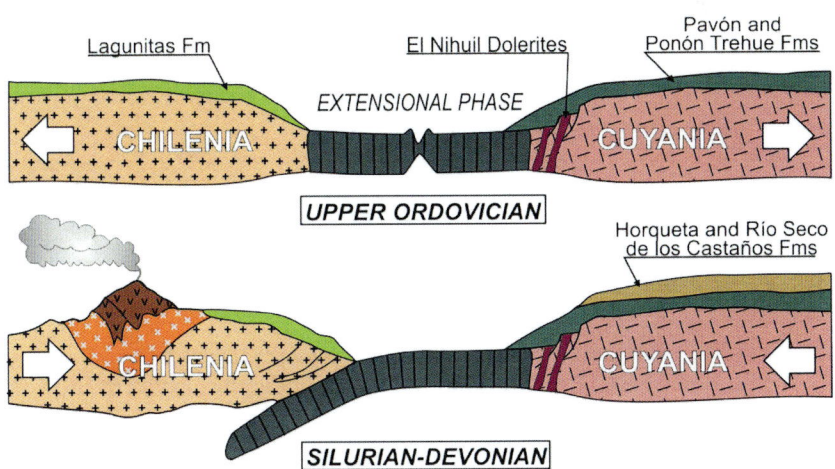

◄**Fig. 11 a** Location of the western Precordillera mafic belt in the context of the Cuyania terrane, **b** tectonic diagram showing the *Lower* Paleozoic extensional phase and intrusion of the El Nihuil Dolerites in a western continental margin of Cuyania. Ordovician sedimentary sequences are exposed over Cuyania (Ponón Trehué and Pavón Fms) and Lagunitas Fm in the eastern border of Chilenia terrane, and **c** during Silurian-Devonian time, west vergence subduction is developed generated a magmatic arc in Chilenia terrane before terranes collision during the Chanic tectonic phase

('Occidentalia terrane' after Dalla Salda et al. 1998) during the Middle to Late Ordovician. Many authors (Alonso et al. 2008 and references therein) described an extensional regime developing in a passive margin environment during the Ordovician in the western margin of Precordillera. These data were discussed in several geotectonic models proposed by different authors such as Ramos et al. (1986), Dalla Salda et al. (1992), Rapela et al. (1998), Davis et al. (2000), Gerbi et al. (2002), Ramos (2004), Thomas et al. (2012), González Menéndez et al. (2013) and Boedo et al. (2013).

The El Nihuil Dolerites record a geochemical and isotopic signature similar to that exposed along the western Precordillera belt (Boedo et al. 2013) as a subduction unrelated type at continental margin (Dilek and Furnes 2011). Figure 11 summarized the tectonic scenario during extensional and compressional phases.

It is noteworthy that the El Nihuil mafic rocks record two different tectonic settings: a Mesoproterozoic crustal fragment (represented by gabbros, amphibolites and tonalites) and Lower Paleozoic intrusive MORB-dolerite dykes, such as the two sections defined by Davis et al. (1999) for the mafic belt of western Precordillera.

7 Final Remarks

1. The undeformed porphyritic dolerites composed of sills and dykes are the predominating type rocks in the 'El Nihuil Mafic Unit.'
2. The country rocks for the dolerites are the high deformed gabbros, amphibolites, and tonalites interpreted as part of the Mesoproterozoic continental crust, exposed mainly at the 'Loma del Petiso shear zone' and Lomas Orientales along the northern sector of the 'El Nihuil Mafic Unit.'
3. The highly deformed mafic rocks and dolerite dykes are in unconformity with the Río Seco de los Castaños Fm as a Silurian-Devonian marine-deltaic system which preserved primary sedimentary structures.
4. The dolerites are geochemically characterized as tholeiite type rocks, and plot within the MORB field in tectonic discrimination diagrams. The trace and REE patterns are coherent with this origin.
5. Geochemical data from the El Nihuil Dolerites record a less evolved E-MORB-type signature than the mafic-ultramafic belt known in western Precordillera region.
6. The K–Ar (WR ages) indicate that the dolerite dykes are a unique Lower Paleozoic mafic event exposed within the entire SRB.
7. The juvenile depleted mantle origin is therefore confirmed by positive εNd values ranging from +4.27 to +12.42 at 450 Ma and Nd model ages (second

stage) from 0.51 to 0.80 Ga. The obtained $^{87}Sr/^{86}Sr$ ratios varying from 0.7032 to 0.7050 are in agreement with ocean ridge tholeiites.

8. The El Nihuil dolerite rocks (dykes and sills with chilled margins) formed a non-subduction related ophiolite within a margin facing a shallow ocean basin and could be interpreted as intruded into pre-Lower Silurian sedimentary units at marine and slope environments.
9. These dolerite rocks are interpreted as a shallow section of the dismembered 'Famatinian ophiolite belt' emplaced during the Upper Ordovician-Lower Silurian extensional event, within a thinned continental crust environment on the western side of the Cuyania terrane.

Acknowledgements This research was partially funded by CONICET (PIP 199) and by grants from the University of La Plata, Argentina. Many thanks to our colleagues Leandro Ortiz and Diego Licitra for their help during fieldworks and laboratory sample preparation. We appreciate the suggestions and guidance during laboratory work by Prof. Koji Kawashita (Brazil) and Prof. Ricardo Varela (CIG). We thank Gabriela Coelho dos Santos for her technical collaboration in petrographical descriptions.

References

Abre P, Cingolani C, Zimmermann U, Cairncross B (2009) Detrital chromian spinels from Upper Ordovician deposits in the Precordillera terrane, Argentine: a mafic crust input. J South America Earth Sci, 28:407–418 (Special Issue on Mafic and Ultramafic complexes in South America and the Caribbean)

Abre P, Cingolani CA, Manassero MJ (this volume) The Pavón Formation as the Upper Ordovician unit developed in a turbidite sand-rich ramp. San Rafael Block, Mendoza, Argentina. In: Cingolani CA (Ed.) The pre-Carboniferous evolution of the San Rafael Block, Argentina. Implications in the SW Gondwana margin, Chapter, 5. Springer

Alonso JI, Gallastegui J, García-Sansegundo J, Farias P, Rodríguez Fernández LR, Ramos VA (2008) Extensional tectonics and gravitational collapse in an Ordovician passive margin: the Western Argentine Precordillera. Gondwana Res 13:204–215

Arévalo Jr R, McDonough WF (2010) Chemical variations and regional diversity observed in MORB. Chem Geol 271:70–85

Boedo EL, Vujovich GI, Kay SM, Ariza JP, Pérez Luján SB (2013) The E-MORB like geochemical features of the Early Paleozoic mafic-ultramafic belt of the Cuyania terrane, western Argentina. J S Am Earth Sci 48:73–84

Borrello AV (1963) Elementos del magmatismo simaico inicial en la evolución de la secuencia geosinclinal de la Precordillera. Instituto Nac. Inv. Ciencias Naturales y Museo Argentino de Ciencias Naturales "B. Rivadavia". Revista, Serie Ciencias Geológicas 1:1–19. Buenos Aires

Borrello AV (1969) Los geosinclinales de la Argentina. Dirección Nacional de Geología y Minería, Anales, v.14:1–188. Buenos Aires

Cingolani CA, Llambías EJ, Ortiz L (2000) El magmatismo básico pre-Carbónico del Nihuil, Bloque de San Rafael, Provincia de Mendoza, Argentina. 9° Congreso Geológico Chileno, 2:717–721. Puerto Varas, Chile

Cingolani CA, Uriz NJ, Marques J, Pimentel M (2012) The Mesoproterozoic U–Pb (LA-ICP-MS) age of the Loma Alta gneissic rocks: Basement remnant of the San Rafael Block, Cuyania terrane, Argentina. VIII South American Symposium on Isotope Geology, Medellin, Colombia, 1p. abstracts book

Cingolani CA, Basei MAS, Varela R, Llambías EJ, Chemale Jr F., Abre P, Uriz NJ, Marques, JC (this volume) The Mesoproterozoic Basement at the San Rafael Block, Mendoza Province

(Argentina): geochemical and isotopic age constraints. In: Cingolani CA (ed) The pre-Carboniferous evolution of the San Rafael Block, Argentina. Implications in the SW Gondwana margin, Chapter, 2 Springer

Criado Roqué P, Ibáñez G (1979) Provincia sanrafaelino-pampeana. In: Turner JCM (ed) Simposio de Geología Regional Argentina, No 2, vol 1. Academia Nacional de Ciencias, Córdoba, pp 837–869

Dalla Salda LH, Cingolani CA, Varela R (1992) Early Paleozoic orogenic belt of the Andes in southeastern South America: result of Laurentia-Gondwana collision? Geology 20:617–620

Dalla Salda LH, López de Luchi MG, Cingolani CA (1998) Laurentia-Gondwana collision: the origin of the Famatinian-Appalachian Orogenic Belt (A review). In: Pankhurst RJ, Rapela CW (eds) The Proto-Andean Margin of Gondwana, vol 142. Geological Society, London, Special Publications, pp 219–234

Davicino RE, Sabalúa JC (1990) El Cuerpo Básico de El Nihuil, Dto. San Rafael, Pcia. de Mendoza, Rep. Argentina. In Congreso Geológico Argentino, No. 11, Actas I. San Juan, pp 43–47

Davis J, Roeske S, McClelland W, Snee L (1999) Closing an ocean between the Precordillera terrane and Chilenia; early Devonian ophiolite emplacement and deformation in the southwest Precordillera. In: Ramos VA, Keppie JD (eds) Laurentia-Gondwana connections before Pangea, vol 336. Geological Society of America, Special Publication, pp 115–138

Davis J, Roeske S, McClelland W, Kay SM (2000) Mafic and ultramafic crustal fragments of the southwestern Precordillera terrane and their bearing on tectonic models of the early Paleozoic in western Argentina. Geology 28(2):171–174

DePaolo DJ (1981) Neodymium isotopes in the Colorado Front Range and crust–mantle evolution in the Proterozoic. Nature 291:193–196

Dessanti RN (1956) Descripción geológica de la Hoja 27-C Cerro Diamante (Provincia de Mendoza). Dirección Nacional de Geología y Minería, Boletín, vol 85. Buenos Aires, pp 1–79

Dilek Y, Furnes H (2011) Ophiolite genesis and global tectonics: geochemical and tectonic fingerprinting of ancient oceanic lithosphere. Geol. Soc. Am. Bull. 123:387–411

Gale A, Dalton C, Langmuir CH, Su Y, Schilling JG (2013) The mean composition of ocean ridge basalts. Geochem. Geophys. Geosyst. 14:489–517

Gerbi C, Roeske SM, Davis JS (2002) Geology and structural history of the southwestern Precordillera margin, northern Mendoza province, Argentina. J S Am Earth Sci 14:821–835

González-Menéndez L, Gallastegui G, Cuesta A, Heredia N, Rubio-Ordoñez A (2013) Petrogénesis of Early Paleozoic basalts and gabros in the western Cuyania terrane: constraints on the tectonic setting of the southwestern Gondwana margin (Sierra del Tigre, Andean Argentine Precordillera). Gondwana Res 24(1):359–376

González Díaz EF (1981) Nuevos argumentos en favor del desdoblamiento de la denominada "Serie de la Horqueta", del Bloque de San Rafael, Provincia de Mendoza. In 8° Congreso Geológico Argentino, Actas 3:241–256. San Luis

Haller MJ, Ramos VA (1984). Las ofiolitas famatinianas (Eopaleozoico) de las Provincias de San Juan y Mendoza. 9° Congreso Geológico Argentino, Actas 3:66–83. S.C. Bariloche

Haller MJ, Ramos VA (1993). Las ofiolitas y otras rocas afines. In: Ramos, V.A. (Ed.) Geología y Recursos Naturales de Mendoza, Relatorio 12° Congreso Geológico Argentino, pp 31–39

Irvine TN, Baragar WRA (1971) A guide to the chemical classification of the common volcanic rocks. Can J Earth Sci 8:523–548

Le Maitre RW (1989) A Classification of Igneous Rocks and Glossary of Terms. In Blackwell. Oxford, 193 pp

Liew TC, Hoffmann AW (1988) Precambrian crustal components, plutonic associations, plate environments of Hercynian Fold Belt of Central Europe: indications from Nd and Sr isotopic study. Contrib Mineral Petrol, 98:129–138

Linares E, Parica C, Parica P (1987) Catálogo de edades radimétricas determinadas para la República Argentina (IV años 1979–1980 realizadas por INGEIS y sin publicar y V años 1981–1982 publicadas). Publicaciones Especiales, Asociación Geológica Argentina, Serie B (Didáctica y Complementaria) n. 15:1–49

Malhburg Kay S, Ramos VA, Kay RW (1984) Elementos mayoritarios y trazas de las vulcanitas ordovícicas de la Precordillera Occidental: basaltos de rift oceánico temprano (¿) próximos al margen continental. 9° Congreso Geológico Argentino, Actas 2:48–65. S.C. de Bariloche

Nuñez E (1976) Descripción geológica de la Hoja 28-C "Nihuil". Unpublished report, Provincia de Mendoza. Servicio Geológico Nacional. Buenos Aires

Padula E (1949) Descripción geológica de la Hoja 28-C "El Nihuil". Unpublished report, Provincia de Mendoza. YPF.

Pearce JA, Cann JR (1973) Tectonic setting of basic volcanic rocks determined using trace element analyses. Earth Planet Sci Lett 19:290–300

Ramos VA (1984) Patagonia: un continente paleozoico a la deriva? IX Congreso Geológico Argentino (San Carlos de Bariloche). Actas 2:311–325

Ramos VA, Jordan T, Allmendinger R, Kay S, Cortés J, Palma M (1984) Chilenia: un terreno alóctono en la evolución paleozoica de los Andes Centrales. 10° Congreso Geológico Argentino, Actas 2:84–106

Ramos VA, Jordan T, Allmendinger R, Mpodozis C, Kay S, Cortés J, Palma M (1986) Paleozoic terranes of the central Argentine-Chilean Andes. Tectonics 5:855–888

Ramos VA, Dallmeyer RD, Vujovich G (1999) Time constraints on the Early Palaeozoic docking of the Precordillera, central Argentina. In Pankhurst RJ, Rapela CW (eds) The Proto-Andean Margin of Gondwana, vol 142. Geological Society, Special Publications. London, pp 143–158

Ramos VA, Escayola M, Mutti DI, Vujovich GI (2000) Proterozoic-early Paleozoic ophiolites of the Andean basement of southern South America. In: Dilek Y, Moores EM, Elthon D, Nicolas A (eds.) Ophiolites and Oceanic crust: New insights from Field Studies and the Ocean Drilling Program: Boulder, Colorado, vol 349. Geological Society of America, Special Paper, pp 331–349

Ramos VA (2004) Cuyania, an exotic block to Gondwana: Review of a historical success and the present problems. Gondwana Res 7:1009–1026

Rapela CW, Pankhurst RJ, Casquet C, Baldo E, Saavedra J, Galindo C (1998) Early evolution of the Proto-Andean margin of South America. Geology 26:707–710

Sepúlveda E (1999) Descripción geológica preliminar de la Hoja Embalse El Nihuil, provincia de Mendoza, escala 1:250.000. SEGEMAR, Boletín, vol 268. Buenos Aires, pp 84

Sepúlveda E, Carpio FW, Regairaz MC, Zárate M, Zanettini JCM (2007) Hoja Geológica 3569-II (San Rafael, Provincia de Mendoza). Servicio Geológico Minero Argentino, Boletín 321, Buenos Aires, 59 pp

Taylor SR, McLennan SM (1985) The continental crust. Its composition and evolution. Blackwell, London 312 pp

Thomas WA, Tucker RD, Astini RA, Denison RE (2012) Ages of pre-rift basement and synrift rocks along the conjugate rift and transform margins of the Argentine Precordillera and Laurentia. Geosphere 8(6):1366–1383

Winchester JA, Floyd PA (1976) Geochemical magma type discrimination, application to altered and metamorphosed basic igneous rocks. Earth Planet Sci Lett 28(3):325–343

Winchester JA, Floyd PA (1977) Geochemical discrimination of different magma series and their differentiation products using immobile elements. Chem Geol, 20:325–343

Wood DA, Joron JL, Treuil M (1979) A reappraisal of the use of trace elements to classify and discriminate between magma series erupted in different tectonic settings. Earth Planet Sci Lett 45(2):326–336

Wichmann R (1928) Reconocimiento geológico de la región de El Nihuil, especialmente relacionado con el proyectado dique de embalse del Río Atuel. Dirección Nacional de Minería (unpublished report), Buenos Aires

Magnetic Fabrics and Paleomagnetism of the El Nihuil Mafic Unit, San Rafael Block, Mendoza, Argentina

Augusto E. Rapalini, Carlos A. Cingolani and Ana María Walther

Abstract A reconnaissance magnetic fabric and paleomagnetic study was carried out on the El Nihuil mafic unit exposed in the San Rafael Block, Mendoza Province, Argentina. Sampling comprised eight sites on three different lithologies. While sites N1 and N2 corresponded to gneisses, N3 to N6 were located on dolerite outcrops. Sites N7 and N8 were situated on the Loma Alta gabbro, although site N8 corresponded to a dolerite intruding the gabbro. The magnetic fabric shows that the bulk susceptibility ranges from 4.9×10^{-4} SI to 1.7×10^{-2} SI, with the gabbro showing the lowest values and the dolerite the highest. Despite the different lithologies all sites showed a low anisotropy degree and an oblate fabric. Distribution of foliation planes shows a dominant NW trending describing an antiformal shape for the whole unit. Lineations were in most cases of low angle. Most samples showed well-defined but scattered remanence directions which indicate that no significant paleomagnetic data can be obtained from El Nihuil mafic unit.

Keywords Mesoproterozoic basement · Ordovician dolerites · Paleomagnetic data

---

A.E. Rapalini (✉) · A.M. Walther
Departamento de Ciencias Geológicas, Facultad de Ciencias Exactas y Naturales,
Instituto de Geociencias Básicas, Aplicadas y Ambientales de Buenos Aires (IGEBA),
Universidad de Buenos Aires CONICET, Buenos Aires, Argentina
e-mail: rapalini@gl.fcen.uba.ar

A.M. Walther
e-mail: walther@gl.fcen.uba.ar

C.A. Cingolani
CONICET-UNLP, Centro de Investigaciones Geológicas,
Diag. 113 n. 275, CP1904 La Plata, Argentina
e-mail: carloscingolani@yahoo.com

© Springer International Publishing AG 2017
C.A. Cingolani (ed.), *Pre-Carboniferous Evolution of the San Rafael Block, Argentina*, Springer Earth System Sciences,
DOI 10.1007/978-3-319-50153-6_7

1 Introduction

The San Rafael Block is located in the southwestern corner of the Mendoza Province, Argentina. It presents small outcrops of Mesoproterozoic rocks covered by remnants of a Ordovician carbonate–siliciclastic platform that have permitted its correlation with the Precordillera region in northern Mendoza, San Juan, and southern La Rioja Provinces. Ramos (2004 and references there in) proposed that it integrated the composite Cuyania allochthonous terrane, of Laurentian affinities, which was accreted to the southwestern margin of Gondwana in late Ordovician times (e.g., Astini et al. 1995; Thomas and Astini 1996, among many others). The El Nihuil mafic unit (Dessanti 1956; Cingolani et al. 2000, 2011 and references therein) is a singular composite body, elongated in a NNE-SSW direction. The extension of its outcrops is of around 73 km^2 (Fig. 1). Its central area, known as Loma Alta, is located at 34° 58′ 50″S, 68° 45′ 00″W. The unit is in contact with the surrounding lithologies by unconformities or faults which limit significantly any stratigraphic inference on its age. It is mainly composed of medium to fine-grained

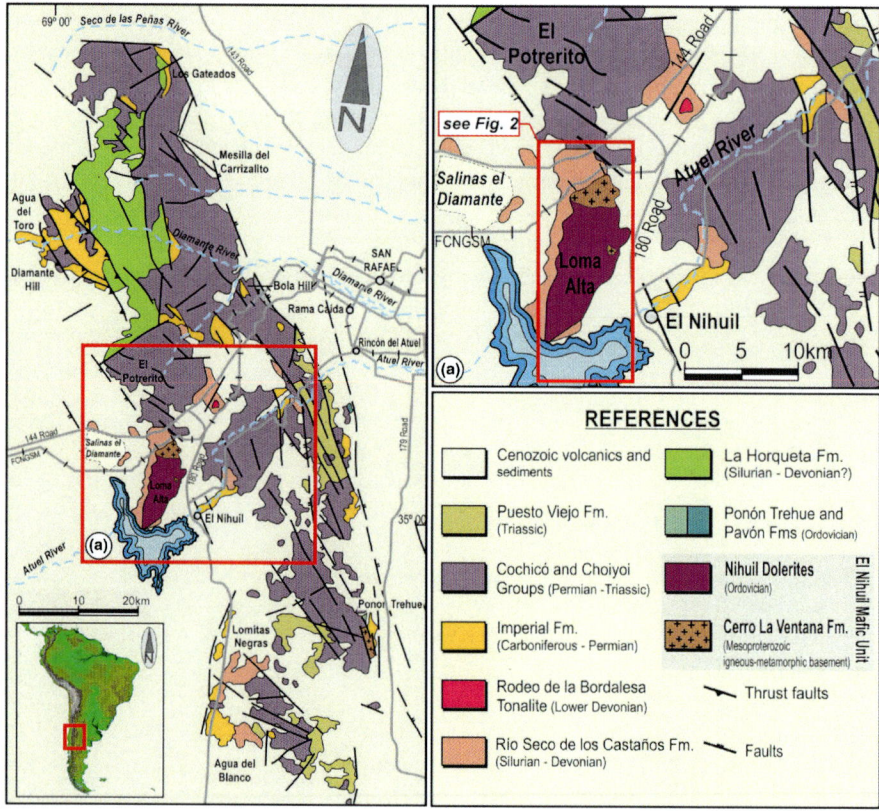

Fig. 1 Location map and geologic sketch of the El Nihuil mafic unit

deformed mafic rocks (gabbros, amphibolites) that are frequently intruded by un-deformed dolerite rocks. To the eastern margin of the unit, small outcrops of ortogneiss of intermediate composition have been observed. In many areas the rocks are significantly altered or covered. The gneiss has been recently dated (Cingolani et al. 2012) by U-Pb (LA-ICP-MS) in zircons as 1266 ± 15 Ma which is very similar to ages obtained for the Cuyania basement in other localities. K-Ar ages on the dolerites (470, 480, and 483 Ma Cingolani et al. 2000) point to a lower Middle Ordovician age and allows correlation with the so-called "Famatinian ophiolites" that mark the western margin of the Cuyania terrane. Geochemical characterization of these dolerites (Cingolani et al. 2000) indicates tholeiitic and MORB origin.

2 Sampling and Laboratory Procedures

In order to perform a preliminary paleomagnetic and magnetic fabric study, fifty-seven cores were drilled at eight different sites on the composite El Nihuil Mafic Unit (Fig. 2). Two sites (N1, N2) were located on the gneisses exposed in the Lomas Orientales. Four sites (N3, N4, N5, and N6) were located on outcrops of the early Paleozoic dolerites in the central and southern areas of the unit. The remaining two sites (N7, N8) were located in the Mesoproterozoic 'Loma del Petiso shear zone' gabbro-amphibolite rocks in the northern exposures of the mafic unit. However, only N7 belongs to that unit while site N8 corresponds to a dolerite intruding the gabbro. Figure 2 shows the distribution of the sampling sites (Table 1). Samples were collected with a portable gasoline-powered drill with non-magnetic bits. They were oriented respect to the magnetic north with a magnetic compass, the geographic north with a sun compass, and the vertical with a clinometer. Samples were cylinders of 2.54 cm in diameter and 5–9 cm long. No systematic deviations were observed between the magnetic and sun compass readings. Samples were sliced in laboratory into 2.2-cm-high specimens. One to three specimens were obtained from each sample (Table 1).

Magnetic fabrics were measured by means of the anisotropy of magnetic susceptibility (AMS). A multi-frequency susceptometer MFK1-A (Agico) was used to measure the bulk susceptibility and the anisotropy parameters (Jelinek 1978). Results from all samples for each site were used to compute the mean site susceptibility ellipsoid and its directional and scalar parameters (Table 2 and Fig. 3).

The paleomagnetic study was carried out with a 2G DC squid cryogenic magnetometer, with a static 3 axis degausser attached to the magnetometer and a dual-chamber ASC thermal demagnetizer. All studies were carried out at the laboratory of Paleomagnetism 'Daniel A. Valencio' (IGEBA, Universidad de Buenos Aires, Argentina).

Fig. 2 Distribution of sampling sites. The paleomagnetic field work is shown in the *inset*

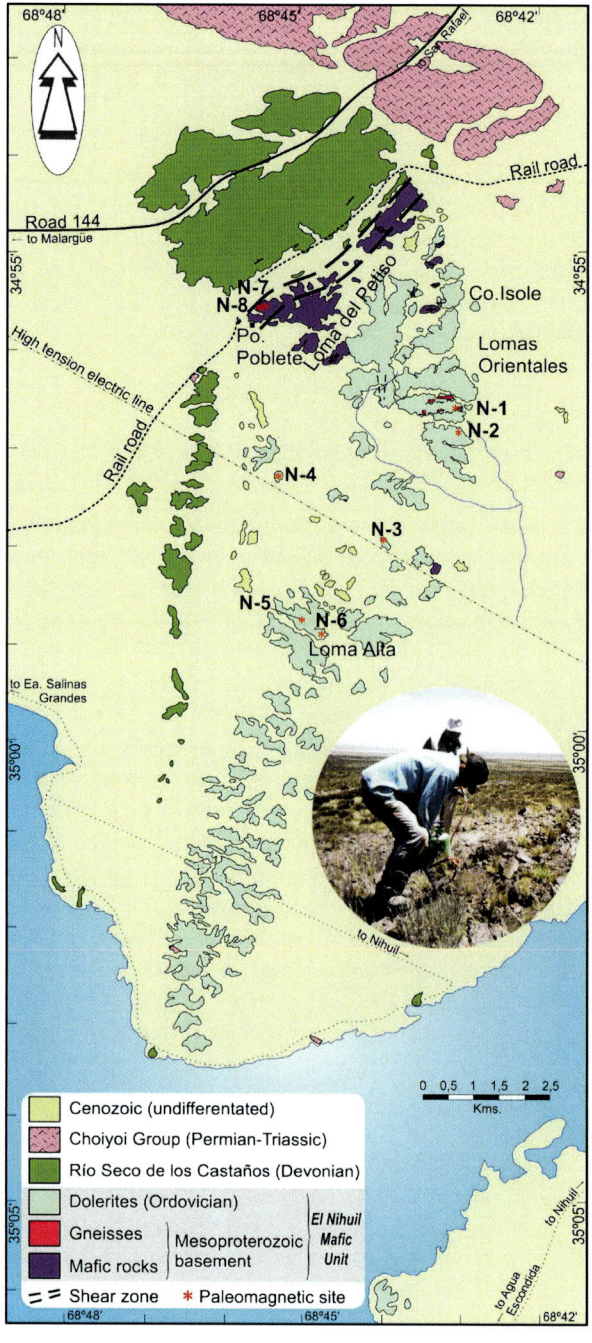

Table 1 GPS coordinates of sampling paleomagnetic sites

Site	Location (GPS)
N1	34° 56' 50"–68° 43' 03"
N2	34° 56' 56.3"–68° 42' 59"
N3	34° 58' 02.6"–68° 43' 55.3"
N4	34° 57' 16.6"–68° 45' 23.8"
N5	34° 58' 49.5"–68° 45' 0.7"
N6	34° 58' 55.6"–68° 44' 48.4"
N7	34° 55' 35.6"–68° 45' 21.6"
N8	34° 55' 35.4"–68° 45' 22.1"

Table 2 Mean site AMS data for the El Nihuil composite unit

Site	n	k1		k3		k	L	F	P_j	T
		D (°)	I (°)	D (°)	I (°)					
N1	7	296.3	74	48.2	6.1	1.19E−03	1.003	1.007	1.010	0.41
N2	5	327.9	29.4	227.7	17.5	1.37E−03	1.006	1.004	1.011	−0.18
N3	6	324.9	9.6	228.9	31.6	3.24E−03	1.008	1.007	1.015	−0.11
N4	11	309.5	49.7	95.6	35.2	9.91E−04	1.002	1.015	1.019	0.75
N5	8	302.5	8.6	107.3	81.1	3.00E−03	1.010	1.018	1.028	0.29
N6	9	299.6	5.9	75.1	81.8	1.69E−02	1.020	1.018	1.039	−0.62
N7	9	46.7	15	161	56.6	4.87E−04	1.001	1.012	1.015	0.67
N8	8	315.4	66.2	141	237	1.09E−03	1.004	1.008	1.012	0.32

n number of samples per site; *k* bulk susceptibility; *L* lineation; *F* foliation; P_j Jelinek's corrected anisotropy degree; *T* Jelinek's shape parameter

3 Magnetic Fabric Results

Mean site AMS parameters are presented in Table 2 and illustrated in Fig. 3. Site mean bulk susceptibility (k) values range between 4.9×10^{-4} SI and 1.7×10^{-2} SI. The lowest value corresponds to the gabbro-amphibolite sampled at site N7, while the dolerite shows the highest values. Both sites on the gneiss (N1 and N2) show very similar intermediate values of 1.1 to 1.4×10^{-3} SI. In all cases the sampled rocks showed a well-defined magnetic fabric, characterized by a low anisotropy degree and a dominance of an oblate shape for the AMS ellipsoid.

Figure 4 illustrates the distribution of the site lineation and foliation plane. Lineations tend to be horizontal or of low angles in the gneiss and dolerite, and of high angle in the gabbro, but they show a systematic trend toward the NNW. On the other hand foliation planes describe an antiformal shape, with high-angle foliations dipping toward the NE on the eastern side of the unit (gneisses and dolerites) and toward the west on the western side (dolerite, N4), sites N5 and N6 correspondent to dolerite in the central areas of the unit show sub-horizontal magnetic foliations.

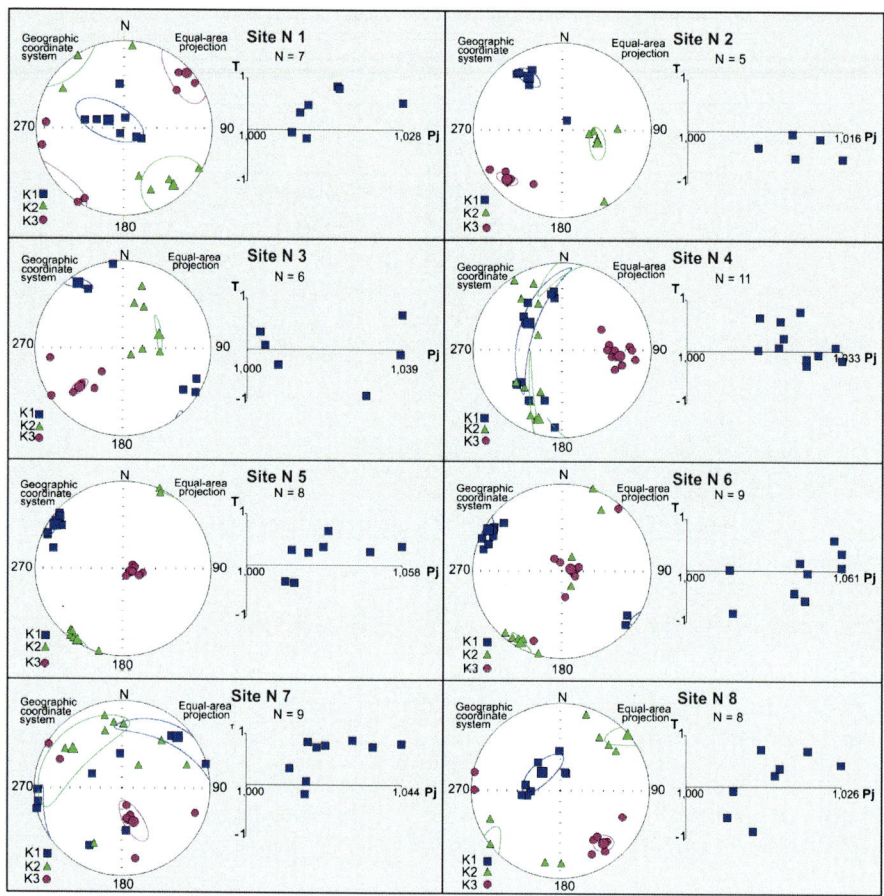

Fig. 3 Equal area stereographic projections of main axes of the anisotropy of magnetic susceptibility ellipsoid for all samples of each site of the El Nihuil mafic unit. *Pj* corrected anisotropy parameter and *T* shape parameter from Jelinek (1978, 1981)

Sites N7 and N8 on the extreme north present a significantly different orientations of the magnetic foliation. This trends NE to ENE with low (N7) and high (N8) dips toward NNW. At first sight they tend to parallel the northern boundary of the unit limited by a major fault. Lineation of site N8 also trends toward the NW but it is of high angle. Despite the small number of sites investigated, the data obtained suggest a dominant NW structural fabric of the El Nihuil mafic unit. The antiformal shape made apparent by the distribution of the foliation planes should be investigated with a more detailed sampling and observations as it may shed light on emplacement and deformational mechanisms.

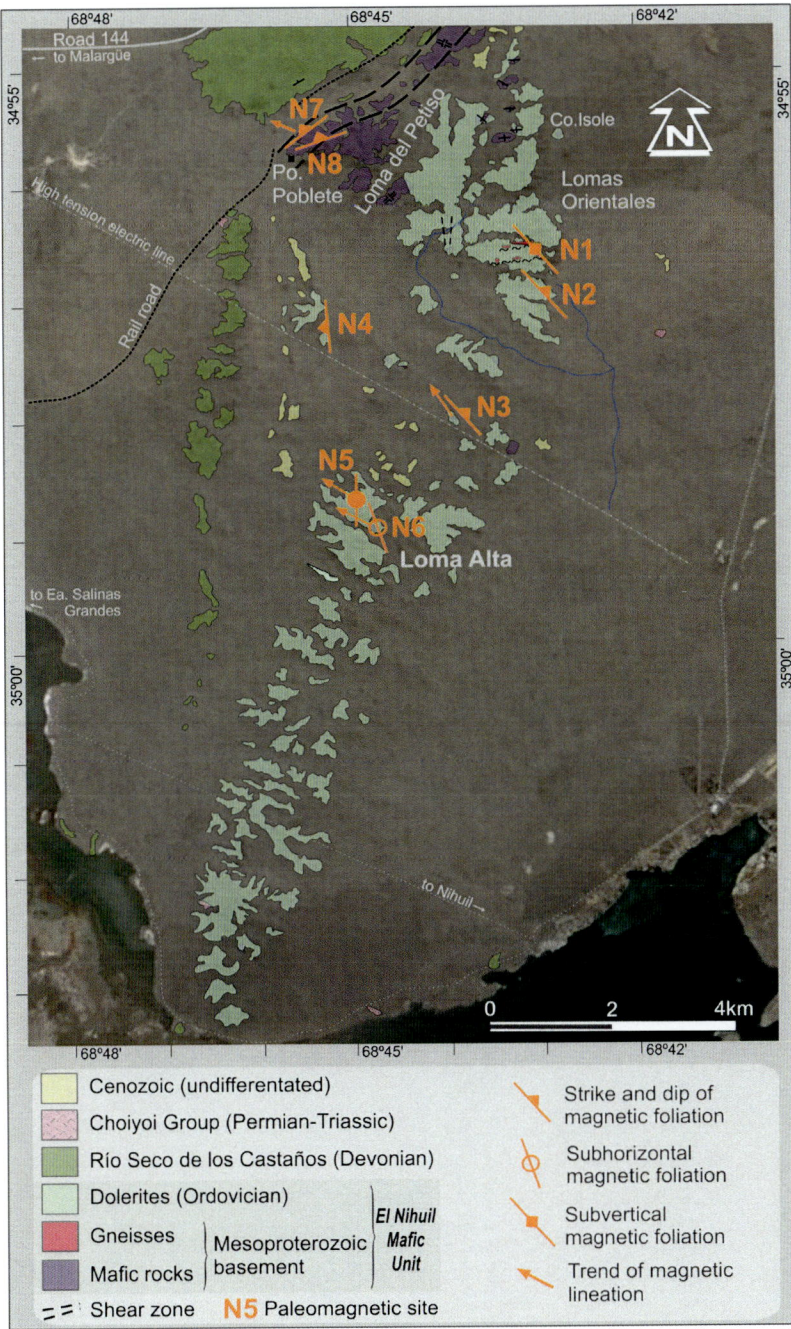

Fig. 4 Distribution of foliation planes and lineations on the El Nihuil mafic unit

4 Paleomagnetic Results

Four to five specimens per site were submitted to alternating field (AF) or thermal stepwise demagnetization. Stages were of 2, 4, 6, 8, 10, 12, 15, 20, 25, 30, 40, 50, 60, 80, 100, and 120 mT or 100, 150, 200, 250, 300, 350, 400, 450, 500, 540, and 580 °C. In most cases samples showed a linear decay toward the origin of coordinates which permitted to determine a characteristic magnetic component. Coercive forces were generally under 60 mT, step at which most specimens lost over 90 % of their original remanence (Fig. 5).

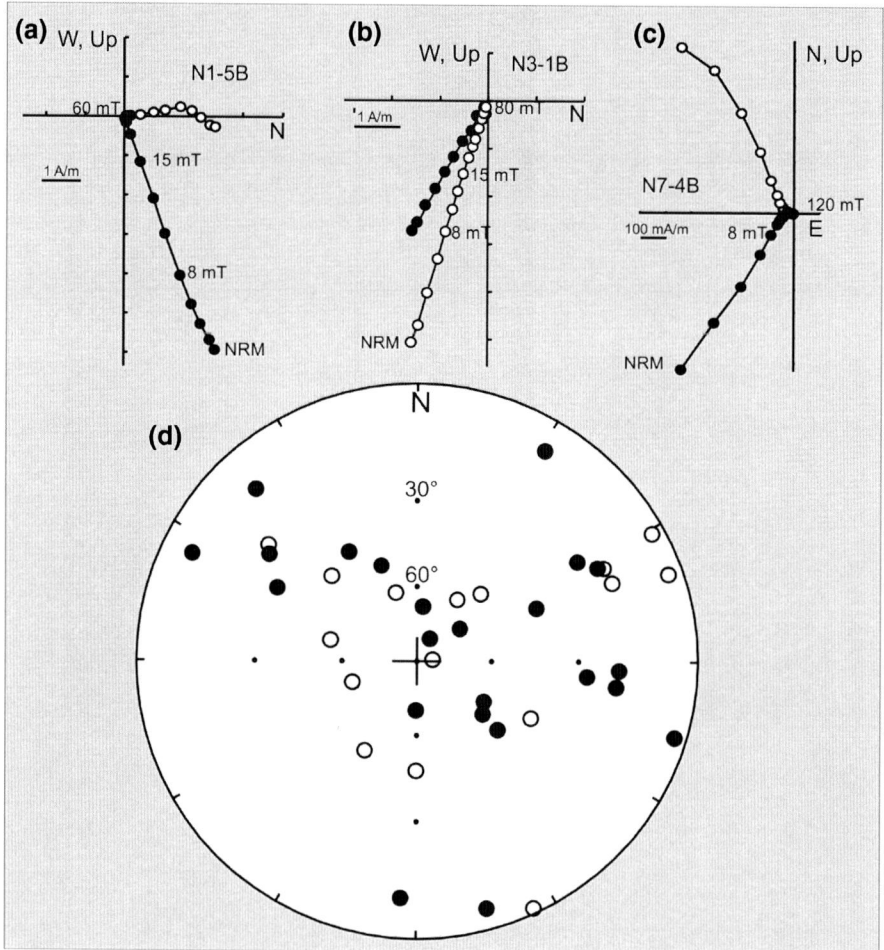

Fig. 5 Representative magnetic behavior of samples of the El Nihuil mafic unit submitted to stepwise demagnetization. **a** A sample from the gneiss; **b** ibidem from the dolerites; **c** from the gabbro; **d** magnetic components isolated from the samples of the El Nihuil mafic unit. In the Zijderveld plots (**a**, **b**, **c**) full (*open*) symbols represent end vector projections in the *horizontal* (*vertical*) plane. In the stereonet, full (*open*) symbols correspond to downward (*upward*) directions

In many cases a soft component with similar or different directions than that of the high-coercivity fraction could be erased with fields of 10–15 mT. This soft component generally represented well over 50 % of the original remanence. Unblocking temperatures were always under 580 °C. Components could be easily determined by principal component analysis (Kirschvink 1980) with MAD under 13° (95 % of components determined with MAD <8°). However, no directional consistency was found either within or between sites (Fig. 5d). This renders useless the paleomagnetic study of this unit. For these reasons processing of the whole collection was not accomplished.

Acknowledgements This study was partially financed by grant projects PIP-CONICET (201101-00294)-647/119 and UBACyT (20020130100465BA). We also acknowledge Leandro Ortiz for his assistance during field work.

References

Astini RA, Benedetto JL, Vaccari NE (1995) The early Paleozoic evolution of the Argentine Precordillera as a laurentian rifted, drifted and collided terrane: a geodynamic model. Geol Soc Am Bull 107:253–273

Cingolani CA, Llambías EJ, Ortiz LR (2000). Magmatismo básico pre-Carbónico del Nihuil, Bloque de San Rafael, Provincia de Mendoza, Argentina. IX Congreso Geol Chileno Puerto Varas 2:717–721

Cingolani CA, Varela R, Abre P, Manassero M, Tickyj H, Uriz N (2011) Insights on the pre-Carboniferous tectonic evolution of the San Rafael Block, Mendoza, Argentina. XVIII Congr Geol Argent, S2, Tectónica Preandina, Neuquén

Cingolani CA, Uriz N, Marques J, Pimentel M (2012) The Mesoproterozoic U-Pb (LA-ICP-MS) age of the Loma Alta gneissic rocks: basement remnant of San Rafael Block, Cuyania terrane, Argentina. In: 7th South American Symposium of Geochronology, Medellín, Colombia, Abstracts, p 140

Dessanti R (1956) Descripción geológica de la Hoja 27c-Cerro Diamante, Provincia de Mendoza. Dir Nac de Min, Boletín 85:79. Buenos Aires

Jelinek JV (1978) Statistical processing of magnetic susceptibility measured in groups of specimens. Stud Geophys Geod 22:50–62

Jelinek JV (1981) Characterization of the magnetic fabrics of rocks. Tectonophysics 79:63–67

Kirschvink JL (1980) The least-squares and plane and the analysis of paleomagnetic data. Geophys J Roy Astron Soc 67:699–718

Ramos VA (2004) Cuyania, an exotic block to Gondwana: review of a historical success and the present problems. Gondwana Res 7:1009–1026

Thomas W, Astini RA (1996) The Argentine Precordillera: a traveler from the Ouachita embayment of North American Laurentia. Science 273:752–757

Low-Grade Metamorphic Conditions and Isotopic Age Constraints of the La Horqueta Pre-Carboniferous Sequence, Argentinian San Rafael Block

Hugo Tickyj, Carlos A. Cingolani, Ricardo Varela and Farid Chemale Jr.

Abstract The San Rafael block is a pre-Andean geological entity situated in central-western Mendoza Province, Argentina. It is part of the Cuyania composite terrane. In this terrane it is important to consider the units exposed at the 'pre-Carboniferous' outcrops because they record geological events before the accretion to Gondwana. One of which is the siliciclastic La Horqueta Formation of an uncertain Lower-Middle Paleozoic sedimentary age. It is characterized by asymmetric, open to similar folds, with southeast vergence. KI values obtained from the La Horqueta Formation in its type area vary from 0.24 to 0.33 $\Delta°2\theta$, indicating very low-grade (high anchizonal) to low-grade (epizonal) metamorphic conditions, that increase slightly from south to north. The white mica b-parameter measured (9.016 ± 0.007 Å) suggest an intermediate pressure regime. Whole rock Rb-Sr isochronic ages were obtained on metapelites from the key outcrops in the La Horqueta area (379 ± 15 Ma, MSWD: 1.4) and in the Los Gateados area (371 ± 62 Ma, MSWD: 3.7), indicating that metamorphism and deformation occurred during the Devonian Chanic Orogenic phase, probably related to Chilenia

---

Electronic supplementary material The online version of this chapter (doi:10.1007/978-3-319-50153-6_8) contains supplementary material, which is available to authorized users.

H. Tickyj (✉)
Facultad de Ciencias Exactas y Naturales, Universidad Nacional de La Pampa,
Av. Uruguay 151, L6300CLB Santa Rosa, La Pampa, Argentina
e-mail: htickyj@exactas.unlpam.edu.ar; htickyj@yahoo.com

C.A. Cingolani · R. Varela
CONICET-UNLP, Centro de Investigaciones Geológicas, Diag. 113 N° 275,
CP1904 La Plata, Argentina
e-mail: carloscingolani@yahoo.com

R. Varela
e-mail: ricardovarela4747@gmail.com

F. Chemale Jr.
Programa de Pós-Graduação Em Geologia, Universidade do Vale do Rio dos Sinos,
93.022-000 São Leopoldo-RS, Brazil
e-mail: faridcj@unisinos.br; faridchemale@gmail.com

terrane collision. U-Pb LA-MC-ICPMS detrital zircon ages patterns suggest that the La Horqueta Formation received a dominant sedimentary input from Mesoproterozoic sources, minor contributions from cratonic environments of Paleoproterozoic and Neoarchean ages, and finally a younger input from Pampean and Famatinian orogenic belts. U-Pb detrital zircon ages indicate a maximum sedimentation age close to the Silurian-Devonian limit for the La Horqueta Formation.

Keywords Metamorphism · Tectonic vergence · Rb-Sr age · Detrital zircons · Silurian-Devonian

1 Introduction

The geological evolution of the proto-Andean Gondwana margin in southern South America during the Paleozoic has been related in several studies to the accretion of allochthonous or displaced terranes (Ramos et al. 1986; Dalla Salda et al. 1992; Astini et al. 1996; Aceñolaza et al. 2002). One of the better studied areas is the Cuyania terrane that was accreted to the active margin of South America contemporaneously with the development of a major orogenic episode—the Famatinian cycle—(Aceñolaza and Toselli 1973; Pankhurst and Rapela 1998). During Upper Devonian-Lower Carboniferous times, the Chilenia terrane was accreted (Ramos 1988; Ramos et al. 1998) along the western side of Cuyania terrane and coetaneous with the Chanic tectonic phase.

The San Rafael Block, located in western Argentina, constitutes, with the Las Matras Block, the southern part of the Cuyania terrane (Fig. 1). It is mainly composed of 'pre-Carboniferous units' (Mesoproterozoic to Devonian), Upper Paleozoic sedimentary and volcaniclastic rocks, superimposed by Gondwanic (Permian-Triassic) magmatism and a widespread Cenozoic volcanism (Dessanti 1956; Polanski 1964; González Díaz 1972; Cuerda and Cingolani 1998; Cingolani et al. 2001).

The 'pre-Carboniferous units' (Fig. 2) include a Mesoproterozoic igneous-metamorphic basement (Cerro La Ventana Formation), Ordovician mafic rocks (El Nihuil complex), Ordovician fossiliferous carbonates and siliciclastic sedimentary rocks of the Ponón Trehué and Pavón Formations, marine metasedimentary rocks known as the Río Seco de los Castaños Formation (Upper Silurian-Lower Devonian) and the La Horqueta Formation which is the subject of this study.

The aim of this paper is to constrain the age and the P-T conditions of the low-grade metamorphic event and the concomitant deformational episode that affected the La Horqueta Formation using white mica b-parameter and Kübler index values, Rb-Sr whole rock isotopic data, and a U-Pb geochronological provenance

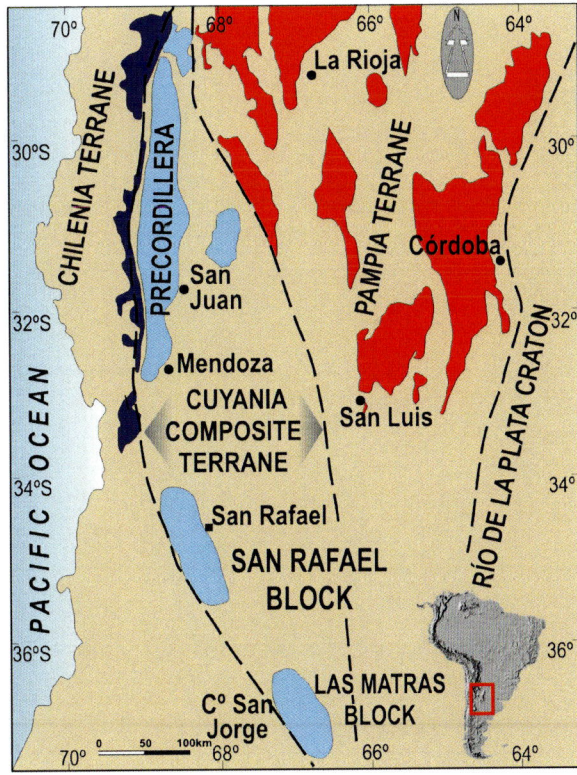

Fig. 1 Different terranes from central-western Argentina and location of the study area in the San Rafael Block, as a southern part of the Cuyania composite terrane (Ramos et al. 1998; Manassero et al. 2009)

study of detrital zircons. The obtained results are discussed in the context of the geological evolution of the SW Gondwana margin.

2 Geological Setting

This sedimentary unit was originally mapped and described by Dessanti (1945, 1956) who called it 'Serie de la Horqueta'. Later, it was renamed as La Horqueta Group (Dessanti and Caminos 1967), and then considered as a Formation by several authors (e.g. González Díaz 1981; Criado Roqué and Ibañez 1979). It is a sandy-dominated meta-sedimentary sequence deposited in a marine environment. The base of the sequence is not exposed. The metamorphic conditions were estimated to range from very low grade in the southernmost outcrops (González Díaz 1972; Criado Roque and Ibáñez 1979; Tickyj and Cingolani 2000) to amphibolite facies in the northern area (Polanski 1964). The sequence was affected by deformational events that developed folding with cleavage. In the area crossed by the Diamante River, Dessanti (1956) described a tight folding with similar, recumbent to asymmetric folds. The northernmost outcrops show folded rocks characterized by

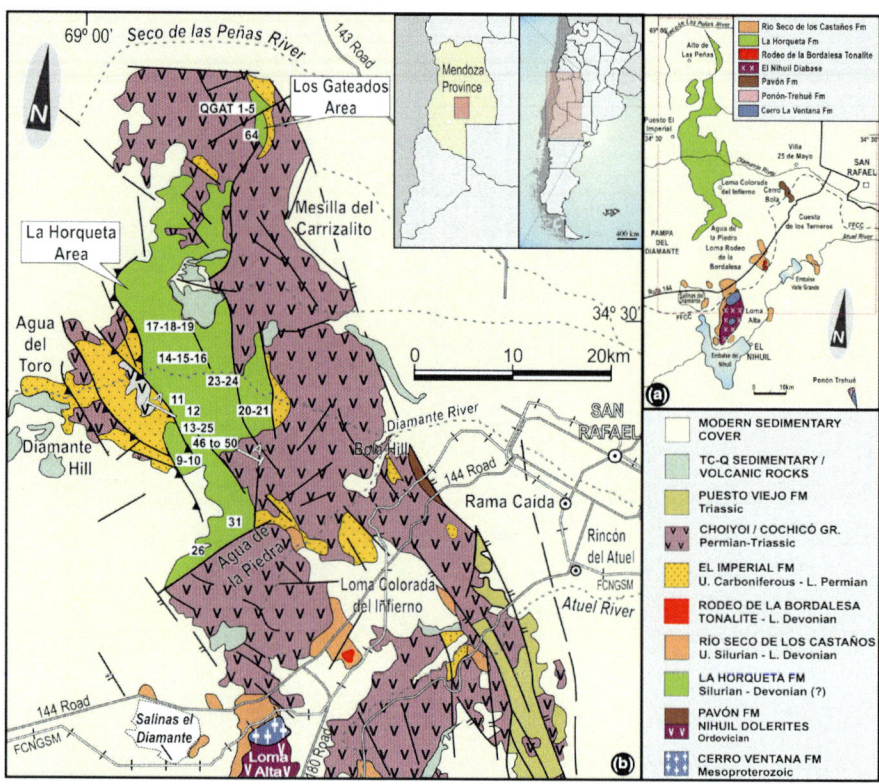

Fig. 2 a Sketch map of the 'pre-Carboniferous' units in the San Rafael Block (modified from González Díaz 1981). **b** Outcrop distributions of the main geological units within the study area (Dessanti 1956). Numbers correspond to sample locations from the La Horqueta Formation. A–A': Structural section shown on Fig. 5

tight to isoclinal gently plunging, upright folds with N-S trending axial planes and rare recumbent folds (Polanski 1964). The La Horqueta Formation was affected by a tectonic phase that put the metasedimentary sequence in contact with the Carboniferous continental to shallow marine (glacial) deposits of the El Imperial Formation. Some faults could have been reactivated during the Cenozoic Andean Orogeny (Moreno Peral and Salvarredi 1984; Cortés and Kleiman 1999; Japas and Kleiman 2004).

It is important to note that the "La Horqueta" unit initially comprised all 'pre-Carboniferous' sedimentary rocks of the San Rafael Block, exposed between the Los Gateados area and the Lomitas Negras and Agua del Blanco localities. Due to the lack of diagnostic fossils, an uncertain Precambrian to Devonian age was assigned (Dessanti 1956; Polanski 1964). At a later date, a fossil record including a Devonian coral similar to *Pleurodyctium* in Agua del Blanco exposures (Di Persia 1972), microfossils (acritarchs) of the Upper Silurian age in outcrops near the 144 road and ichnofossils like *Nereites-Mermia facies* in several outcrops (Rubinstein 1997; Poiré et al. 2002) were mentioned. More recently Morel et al. (2006) found herbaceous *Lycophytes* in the Atuel River section. These data support an Upper Silurian-Lower Devonian sedimentation age for part of the rocks included originally in the "La Horqueta" unit, now assigned to the Río Seco de los Castaños Formation. According to Manassero et al. (2009) this formation was deposited in a marine platform-deltaic system, the dominant sedimentary processes were wave and storm action, whereas source areas were located mainly to the east. However, K-Ar geochronological data of two magmatic complexes (originally described as intrusive bodies) yielded Lower Paleozoic ages and suggested that the "La Horqueta" unit could be Lower Paleozoic in age (González Díaz 1981). U-Pb age on zircons of 401 ± 4 Ma was obtained for the intrusive Rodeo de la Bordalesa Tonalite (Cingolani et al. 2003a), which is in according with mentioned fossil record, at least for a part of the unit now called Río Seco de los Castaños Formation.

At this point it is important to mention some stratigraphic changes (a) González Díaz (1981) splitted up the La Horqueta unit in the sense of Dessanti (1956) into two units: the La Horqueta and Río Seco de los Castaños formations. The latter lacks the regional metamorphic overprint as well as the mafic rock mentioned by Dessanti (1956) in the La Horqueta Formation. Furthermore, the Río Seco de los Castaños Formation preserved some diagnostic fossils as we mentioned before (acritarchs, lycophytes, coral). This suggestion was followed by Cuerda and Cingolani (1998), Cingolani et al. (2005) and Manassero et al. (2009) who also included in the Río Seco de los Castaños Formation the outcrops placed near road 144 (Fig. 2a) where Rubinstein (1997) found Upper Silurian microfossils (acritarchs), and Rodeo de la Bordalesa section with the intrusive tonalite; and (b) the Caradocian graptolite-rich sedimentary rocks located on the eastern slope of the Cerro Bola and originally comprising the "La Horqueta" unit, are now know as the siliciclastic Pavón Formation (Holmberg 1948 *emend*; Cuerda and Cingolani 1998).

Summarizing, we agree with the suggestions of Cuerda and Cingolani (1998) and Cingolani et al. (2003b) that the La Horqueta Formation (sensu stricto) should be restricted to the outcrops located on a strip reaching from the Seco de las Peñas River in the North to the Agua de la Piedra creek in the South. Where the best section is exposed—at the Diamante river area (Fig. 2)—these outcrops are 12 km wide. The La Horqueta Formation is bounded by reverse faults that bring this unit in contact with the Carboniferous El Imperial sedimentary sequence (Dessanti 1956; Giudici 1971) but in some outcrops like at Punta del Agua area, the Carboniferous rocks overlay the La Horqueta Formation separated by an angular unconformity (Fig. 3). The La Horqueta folded metasedimentary sequence is intruded by the Permian granitic stocks like Agua de la Chilena (Cingolani et al. 2005). All these rocks are overlain by Permian-Triassic volcano-sedimentary sequences related to the Choiyoi Gondwanian magmatism (Llambías 1999; Rocha Campos et al. 2011) (Fig. 2b).

Previous geochronological data of the metamorphic event that affected the "La Horqueta Formation" are K-Ar whole rock ages of 320 ± 20, 390 ± 15 and 395 ± 15 Ma (Toubes and Spikermann 1976, 1979; Linares and González 1990).

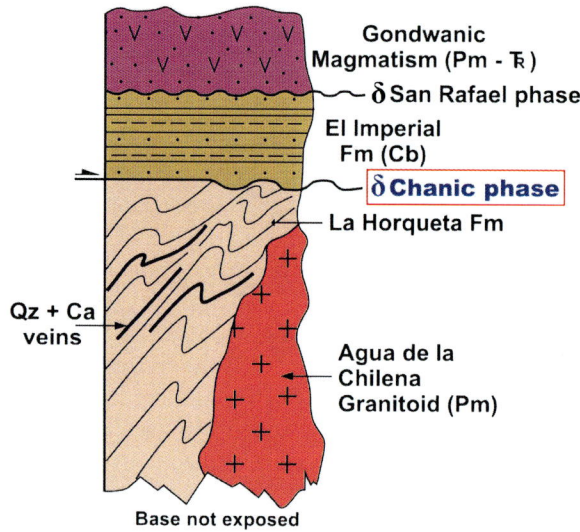

Fig. 3 Schematic main stratigraphical relationships of the La Horqueta Formation in the San Rafael Block

3 Results on key sections of the La Horqueta Formation

Two separate and very well exposed areas—as previously stated—were selected to perform metamorphic and isotopic studies: La Horqueta type section and Los Gateados area (Fig. 2).

Sedimentological and petrographical aspects: At the La Horqueta type area the sequence consists of alternate beds of metawackes, metasiltstones, metapelites, and rare metaconglomerates, deposited in a marine environment (Fig. 4a, b). The metasandstones are the commonest rock type. They show tabular layers of variable thickness—between 0.1 and 6 m—which usually preserve sedimentary structures such as graded bedding, lamination and cross-bedding. The meta-sandstones show metaclastic textures with a matrix recrystallized into chlorite, illite, quartz, albite and minor smectite. The original texture has been modified to variable degrees. Thick layers usually present rough foliation with recrystallized matrix (Fig. 4c, d), while other layers show penetrative foliation with ductile deformed clasts and pseudo-matrix development (Fig. 4e, f). In less deformed metawackes, clasts are mainly composed of quartz (mono and polycrystals), and sedimentary and metasedimentary, with scarce volcanic and limestone lithoclasts, and minor feldspars. The presence of carbonaceous material (0.5–2%) and authigenic pyrite is common.

Fine grained sediments are less abundant and they have been mostly metamorphosed to phyllites. They are mainly composed of well oriented crystals of illite and chlorite (up to 10μ wide) with a minor proportion of quartz and feldspar. Abundant thin veins of quartz and calcite cut the phyllites. The meta-conglomerates are scarce and usually appear at the base of graded sandy layers.

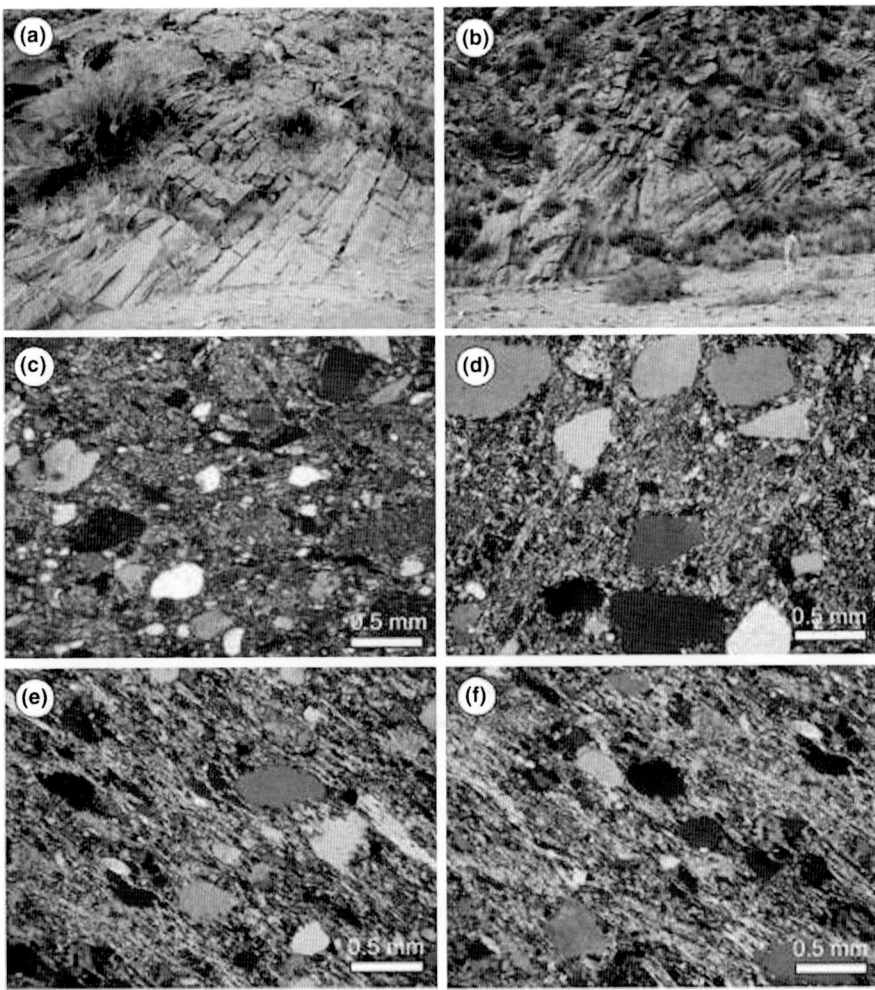

Fig. 4 **a** General view of the La Horqueta unit outcrop at the La Horqueta type section. **b** Similar folding, east vergence **c** and **d** Photomicrographs of metawackes showing a slight modification of the clastic texture. **e** and **f** strong foliated metasandstones (crossed nicols)

At the Los Gateados river area, the unit consists of intercalated layers of muscovite-biotite schists and quartzitic schists. They have granolepidoblastic textures with a typical mineral association of chlorite + muscovite + quartz ± biotite, with accessory tourmaline, zircon and opaque minerals. Its structure is characterized by a continuous penetrative foliation, with a NNE trend and dips of 35–40° to the East.

Structural characteristics: A structural profile was described between the Puesto La Horqueta and Loma Colorada del Infierno at the La Horqueta River area (Fig. 5). In this section the La Horqueta Formation is in tectonic contact by a

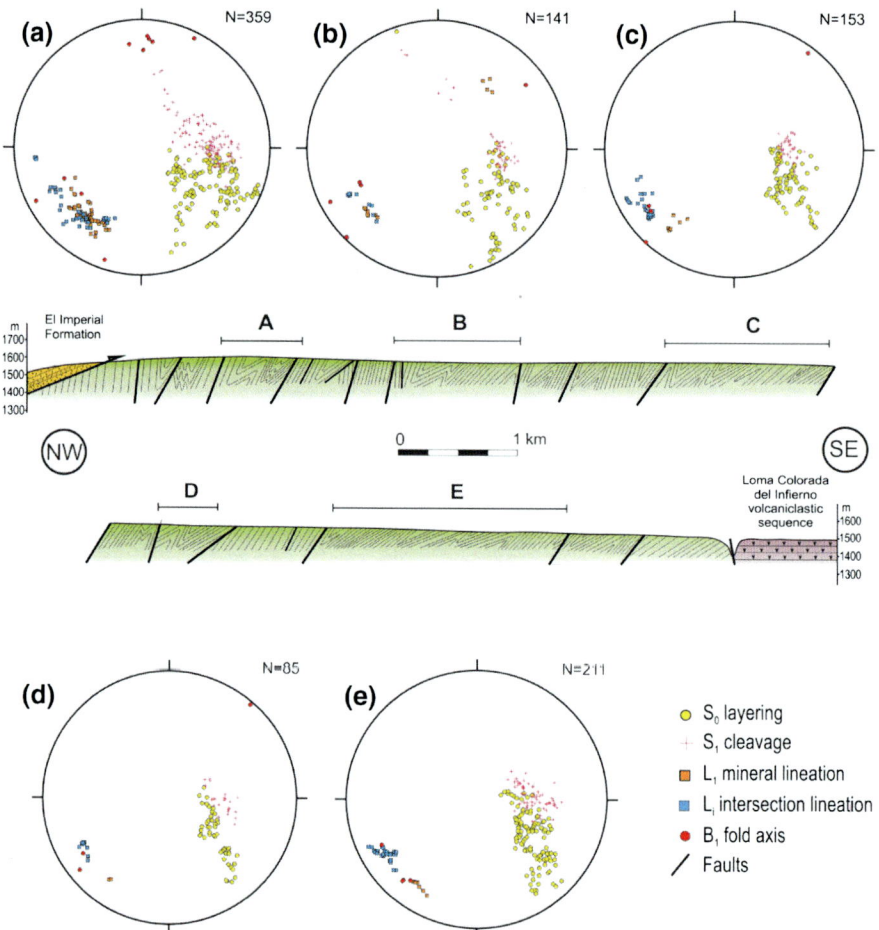

Fig. 5 Geological cross-section through the La Horqueta Formation. For location see A–A' profile in Fig. 2. Structural data plotted in lower hemisphere equal area stereonets

reverse faulting with the mainly Carboniferous El Imperial Formation at the Northwestern tip of the profile. In the SE outcrops the La Horqueta Formation is covered by the Loma Colorada del Infierno sub-volcanic rocks (Dessanti 1956; Giudici 1971; Rubinstein et al. 2013). The whole sequence is characterized by asymmetric, open to similar folds, with straight limbs and rounded hinges. These folds have axial planes striking to the NE and dipping to the NW, and axes plunging a few degrees to the NE or SW (Figs. 4b and 5). The fold vergence of the whole unit is towards SE. The main mesoscopic structure is a secondary foliation S_1, usually defined by aligned illite and chlorite. It has a consistent orientation with a north strike and moderate dip to the west. The S_1 foliation is continuous in metapelites, whereas it is anastomosed and spaced in metasandstones. Two types of lineations have been recognized linked to the folding. On S_1-planes a first mineral lineation is indicated by aligned illite + chlorite and tails of quartz on clasts, whereas another lineation is determined by the intersection of bedding planes and cleavage surfaces (Fig. 5).

Several faults have been recognized, some of them are in the limbs of large folds, suggesting that they could be reverse faults related to the asymmetric folding. However, they may have been reactivated, or even generated, by post-Carboniferous tectonic events.

These folds have axial planes striking to the NE and dipping to the NW, and axes plunging a few degrees to the NE or SW (Figs. 4b and 5). The fold vergence of the whole unit is towards SE.

Metamorphic conditions (clay minerals): Thirty two samples of fine to medium grain (shale to fine sandstone) were collected from several outcrops covering a wide area (Tickyj and Cingolani 2000) (Fig. 2). Clay minerals assemblages and Kübler index (Kübler 1968; Guggenheim et al. 2002) were determined in all of them. Thirteen samples from the La Horqueta type section were selected to determine the white mica b-parameter.

Methodology: Whole rock samples were crushed using an agate-mill and sieved repeatedly#30 sieves to avoid over grinding. Organic matter and carbonates were eliminated using acid treatment. Twenty grams of each sample were dispersed with an ultrasound equipment during 25 min and the <2 μm fractions were separated by centrifugation. Clay mineral identification and Kübler index measurements were performed in the Centro de Investigaciones Geológicas (La Plata, Argentina). The <2 μm fractions were sedimented onto glass slides from which X-ray diffractions patterns were obtained using a Philips PW 2233/20 X-ray diffractometer, CuKα radiation, Ni filter, γ 1.54 Å and 36 kV/18 mA. Samples were scanned over the range 2θ = 2–32° at 2° 2θ/min, time constant = 1 s, and using divergent slits of 1° and receiving slits of 0.2 mm. Samples for qualitative and semi-quantitative clay mineral identification were first air dried and then treated with ethylene glycol and heated to 550 °C during 2 h (Brindley 1980; Moore and Reynolds 1989).

The obtained Kübler index values (KI_{CIG}) from the width of the (001) white mica peak at half height on air-dried samples, expressed in terms of $\Delta°2\theta$. KI_{CIG}, were converted to standards values, using the international parameters after Warr and Rice (1994). The equation used was: $KI = 1.0999 \, KI_{CIG} - 0.1548$, $R^2 = 0.9755$.

Values used to delimitate diagenetic, low anchizonal, high anchizonal and epizonal conditions were $KI > 0.52 \, \Delta°2\theta$, KI: $0.52–0.42 \, \Delta°2\theta$, KI: $0.42–0.32 \, \Delta°2\theta$ and $KI < 0.32 \, \Delta°2\theta$, respectively (Warr and Ferreiro Mählmann 2015). KI values in the La Horqueta Formation vary from 0.24 to 0.33 $\Delta°2\theta$. Almost all samples belong to the epizone, except for a sample from the central area (HOR 12) that underwent high anchizonal metamorphic conditions.

Measurements of the white mica b-parameter were performed at the Centro de Tecnología de Recursos Minerales y Cerámica (CETMIC, La Plata, Argentina). Randomly-oriented <2 μm powdered samples were analyzed in the range 59–

64° 2θ, using routine parameters indicated in Padan et al. (1982). The results were statistically analyzed by mean values and the relative standard deviation and are presented as cumulative frequency curves.

Results: The clay mineral assemblages are fairly homogeneous in the 32 analyzed samples. They mainly consist of illite (56–86%) and chlorite (14–44%). A few samples have small quantities of smectite and kaolinite (Table 1). The presense of smectite in high anchizone-epizone could be by retrograde phase.

Illite was identified by a strong reflection at 10 Å and weak reflections at 5 and 3.33 Å, which were modified neither in ethylene-glycol nor in heated samples. The presence of chlorite was determined by reflections at 14.2 and 7.1 Å and 3.5 Å on untreated samples; when the samples were heated the first reflection increase its intensity while the second almost disappear.

In two rocks from the Puesto Imperial area significant proportion of smectite, were identified, because the XRD trace of the ethylene-glycol treated samples show reflections at 16.6–16.8 Å (Figs. 2 and 6b, c; Table 1). Other samples show weak reflections in the range 17–18 Å, which probably represent small quantities of smectite. Kaolinite was identified for its reflection at 3.58 Å, close to the (004) peak of chlorite (Moore and Reynolds 1989).

KI values in the La Horqueta Formation vary from 0.24 to 0.33 Δ°2θ. Almost all samples belong to the epizone, except for a sample from the central area (HOR 12) that underwent high anchizonal metamorphic conditions.

The whole data set shows a slight increase of metamorphic grade from south to north, as mentioned by Polanski (1964) in its regional study.

Natural K-white micas are a solid solution between the ideal muscovite and celadonite end-members (Guidotti 1984). The content of celadonite is controlled by the Tschermak substitution [(Mg, Fe^{2+})^{VI}Si^{IV} = Al^{VI}Al^{IV}], which is particularly sensitive to pressure at low-grade metamorphic conditions (Guidotti et al. 1989). The increase of celadonite content in white micas has been widely used to estimate geobaric conditions in low- and very low-grade metamorphic terranes (Padan et al. 1982).

The b-parameter values obtained for the La Horqueta Formation range from 9.004 to 9.029 Å, with an average of 9.016 Å (σ: 0.007 Å), indicating a low-intermediate pressure regime as they plot between the curves corresponding to the New Hampshire and the Ryoke terranes in the cumulative frequency plot (Fig. 7) proposed by Sassi and Scolari (1974).

Rb-Sr whole-rock data: To constrain the age of the main deformational event of the La Horqueta Formation we applied the whole rock Rb-Sr method (Tickyj et al. 2001). The determinations of Rb and Sr contents were performed by XRF whereas the isotopic composition on natural Sr was analyzed by mass spectrometry. Sample preparation, chemical attack and extraction of natural Sr using cation exchange resin, were carried out at the clean laboratory of the Centro de Investigaciones Geológicas (CIG, University of La Plata, Argentina). Mass spectrometry (TIMS) measurements were developed at the Laboratorio de Geología Isotópica, Porto Alegre, Brazil. The results were plotted on isochronic diagrams, using the Isoplot software after Ludwig (1998).

Table 1 Mineral composition of <2 μ fraction and KI (Kübler index) values of the La Horqueta Formation (location of samples in Fig. 2)

Sample	Location		Illite-Mus. (%)	Chlorite (%)	Smectite (%)	Kaolinite (%)	KI
Hor 9	La Horqueta area	34° 38′ 16.53″S	67	27	2	4	0.28
		68° 53′ 11.11″W					
Hor 10	La Horqueta area	34° 38′ 16.53″S	63	37	–	–	0.27
		68° 53′ 11.11″W					
Hor 11	La Horqueta area	34° 33′ 57.64″S	75	25	–	–	0.28
		68° 54′ 05.05″W					
Hor 12	La Horqueta area	34° 35′ 27.18″S	71	29	–	–	0.33
		68° 53′ 29.63″W					
Hor 13	La Horqueta area	34° 36′ 46.12″S	74	22	–	3	0.31
		68° 50′ 41.23″W					
Hor 14	El Baqueano	34° 33′ 02.56″S	84	16	–	–	0.25
		68° 51′ 04.83″W					
Hor 15	Puesto El Chacay	34° 33′ 02.56″S	76	24	–	–	0.3
		68° 51′ 04.83″W					
Hor 16	El Baqueano	34° 33′ 02.56″S	76	20	3	1	0.24
		68° 51′ 04.83″W					
Hor 17	Puesto Imperial	34° 30′ 04.40″S	72	19	9	–	0.28
		68° 55′ 07.01″W					
Hor 18	Puesto Imperial	34° 30′ 04.40″S	64	19	15	2	0.3
		68° 55′ 07.01″W					
Hor 19	Puesto Imperial	34° 30′ 04.40″S	58	42	–	–	0.28
		68° 55′ 07.01″W					
Hor 20	Los Reyunos	34° 34′ 25.96″S	60	40	–	–	0.28
		68° 46′ 40.08″W					

(continued)

Table 1 (continued)

Sample	Location		Illite-Mus. (%)	Chlorite (%)	Smectite (%)	Kaolinite (%)	KI
Hor 21	Los Reyunos	34° 34' 25.96"S	56	44	–	–	0.28
		68° 46' 40.08"W					
Hor 23	La Picaza	34° 33' 53.22"S	81	19	–	–	0.28
		68° 49' 35.48"W					
Hor 24	La Picaza	34° 33' 53.22"S	81	17	1	–	0.27
		68° 49' 35.48"W					
Hor 25	La Horqueta area	34° 38' 02.00"S	86	14	–	–	0.29
		68° 50' 49.31"W					
Hor 26	Agua de la Piedra	34° 44' 31.69"S	68	32	–	–	0.26
		68° 49' 34.90"W					
Hor 31	Agua de la Piedra	34° 43' 04.69"S	70	25	5	–	0.26
		68° 46' 53.81"W					
Hor 46	La Horqueta area	34° 35' 52.09"S	69	31	–	–	0.25
		68° 52' 28.63"W					
Hor 47	La Horqueta area	34° 36' 10.59"S	62	38	–	–	0.25
		68° 51' 54.36"W					
Hor 48	La Horqueta area	34° 36' 24.78"S	78	22	–	–	0.25
		68° 51' 26.53"W					
Hor 49	La Horqueta area	34° 37' 33.36"S	67	33	–	–	0.27
		68° 49' 50.92"W					
Hor 50	La Horqueta area	34° 38' 08.78"S	84	16	–	–	0.24
		68° 48' 49.57"W					

Fig. 6 Powder X-ray diffraction patterns of oriented mounts of the <2 μm fraction of representative samples of La Horqueta Formation

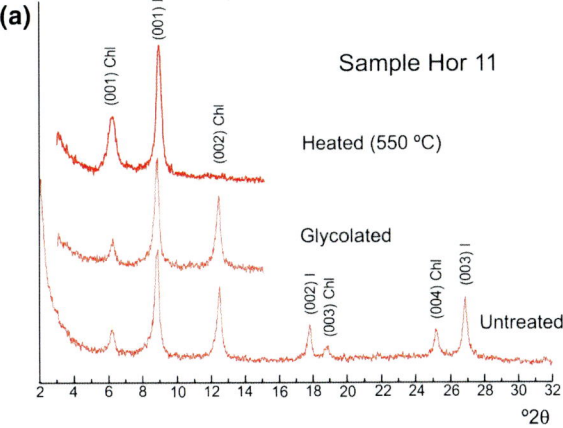

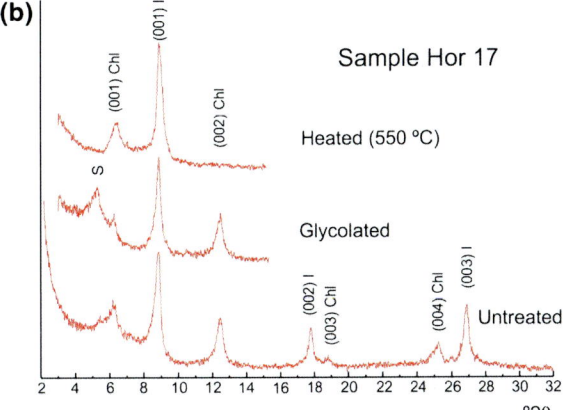

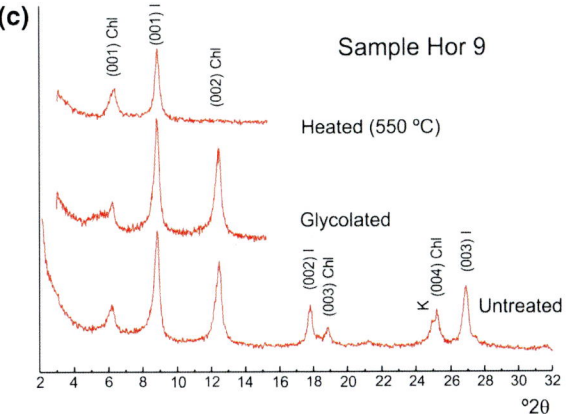

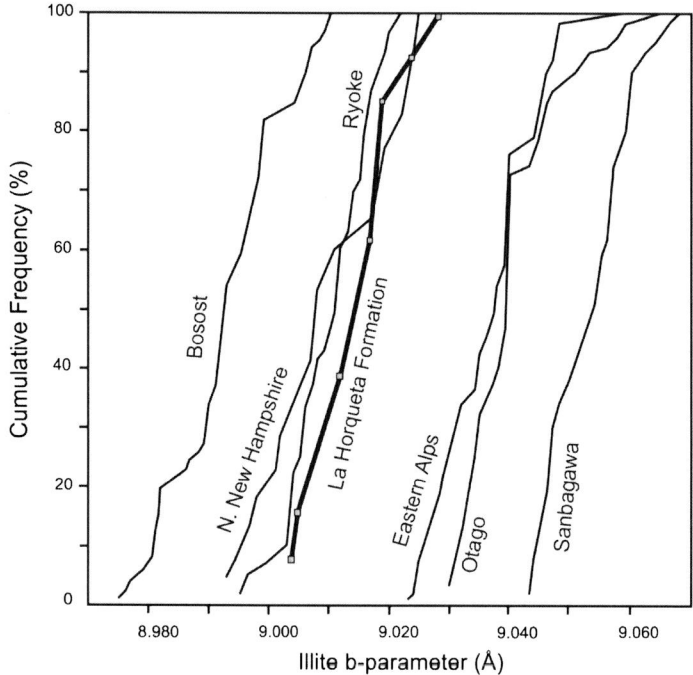

Fig. 7 Cumulative frequency-white mica b parameter plots for samples of the La Horqueta Formation (this study) and from other regional metamorphic terrains (Sassi and Scolari 1974)

Results: Six samples of micaschists from the Los Gateados section were analyzed. The Rb contents vary between 57 and 151 ppm, while the Sr contents vary from 50 to 83 ppm (Table 2). The age obtained from the isochron calculated with using Isoplot/Ex Model 1 (Ludwig 1998) is 371 ± 62 Ma, initial $^{87}Sr/^{86}Sr$ 0.7165 ± 0.0034 and MSWD: 3.7 (Fig. 8a). Furthermore, seven samples of metapelites from the La Horqueta type section were analyzed. The Rb contents vary between 116 and 290 ppm, whereas the Sr contents from 29 to 57 ppm (Table 2). The Rb-Sr isochron calculated with Isoplot/Ex Model 1 (Ludwig 1998) yielded an age of 379 ± 15 Ma, with IR: 0.7151 ± 0.0026, and MSWD: 1.4 (Fig. 8b) (Tickyj et al. 2001).

The Rb-Sr data pointed out that the low-grade metamorphism and folding events of the La Horqueta Formation are Late Devonian.

These data agree with previous K-Ar ages reported by Linares and González (1979). Similar data were obtained on low-grade metamorphic units from the western and south-western sections of the Precordillera (Cucchi 1971; Buggisch et al. 1994; Gerbi et al. 2002). This geochronological data let infer a Devonian age for the synmetamorphic Ductile deformation in the western side of the Cuyania terrane, probably connected with the accretion of the Chileniaterrane (Ramos 1988). This hypothesis is also supported by $^{40}Ar/^{39}Ar$ plateau data on white micas

Table 2 Rb-Sr analytical data

Lab. Number	Field sample	Rb	Sr	$^{87}Rb/^{86}Sr$	Error	$^{87}Sr/^{86}Sr$	Error
La Horqueta formation							
CIG 1205	QGAT 1	74.2	82.6	2.6063	0.0521	0.730473	0.000037
CIG 1206	QGAT 2	83.4	62.3	3.8869	0.0777	0.738029	0.000059
CIG 1207	QGAT 3	85.3	56.9	4.3530	0.0871	0.738759	0.000051
CIG 1208	QGAT 4	70.3	61.2	3.3341	0.0667	0.734435	0.000066
CIG 1209	QGAT 5	57.4	50.5	3.2987	0.0660	0.733119	0.000029
CIG 1210	HOR 64	151.5	73.7	5.9745	0.1195	0.748189	0.000120
CIG 1232	HOR 11	258.7	53.6	14.0907	0.2818	0.794198	0.000008
CIG 1233	HOR 12	247.2	35.3	20.5050	0.4101	0.824753	0.000009
CIG 1234	HOR 24	289.7	28.7	29.6940	0.5939	0.872864	0.000010
CIG 1235	HOR 25	253.6	57.5	12.8633	0.2573	0.784016	0.000008
CIG 1236	HOR 46	205.8	40.2	14.9486	0.2990	0.796088	0.000007
CIG 1237	HOR 47	116.2	38.5	8.7833	0.1757	0.761234	0.000008
CIG 1238	HOR 50	142.9	53.0	7.8438	0.1569	0.757868	0.000010

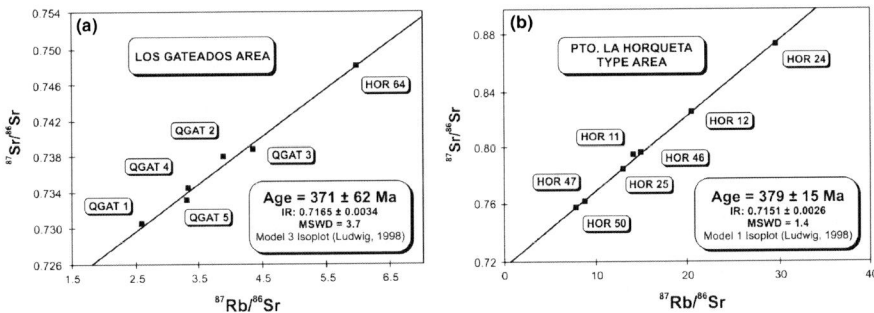

Fig. 8 Rb-Sr isochronic plot of samples from Los Gateados region (**a**) and from Puesto La Horqueta type region (**b**) (after Tickyj et al. 2001)

(384 ± 0.5 and 378 ± 0.5 Ma) from low-grade metamorphic rocks from Bonilla and Portillo areas obtained by Davis et al. (1999).

U-Pb geochronology: U-Pb dating on detrital zircons of six metasandstone was performed in order to estimate maximum age of deposition and to accomplish a geochronological provenance study of the La Horqueta unit. In situ U–Pb zircon dating was carried out at the Isotope Geology Laboratory of the Federal University of Rio Grande do Sul, Porto Alegre, Brazil, by means of the LA-MC-ICPMS technique (Jackson et al. 2004); preliminary results were presented by Cingolani et al. (2008). In this technique all zircon grains and a sample of the GJ-1 (GEMOC ARC Nat. Key Center) standard zircon were mounted in 2.5 cm-diameter circular epoxy and polished until the zircons were revealed. Back-scattered electron images of zircons were obtained using a Jeol JSM 5800 electron microscope in order to analyze the internal morphology of grains and select the spot areas to be dated with a New Wave UP213 laser ablation microprobe coupled to a Neptune MC-ICPMS, with collector configuration for simultaneous measurements of Th, U, Pb and Hg isotopes. The isotope ratios and inter-element fractionation data were evaluated in comparison with the GJ-1 standard at every set of 4–10 zircon spots, and used to estimate the necessary corrections and internal instrumental fractionation. The laser spot size was 25 μm. For each standard and spot run, a blank sample was also run and its values subtracted from all individual cycle measurements. The ^{204}Pb value was corrected for ^{204}Hg, assuming the ^{202}Hg/^{204}Hg ratio to be 4.355. The necessary correction for common ^{204}Pb, after Hg correction based on the simultaneously measured ^{202}Hg was insignificant in most cases and the Pb isotopic composition assumed to follow the isotopic evolution proposed by Stacey and Kramers (1975), which is required to attribute an initial estimated age. After the blank and common Pb corrections, the ratios and their absolute errors (1σ) of ^{206}Pb*/^{238}U, ^{232}Th/^{238}U and ^{206}Pb*/^{207}Pb* were calculated in an Excel spreadsheet. Usually zircon-rims were dated and in some cases also grain-cores, for comparison.

Results: As we depicted in Fig. 9 the detrital zircon population of samples **Hor 21**(n = 61), **Hor 46** (n = 60) and **Hor 10** (n = 60) show patterns dominated by grains of Mesoproterozoic ages, minor peaks corresponding to Neoproterozoic-Lower Paleozoic age and few subordinate peaks in the Paleoproterozoic-Neoarchean.

As it is shown in Fig. 9 the sample **Hor 21** notably record 85% of zircons derived from Mesoproterozoic sources, most of them from Upper Mesoproterozoic ("Grenvillian-age" or M3) in a polymodal detrital zircon-age pattern. The 9% correspond to the Neoproterozoic (Pampean-Brasiliano cycle), 6% of zircon grains were derived from cratonic domains (Paleoproterozoic ages). The sample **Hor 46** could be described as bimodal that shows (Fig. 9) more than 80% of zircons of Mesoproterozoic sources, with 55% that correspond from the"Grenvillian-age" or M3; 11% are from the Pampean-Brasiliano cycle, 9% from cratonic sources and only 3% derived from the Famatinian belt. The sample **Hor 10** also presents main peaks (unimodal age pattern) in the Mesoproterozoic (72%) with 62% from the M3 or "Grenvillian-age". Zircons of the Pampean-Brasiliano cycle are present with 18% and about 10% derived from cratonic sources (Paleoproterozoic).

The samples **Hor 15** (n = 59) and **Hor 81** (n = 56) record a pattern with main peaks in the Neoproterozoic-Lower Paleozoic as well as in the Mesoproterozoic, with minor peaks from Paleoproterozoic to Neoarchean ages (Fig. 10). The detrital zircon age pattern for sample **Hor 15** shows two major groups corresponding to Mesoproterozoic (62% of the grains, with 42% of M3), and Pampean (27% of zircons), with minor contribution from cratonic sources (11%), where 3% were Neoarchean. For the sample **Hor 81** the zircon population is dominated by a strong peak corresponding to Pampean-Brasiliano ages (40%), then 37% from Mesoproterozoic ages (25% of M3), 13% from Paleoproterozoic ages (major percentage of cratonic sources without Neoarchean ages), whereas zircons from the Famatinian cycle represent a 10% of the analyzed grains.

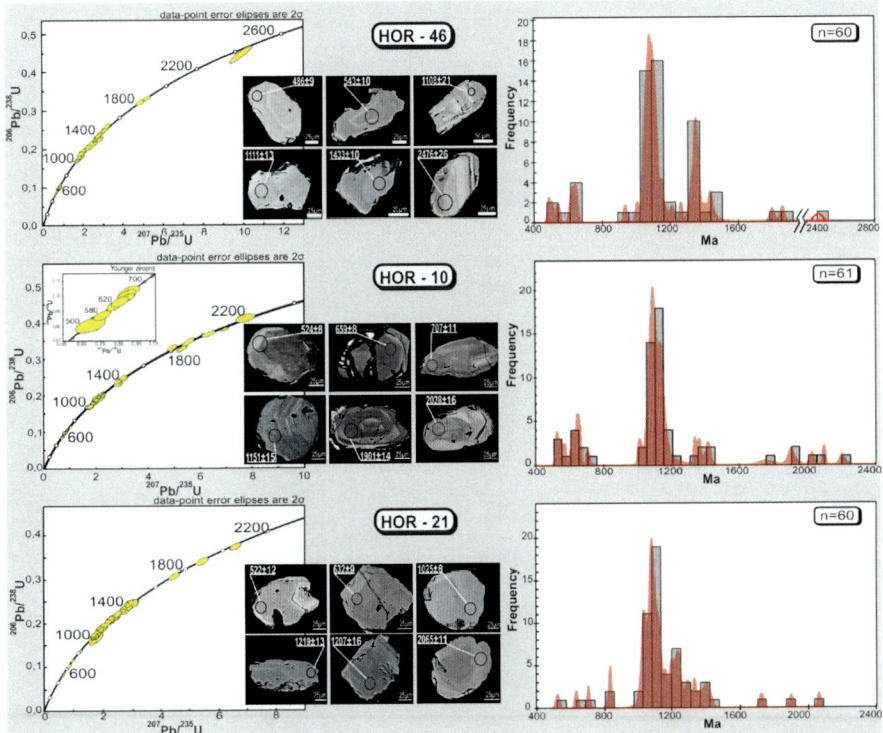

Fig. 9 U-Pb (LA-ICP-MS) concordia diagrams, frequency histograms and some of the electronic microscope (SEM) images of the studied zircon grains from samples Hor 46, Hor 10 and Hor 21. The location of the spots and obtained ages by LA-ICP-MS are in Ma. See Tables 3, 4 and 5 (supplementary material)

The sample **Hor 27** ($n = 64$) show a quite different pattern from other studied samples (Fig. 10). In this sample the detrital zircon age pattern is dominated by 54% of Famatinian zircon grains (23% Silurian and 31% Early Devonian in age), then a 35% derived from a source of Mesoproterozoic age (while 26% from M3), and subordinate contributions from Pampean-Brasiliano sources (8%) and cratonic areas (4%, with 2% from Neoarchean).

Constrains on provenance of the main detrital zircon sources could be as follows, from older to younger age components (Fig. 11):

(1) <u>Archean to Paleoproterozoic</u>: The source rocks of the obtained clusters (4–13%) are probably derived from the erosion of the basement of the Río de la Plata craton located toward the East. (2) <u>Mesoproterozoic</u>: Prominent clusters at M3 or "Grenvillian-age" 1.0–1.2 Ga were registered in all samples (26–62%), the most probable source of zircons of this age is the juvenile basement of Laurentian affinity of Precordillera-Cuyania, outcropping at the Pie de Palo, Umango ranges, Cerro La Ventana Formation at the San Rafael and Las Matras blocks (Sato et al. 2000;

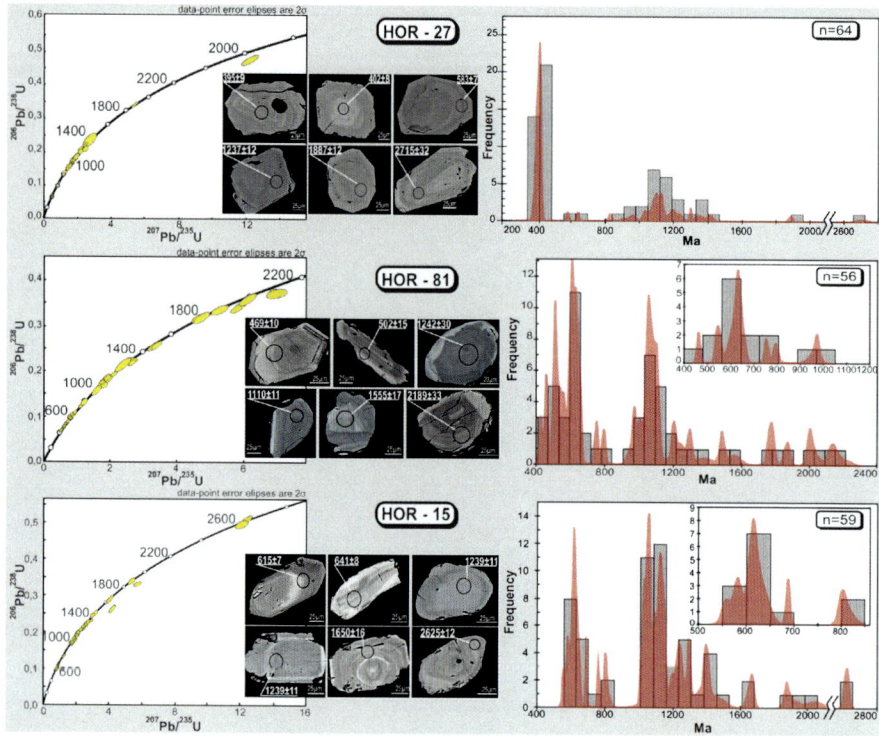

Fig. 10 U-Pb (LA-ICP-MS) concordia diagrams, frequency histograms and some of the electronic microscope (SEM) images of the studied zircon grains from samples Hor 15, Hor 81 and Hor 27. The location of the spots and obtained ages by LA-ICP-MS are Ma. See Tables 6, 7 and 8 (supplementary material)

Varela et al. 2011 and references). (3) Neoproterozoic-Lower Paleozoic: Clusters of these ages were found in all studied samples (8–40%). Zircons of these ages are abundant in southern South America, evidencing the uplift and denudation of the Pampean-Brasiliano orogenic belts. (4) Ordovician-Early Devonian: Zircon grains of these ages are recorded in samples Hor 27 (with more than 50%), Hor 46 and Hor 81 (in between 3 and 10%). These grains probably derived from the erosion of the igneous rocks from the Late Famatinian magmatic arc, well known in western-central Argentina. The Devonian ages are abundantly registered on magmatic zircons from sample Hor 27.

The younger detrital zircon ages (ca. 410 Ma) recorded on the sample Hor 27 allows to constrain the age of deposition of the La Horqueta Formation to the Silurian-Devonian limit.

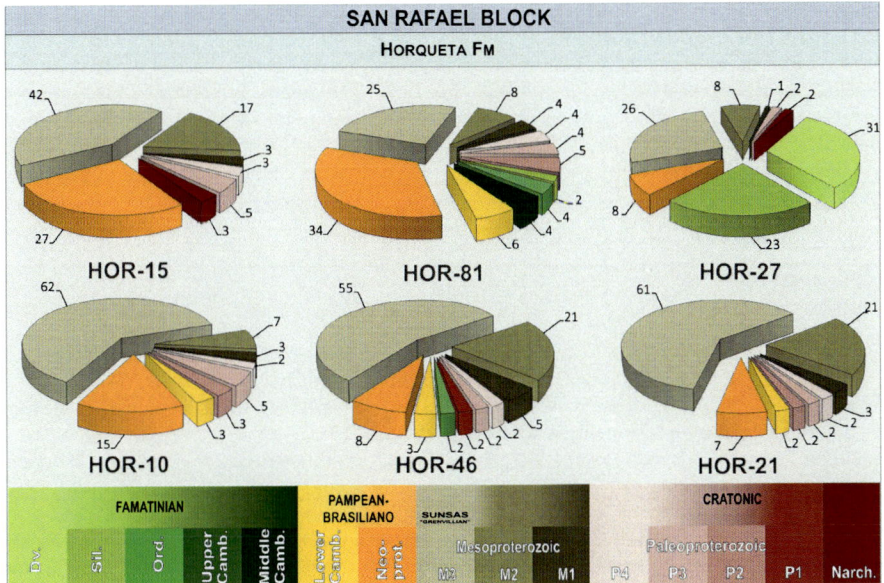

Fig. 11 Percentage representation 'pie diagrams' of the U-Pb detrital zircon ages in terms of the main orogenic cycles defined for South America

4 Concluding Remarks

The La Horqueta Formation is a marine metasedimentary sequence that has been folded, cleaved and faulted by a main regional deformational event with southeast vergence that was coeval with a regional metamorphic event of very low to low-grade. Studies performed on samples from La Horqueta Formation in its type section area indicate that the regional metamorphism affecting this unit attained high anchizone-epizone and intermediate P facies series.

The Rb-Sr whole-rock age obtained from Los Gateados sector is 371 ± 62 Ma, with IR: 0.7165 ± 0.0034, while from La Horqueta type section yielded an age of 379 ± 15 Ma, with IR: 0.7151 ± 0.0026. These ages constrain the metamorphic and the main deformational event to the Devonian and linked it to the Chanic Orogenic phase.

Based on U-Pb isotopic data the La Horqueta Formation received a minor sedimentary input from cratonic sources such as the Rio de la Plata craton, which contribute with the older ages, then an important source from Cuyania terrane basement (Mesoproterozoic ages) and finally a younger source from Pampean and Famatinian orogenic belts. The U-Pb detrital zircon ages constrain the age of sedimentation to the Silurian-Devonian limit.

Acknowledgements We thank Patricia Zalba (CETMIC), Daniel Poiré and Jorge Maggi (CIG) for access to XRD facilities. To Paulina Abre for helpful assistance in improving the early version of the paper. Funding for fieldwork and laboratory research was provided by the Argentine CONICET-PIPs 0647-0199 and ANPCYT (PICT 07829) grants. Important laboratory work facilities were also obtained at the Universities of La Pampa and La Plata, Argentina. We are grateful to Norberto Uriz (University of La Plata) for help us in the presentation of some figures and tables. Many thanks to Margarita Do Campo for careful and thoughtful review of this paper.

References

Aceñolaza FG, Toselli AJ (1973) Consideraciones estratigráficas y tectónicas sobre el Paleozoico inferior del Noroeste Argentino. II Congr Latinoam Geol (Caracas), Actas 2:755–763

Aceñolaza FG, Miller H, Toselli AJ (2002) Proterozoic-Early Paleozoic evolution in western South America-a discusión. Tectonophysics, 354:121–137

Astini R, Ramos V, Benedetto JL, Vaccari N, Cañas FL (1996) La Precordillera: Un terreno exótico a Gondwana. 13° Congr Geol Argent y 3° Congr Expl Hidrocarb (Buenos Aires), Actas 5:293–324

Brindley G (1980) Quantitative X-ray diffraction procedures for clay mineral identification. In: Brindley G, Brown G (eds) Crystal Estructures of Clay Minerals and their X-ray identification, Mineral Soc London. Monograph 5, p 411–438. London

Buggisch W, von Gosen W, Henjes-Kunst F, Krumm S (1994) The age of Early Paleozoic deformation and metamorphism in the Argentine Precordillera; evidence from K-Ar data. Zentralblatt fuer Geol und Palaeontologie 1:275–286

Cingolani CA, Llambías EJ, Tickyj H, Manassero M, Abre P (2001) El Pre-carbonífero del Bloque de San Rafael, Mendoza (Argentina): Evolución y correlaciones en el margen proto-andino de Gondwana. XI Congr Latinoam Geol y III Congr Uruguayo Geol Montevideo, R.O. Uruguay. (Version CD) p 8

Cingolani CA, Manassero M, Abre P (2003a) Composition, provenance, and tectonic setting of Ordovician siliciclastic rocks in the San Rafael block: southern extension of the Precordillera crustal fragment. Argentina J South Am Earth Sci 16:91–106

Cingolani CA, Basei MAS, Llambías EJ, Varela R, Chemale Jr F, Siga Jr O, Abre P (2003) The Rodeo Bordalesa Tonalite, San Rafael Block (Argentina): Geochemical and isotopic age constraints. 10° Congr Geol Chileno, Concepción, Octubre 2003. (Versión CD Rom) p 10

Cingolani CA, Varela R, Abre P (2005) Geocronología Rb-Sr del Stock de Agua de la Chilena: magmatismo pérmico del Bloque de San Rafael, Mendoza

Cingolani CA, Tickyj H, Chemale Jr F (2008) Procedencia sedimentaria de la Formación La Horqueta, Bloque de San Rafael, Mendoza (Argentina): primeras edades U-Pb en circones detríticos. XVII Congr Geol Argent, Actas, Tomo III. San Salvador de Jujuy, Argentina, pp 998–999

Cortés JM, Kleiman LE (1999) La Orogenia Sanrafaélica en los Andes de Mendoza. 14° Congr Geol Argent, Actas 1:31

Criado Roque P, Ibáñez G (1979) Provincia Geológica Sanrafaelino Pampeana. In Segundo Simposio de Geología Regional Argentina. Academia Nacional de Ciencias, vol 1. Córdoba, Argentina, pp 837–869

Cucchi RJ (1971) Edades radimétricas y correlación de metamorfitas de la Precordillera, San Juan-Mendoza. Rep Argentina. Rev Asoc Geol Argent 26:503–515

Cuerda AJ, Cingolani CA (1998) El Ordovícico de la región del Cerro Bola en el Bloque de San Rafael, Mendoza: sus faunas graptolíticas. Ameghiniana 35(4):427–448

Dalla Salda LH, Cingolani CA, Varela R (1992) Early Paleozoic Orogenic belt of the Andes in Southwestern South America: results of Laurentia-Gondwana collision?. Geology 20:617–620

Davis JS, Roeske SM, McClelland WC, Snee LW (1999) Closing the ocean between the Precordillera terrane and Chilenia: Early Devonian ophiolite emplacement and deformation in the SW Precordillera. In: Ramos VA, Keppie JD (eds) Laurentia–Gondwana connections before Pangea, Special Paper 336, Geol Soc Am, Boulder, Colorado

Dessanti R (1945) Informe geológico preliminar sobre la Sierra Pintada, Departamento San Rafael, provincia de Mendoza. Direcc Nac Geol y Min. Carpeta 28, Buenos Aires

Dessanti R (1956) Descripción geológica de la Hoja 27c, Cerro Diamante (provincia de Mendoza). Direcc Nac Min, Boletín 85:1–79. Buenos Aires

Dessanti RN, Caminos R (1967) Edades Potasio-Argón y posición estratigráfica de algunas rocas ígneas y metamórficas de la Precordillera, Cordillera Frontal y Sierras de San Rafael, prov. Mendoza. Rev Asoc Geol Argent 22(2):135–162

Di Persia CA (1972) Breve nota sobre la edad de la denominada Serie de La Horqueta, zona Sierra Sierra Pintada, Dpto. San Rafael, Prov. Mendoza. Actas 4ª Jor Geol Argent, 3:29–41

Gerbi C, Roeske SM, Davis JS (2002) Geology and structural history of the southwest Precordillera margin, northern Mendoza Province, Argentina. J South Am Earth Sci 14:821–835

Giudici AR (1971) Geología de las adyacencias del río Diamante al este del cerro homónimo, provincia de Mendoza. República Argentina. Rev Asoc Geol Argent 26(4):439–458

González Díaz EF (1972) Descripción geológica de la Hoja 27d, San Rafael (provincia de Mendoza). Direcc Nac Min, Boletín 132. Buenos Aires

González Díaz EF (1981) Nuevos argumentos a favor del desdoblamiento de la denominada "Serie de La Horqueta" del Bloque de San Rafael, provincia de Mendoza. In Congr Geol Argent, N° 7, Actas 3:241–256. San Luis, Argentina

Guggenheim S, Bain D, Bergaya F, Brigatti M, Drits V, Eberl D, Formoso M, Galán E, Merriman R, Peacor D, Stanjek H, Watanabe T (2002) Report of the Association International pour L'Etude Des Argiles (AIPEA) Nomenclature Committee for 2001; Order, Disorder and Crystallinity in Phillosilicates and the use of the "Crystallinity Index": Clay. Clay Miner 50:406–409

Guidotti CV (1984) Micas in metamorphic rocks. In: Bailey SW (ed) Micas. Reviews in Mineralogy, Mineral Soc Am, 13:357–367

Guidotti CV, Sassi FP, Blencoe JG (1989) Compositional control of the a and b cell dimensions of 2M1 muscovite. Euro J Miner 1:71–84

Holmberg E (1948) Geología del Cerro Bola. Contribución al conocimiento de la tectónica de la Sierra Pintada. Secretaría de Industria y Comercio de la Nación. Direcc Gen Ind y Min. Boletín 69:313–361. Buenos Aires

Jackson SE, Pearson NJ, Griffin WL, Belousova EA (2004) The application of laser ablation-inductively coupled plasma-mass spectrometry to in situ U-Pb zircon geochronology. Chem Geol 211:47–69

Japas MS, Kleiman LE (2004) El Ciclo Choiyoi en el Bloque de San Rafael (Mendoza): de la orogénesis tardía a la relajación mecánica. Asoc Geol Argent, Serie D: Publicación Especial 7:89–100

Kübler B (1968) Evaluation quantitative de metamorphisme par la crystallinité de l'illite. Centre de Recherche de Pau, Societé National des Pétroles d'Aquitaine, Bulletin 2:385–397

Linares E, González RR (1990) Catálogo de edades radimétricas de la República Argentina, años 1957–1987. Serie B (Didáctica y Complementaria) 19. Publicaciones Especiales de la Asoc Geol Argent, pp 630. Buenos Aires

Llambías E (1999) Las Rocas ïgneas Gondwánicas. I. El magmatismo gondwánico durante el Paleozoico superior-Triásico. Geol Argent, Anales SEGEMAR 29(14):349–376. Buenos Aires

Ludwig KR (1998) Using Isoplot/Ex—a geochronological toolkit for Microsoft Excel. Berkeley Geochronology Center, Special Publication, 1, Berkeley

Manassero MJ, Cingolani CA, Abre P (2009) A Silurian-Devonian marine platform-deltaic system in the San Rafael Block, Argentine Precordillera-Cuyania terrane: lithofacies and provenance. In: Königshof P (ed) Devonian change: case studies in palaeogeography and palaeoecology, vol 314. Geol Soc, London, Special Publications, pp 215–240

Moore DM, Reynolds RC (1989) X-ray diffraction and the identification and analysis of Clay minerals. Oxford University Press, p 332. Oxford

Morel EM, Cingolani CA, Ganuza DG, Uriz NJ (2006) El registro de Lycophytas primitivas en la Formación Río Seco de los Castaños, Bloque de San Rafael, Mendoza. 9° Congr Argent Paleont y Bioest, Córdoba. Abstract

Moreno Peral CA, Salvarredi JA (1984) Interpretación del origen de las estructuras anticlinales del Pérmico inferior en el Bloque de San Rafael, provincia de Mendoza. 9° Congr Geol Argent, Actas 2:396–413

Padan A, Kisch HJ, Shagam R (1982) Use of the lattice parameter bo of dioctahedral illite/muscovite for the characterization of P/T gradients of incipient metamorphism. Contrib Mineral Pet 79:85–95

Pankhurst RJ, Rapela CW (1998) The proto-Andean margin of Gondwana: an introduction. In: Pankhurst RJ, Rapela CW (eds) The proto-Andean margin of Gondwana, vol 142. Geol Soc London, Special Publications, pp 1–9

Poiré DG, Cingolani CA, Morel ED (2002) Características sedimentológicas de la Formación Rio Seco de los Castaños en el perfil de Agua del Blanco: Pre-carbonífero del Bloque de San Rafael, Mendoza. XV Congr Geol Argent, Actas 3:129–133. Calafate

Polanski J (1964) Descripción geológica de la Hoja 26-c La Tosca. Provincia de Mendoza. Dirección Nacional de Geología y Minas, Boletín 101:1–86. Buenos Aires

Ramos VA (1988) Late Proterozoic-Early Paleozoic of South America, a collisional history. Episodes 2:168–173

Ramos V, Jordan TE, Allmendinger RW, Mpodozis C, Kay SM, Cortés JM, Palma MA (1986) Paleozoic terranes of the central Argentine-Chilean Andes. Tectonics 5:855–880

Ramos VA, Dallmeyer RD, Vujovich G (1998) Time constraints on the Early Palaeozoic docking on the Precordillera, central Argentina. In: Pankhurst RJ, Rapela CW (eds) The proto-Andean margin of Gondwana vol 142. Geol Soc London, Special Publications, pp 143–158

Rocha Campos AC, Basei MAS, Nutman AP, Kleiman LE, Varela R, Llambías EJ, Canile FM, Da Rosa OCR (2011) 30 million years of Permian volcanism recorded in the Choiyoi igneous province (W Argentina) and their source for younger ash fall deposits in the Paraná Basin: SHRIMP U-Pb zircon geochronology evidence. Gondwana Res 19(2011):509–523

Rubinstein C (1997) Primer registro de palinomorfos silúricos en la Formación La Horqueta, Bloque de San Rafael, provincia de Mendoza. Argentina. Ameghiniana 34(2):163–167

Rubinstein NA, Gómez A, Kleiman L (2013) Caracterización litofacial y geoquímica de las volcanitas del área del distrito minero El Infiernillo, Mendoza. Revista de la Asociación Geológica Argentina, 70(3):382–389

Sassi FP, Scolari A (1974) The bo value of the potassic white mica as a barometric indicator in low-grade metamorphism of politic schists. Contrib Miner Pet 45:143–152

Sato AM, Tickyj H, Llambías EJ, Sato K (2000) The Las Matras tonalitic-trondhjemitic pluton, Central Argentina: Grenvillian age constraints, geochemical characteristics, and regional implications. J South Am Earth Sci 13:587–610

Stacey JS, Kramers JD (1975) Approximation of terrestrial lead isotope evolution by a two-stage model. Earth Planet Sci Lett 26:207–221

Tickyj H, Cingolani CA (2000) Metamorfismo de muy bajo grado de la Formación La Horqueta (Proterozoico-Paleozoico inferior), Bloque de San Rafael (Mendoza), Argentina. IX Congr Geol Chile, Actas 2:539–544

Tickyj H, Cingolani CA, Varela R, Chemale Jr F (2001) Rb-Sr ages from La Horqueta Formation, San Rafael Block, Argentina. III South Am Symp Isotope Geol. Pucón, Chile, p 4

Toubes RO, Spikermann JP (1976) Algunas edades K-Ar para la Sierra Pintada, provincia de Mendoza. Rev Asoc Geol Argent 31(2):118–126

Toubes RO, Spikermann JP (1979) Nuevas edades K-Ar para la Sierra Pintada, provincia de Mendoza. Rev Asoc Geol Argent 34(1):73–79

Varela R, Basei MAS, González PD, Sato AM, Naipauer M, Campos Neto M, Cingolani CA Meira VT (2011) Accretion of Grenvillian terranes to the southwestern border of the Río de la Plata craton, western Argentina. Int J Earth Sci (Geol Rundsch) 100:243–272. Springer Verlag

Warr LN, Ferreiro Mählmann R (2015) Recommendations for Kubler Index standardization. Clay Miner 50(3):283–286

Warr LN, Rice AHN (1994) Interlaboratory standardization and calibration of clay mineral crystallinity and crystallite size data. J Metamorph Geol 12:141–152

La Horqueta Formation: Geochemistry, Isotopic Data, and Provenance Analysis

Paulina Abre, Carlos A. Cingolani, Farid Chemale Jr. and Norberto Javier Uriz

Abstract La Horqueta Formation is developed from the Seco de las Peñas River to Agua de la Piedra creek within the San Rafael block and was deposited in a marine environment. It comprises dominantly metasandstones, although metasiltstones, metapelites, and rare metaconglomerates are also present. The base of the succession is not exposed and it is superposed through unconformity by Upper Carboniferous units. La Horqueta Formation is folded and shows cleavage. Provenance analyses based on whole-rock geochemistry and isotope data is the main focus of the work. Whole-rock geochemical data point to a derivation from unrecycled upper continental crust, based mainly on Th/Sc, Zr/Sc, La/Th, and Th/U ratios and rare earth element (REE) patterns (including Eu anomalies). Sc, Cr, and V concentrations and low Th/Sc ratios are indicative of a source slightly less evolved than the average upper continental crust. The εNd values are within the range of variation of data from the Mesoproterozoic Cerro La Ventana Formation, which is part of the basement of the Cuyania terrane outcropping within the San Rafael block. The Rb-Sr whole-rock data indicate that the low-grade metamorphism and folding events are Devonian in age. U-Pb detrital zircon ages suggest main derivation from the Mesoproterozoic ("Grenvillian-age") basement of the San

---

P. Abre (✉)
Centro Universitario de la Región Este, Universidad de la República,
Ruta 8 Km 282, Treinta y Tres, Uruguay
e-mail: paulinabre@yahoo.com.ar

C.A. Cingolani
Centro de Investigaciones Geológicas, CONICET-Universidad Nacional de La Plata,
Calle 1 no. 644, B1900TAC La Plata, Argentina
e-mail: carloscingolani@yahoo.com

C.A. Cingolani · N.J. Uriz
División Geología, Museo de La Plata, Universidad Nacional de La Plata,
Paseo del Bosque s/n, La Plata, Argentina
e-mail: norjuz@gmail.com

F. Chemale Jr.
Programa de Pós-Graduação Em Geologia, Universidade Do Vale Do Rio Dos Sinos,
São Leopoldo RS, Brazil
e-mail: faridchemale@gmail.com

© Springer International Publishing AG 2017
C.A. Cingolani (ed.), *Pre-Carboniferous Evolution of the San Rafael Block, Argentina*, Springer Earth System Sciences,
DOI 10.1007/978-3-319-50153-6_9

Rafael block and the Pampean–Brasiliano cycle, as well as a detrital input from the Río de la Plata craton and the Famatinian belt. Despite geochemical similarities, Río Seco de los Castaños Formation display different proportions of detrital zircon ages, when compared to La Horqueta Formation.

Keywords Geochemistry · Isotope data · Provenance · La Horqueta Formation · San Rafael Block · Cuyania terrane

1 Introduction and Geological Setting

La Horqueta Formation (Dessanti 1956; González Díaz 1981) crops out on a 12 km-wide strip developed from the Seco de las Peñas River to Agua de la Piedra creek (Fig. 1; Cuerda and Cingolani 1998; Cingolani et al. 2003a), within the San Rafael block. It is in tectonic contact with Carboniferous units, either by reverse faults or by an angular unconformity (Tickyj et al. this volume).

La Horqueta Formation was deposited in a marine environment and comprises dominantly metasandstones, although metasiltstones, metapelites, and rare metaconglomerates are also present. The matrix of the metasandstones was recrystallized into chlorite, illite, quartz, albite, and minor smectite. Foliation is penetrative in some layers; ductile deformed clasts are present as well as pseudomatrix. In less-deformed metawackes, the relictic clasts are mainly composed of monocrystalline and polycrystalline quartz, sedimentary and metasedimentary lithoclasts, with scarce volcanic and limestone lithoclasts, and rare feldspars. The fine-grained levels are metamorphosed to phyllites and they comprise oriented illite and chlorite with scarce quartz and feldspar grains (Tickyj et al. this volume). In several outcrops quartz veins cutting the La Horqueta unit are conspicuous (Fig. 2c, d, e). Toward north, (Los Gateados river; Fig. 1) the unit consists of muscovite–biotite schists interlayered with quartzitic schists showing granolepidoblastic textures.

The base of the succession is not exposed and it is overlaid through unconformity by the Upper Carboniferous marine-glacial-continental unit (El Imperial Formation). La Horqueta Formation was affected by deformational events; it is folded and develops cleavage. The regional metamorphic conditions slightly increase from south to north (Criado Roqué 1972; Criado Roqué and Ibáñez 1979), ranging from very low (anchizone) to low grade (epizone). Maximum Silurian–Devonian depositional age was determined using U-Pb detrital zircon dating (Cingolani et al. 2008; Tickyj et al. this volume).

La Horqueta Formation is intruded by a granitic stock known as Agua de la Chilena, which extends over 5 km^2 of the northwestern part of the San Rafael Block and it is covered by Quaternary volcanic rocks (Cingolani et al. 2005a). The stock is composed of diorites, tonalites and biotitic-horblendiferous, and leucocratic granodiorites; grain size is medium to fine. Xenoliths and enclaves are frequent. Their mineralogical constituents are subhedral to anhedral quartz (27–33%), subhedral to euhedral altered alkaline feldspars (10–20%) and plagioclases (51–59%)

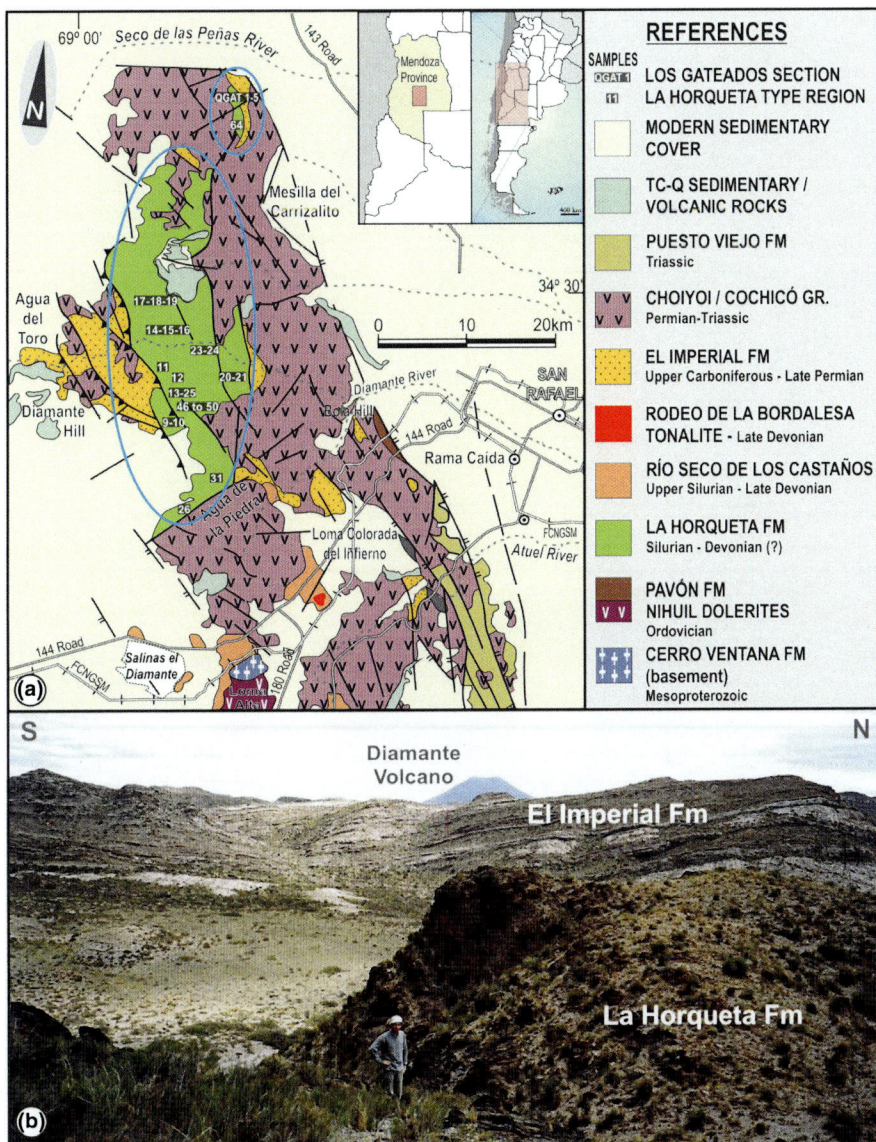

Fig. 1 a Geological sketch map of the studied area within the San Rafael block, where outcrops and sampling zone of La Horqueta Formation are located northward. b Regional view toward the West near La Horqueta type area. It is shown the deformed outcrops of La Horqueta Fm superposed by the *Upper* Paleozoic El Imperial Fm. The Quaternary Diamante Volcano is also shown

Fig. 2 a Google satellite image around Diamante river type section, within the area of the La Picaza old mine and Agua de la Chilena. **b** Outcrops of La Horqueta Formation at Agua de la Piedra section. **c, d, e** Some details of abundant quartz veins crosscutting folded layers of La Horqueta Formation at Agua de la Chilena and Agua de la Piedra outcrops

as well as biotite, amphiboles, and epidotes. Accessory minerals are apatite and zircon, and scarce titanite. The texture is granular hypidiomorphic and locally pegmatitic.

The stock was dated using the Rb-Sr method on whole rock and biotite on one sample, giving an age of 256 ± 2 Ma, with a Ri = 0.7073 ± 0.0001. However, a more accurate age is obtained combining whole-rock Rb-Sr of five samples with data from feldspars and biotite, which assigned a Guadalupian–Lopingian age of 257 ± 3 Ma, with a Ri = 0.7069 ± 0.0003 and MSWD of 8.6 following ISOPLOT model 3. The stock would have been emplaced after the Orogenic San Rafael Phase (Asselian—Sakmarian), and during the latest stages of volcanic activity linked to the Cochicó Group (Cingolani et al. 2005a). Based on the presence of amphibole together with biotite a metaluminous series with calcoalcaline characteristics can be assumed, which are typical of magmatic arcs related to the subduction of the paleopacific plate within the southwestern margin of Gondwana. This Permian magmatism could have originated the mineralization of El Rodeo and Las Picazas sulfides mines (arsenopyrite, pyrite and sphalerite), as well as the hydrothermal hematite of the Alto Molle mine (within the La Horqueta Formation; (Cingolani et al. 2005a).

The present work focus on provenance analyses of La Horqueta Formation based on whole-rock geochemistry and Sm–Nd data, which altogether with the information presented in Tickyj et al. (this volume), particularly regarding detrital zircon dating and Rb-Sr whole-rock data, give insights into source composition and the comparison with Río Seco de los Castaños Formation.

2 Sampling and Analytical Techniques

Sampling was done (see Tickyj et al. this volume) at Los Gateados and La Horqueta type areas (Fig. 1 and Table 1). A total of eighteen samples were selected for chemical analyses done at ACME Labs, Canada. Major elements were obtained by inductively coupled plasma element spectroscopy (ICP-ES) on fusion beads and the loss on ignition (LOI) was calculated by weight after ignition at 1000 °C. Mo, Cu, Pb, Zn, Ni, As, Cd, Sb, Bi, Ag, Au, Hg, Tl, and Se were analyzed by inductively coupled plasma mass spectroscopy (ICP-MS) after leaching each sample with 3 ml 2:2:2 $HCl-HNO_3-H_2O$ at 95 °C for 1 hour and later diluted to 10 ml. Rare earth elements (REE) and certain trace elements (Ba, Be, Co, Cs, Ga, Hf, Nb, Rb, Sn, Sr, Ta, Sc, Th, U, V, W, Zr, Y, La, Ce, Pr, Nd, Sm, Eu, Gd, Tb, Dy, Ho, Er, Tm, Yb, Lu) were analyzed by ICP-MS following lithium metaborate/tetraborate fusion and nitric acid digestion. Detection limits are: 0.01% for major elements, except for Fe_2O_3 which is 0.04%; 0.1 ppm for Mo, Cu, Pb, Cd, Sb, Bi, Ag, Tl, Cs, Hf, Nb, Rb, Ta, U, Zr, Y, La, and Ce; 1 ppm for Zn, Ba, Be, Sn, and Sc; 0.5 ppm for As, Au, Ga, Sr, and W; 0.01 ppm for Hg, Tm, Lu, and Tb; 0.2 ppm for Co and Th; 8 ppm for V; 20 ppm for Ni; 0.002 ppm for Cr; 0.02 ppm for Pr, Eu, and Ho; 0.3 ppm for Nd; 0.05 ppm for Sm, Gd, Dy, and Yb and 0.03 ppm for Er. Data are presented in Tables 2, 3, 4 and 5.

Seven whole-rock samples were used for Sm–Nd determinations; they were spiked with mixed $^{149}Sm-^{150}Nd$ tracer and dissolved in Teflon vial using an HF–HNO_3 mixture and 6 N HCl until complete material dissolution. The cationic resin AG-50 W-X8 (200–400 mesh) were used for column separation of the REE,

Table 1 GPS location of studied samples (after Tickyj et al,. this volume)

Sample	Location
Hor 9	34° 38′ 16.53″S–68° 53′ 11.11″W
Hor 10	34° 38′ 16.53″S–68° 53′ 11.11″W
Hor 11	34° 33′ 57.64″S–68° 54′ 05.05″W
Hor 14	34° 33′ 02.56″S–68° 51′ 04.83″W
Hor 15	34° 33′ 02.56″S–68° 51′ 04.83″W
Hor 16	34° 33′ 02.56″S–68° 51′ 04.83″W
Hor 17	34° 30′ 04.40″S–68° 55′ 07.01″W
Hor 18	34° 30′ 04.40″S–68° 55′ 07.01″W
Hor 20	34° 34′ 25.96″S–68° 46′ 40.08″W
Hor 21	34° 34′ 25.96″S–68° 46′ 40.08″W
Hor 24	34° 33′ 53.22″S–68° 49′ 35.48″W
Hor 27	34° 44′ 31.69″S–68° 49′ 34.90″W
Hor 64	34° 17′ 24.00″S–68° 48′ 49.57″W
Hor 50	34° 38′ 08.78″S–68° 48′ 49.57″W
Hor 53	34° 38′ 08.78″S–68° 48′ 9.57″W
Hor 66	34° 35′ 48.00″S–68° 52′ 32.00″W
Hor 67	34° 35′ 48.00″S–68° 52′ 32.00″W
QGAT1	34° 17′ 24.00″S–68° 48′ 49.57″W
QGAT2	34° 17′ 24.00″S–68° 48′ 49.57″W
QGAT3	34° 17′ 24.00″S–68° 48′ 49.57″W
QGAT4	34° 17′ 24.00″S–68° 48′ 49.57″W
QGAT5	34° 17′ 24.00″S–68° 48′ 49.57″W

followed by Sm and Nd separation using anionic politeflon HDEHP LN-B50-A (100–200 μm) resin according to Patchett and Ruiz (1987). Each sample was dried to a solid and then loaded with 0.25 N H_3PO_4 on appropriated filament (single Ta for Sm and triple Ta–Re–Ta for Nd). Isotopic ratios were measured in static mode with a VG Sector 54 multicollector mass spectrometer at the Laboratorio de Geología Isotópica, Universidade Federal do Rio Grande do Sul (LGI-UFRGS, Porto Alegre, Brazil). 100–120 ratios with a 0.5–1 V ^{144}Nd beam were normally collected. Nd ratios were normalized to $^{146}Nd/^{144}Nd = 0.72190$. All analyses were adjusted for variations instrumental bias due to periodic adjustment of collector positions as monitored by measurements of our internal standards. Measurements for the Spex $^{143}Nd/^{144}Nd$ are 0.511130 ± 0.000010. Correction for blank was insignificant for Nd isotopic compositions and generally insignificant for Sm/Nd ratios. $f_{Sm/Nd}$ is the fractional deviation of the sample $^{147}Sm/^{144}Nd$ from achondritic reference and is calculated as $(^{147}Sm/^{144}Nd)_{sample}/(^{147}Sm/^{144}Nd)_{CHUR} - 1$. The εNd indicates the deviation of the $^{143}Nd/^{144}Nd$ value of the sample from that of CHUR (DePaolo and Wasserburg 1976) and it is calculated as $εNd_{(0)} = \{[(^{143}Nd/^{144}Nd)_{sample(t=0)}/0.512638] - 1\} * 10{,}000$, whereas $εNd_{(t=420 Ma)} = \{[(^{143}Nd/^{144}Nd)_{sample\ (t)}/(^{143}Nd/^{144}Nd)_{CHUR\ (t)}] - 1\} * 10{,}000$. Parameters used are: $(^{147}Sm/^{144}Nd)_{CHUR} = 0.1967$. $(^{143}Nd/^{144}Nd)_{CHUR} = 0.512638$. T_{DM}

Table 2 Major elements (expressed in %) of La Horqueta Formation

	HOR 10	HOR 15	HOR 21	HOR 27	HOR 9	HOR 11	HOR 14	HOR 16	HOR 17	HOR 24	HOR 18	HOR 20	HOR 64	QGAT1	QGAT2	QGAT3	QGAT4	QGAT5	Average
SiO_2	67.85	70.28	69.32	65.23	59.26	52.50	58.79	58.39	49.97	44.69	52.79	45.48	66.27	77.45	76.78	76.71	78.19	83.17	64.06
Al_2O_3	13.62	12.51	12.54	15.43	18.09	21.89	19.78	20.43	24.46	24.48	22.36	24.35	15.48	9.78	10.32	10.45	9.94	7.32	16.29
Fe_2O_3	7.27	6.98	6.95	2.43	8.50	7.77	6.86	6.29	7.21	10.89	7.16	9.19	5.72	4.43	4.51	4.47	4.11	3.33	6.34
MnO	0.07	0.05	0.04	0.02	0.07	0.07	0.04	0.04	0.03	0.10	0.03	0.09	0.04	0.04	0.04	0.04	0.04	0.03	0.05
MgO	2.26	2.30	2.63	1.48	2.62	3.10	2.33	2.19	2.87	5.11	2.70	3.74	1.85	1.52	1.53	1.49	1.27	1.09	2.34
CaO	0.78	0.39	0.60	2.62	0.25	0.47	0.22	0.28	0.29	0.34	0.62	1.56	0.31	0.44	0.39	0.35	0.39	0.30	0.59
Na_2O	1.63	1.63	1.38	2.78	1.40	1.90	1.37	1.04	0.77	0.08	0.66	0.56	1.81	2.16	1.92	1.91	1.97	1.55	1.47
K_2O	2.14	1.86	1.89	2.87	3.51	5.36	4.85	5.59	6.84	5.92	6.01	5.82	3.30	1.71	1.90	2.01	1.72	1.23	3.59
TiO_2	1.06	1.05	1.12	0.74	0.99	1.07	0.86	0.96	1.09	1.20	1.05	1.16	0.83	0.69	0.66	0.65	0.62	0.46	0.90
P_2O_5	0.25	0.20	0.19	0.12	0.16	0.33	0.18	0.17	0.18	0.22	0.39	0.32	0.19	0.20	0.18	0.19	0.17	0.18	0.21
LOI	3.49	3.04	3.39	5.65	4.39	4.13	3.53	3.97	4.99	5.80	4.97	6.45	2.83	1.94	2.11	2.04	1.94	1.54	3.68
TOTAL	100.40	100.29	100.05	99.38	99.24	98.60	98.82	99.36	98.69	98.83	98.73	98.72	98.64	100.36	100.34	100.32	100.37	100.20	99.52
CIA	68	70	70	55	73	69	71	71	73	77	72	71	69	61	64	64	63	62	68

Table 3 Trace elements (expressed in ppm) of La Horqueta Formation

	HOR 10	HOR 15	HOR 21	HOR 27	HOR 9	HOR 11	HOR 14	HOR 16	HOR 17	HOR 24	HOR 18	HOR 20	HOR 64	QGAT1	QGAT2	QGAT3	QGAT4	QGAT5	AVERAGE
Sc	18	16	17	7	20	25	20	21	29	31	26	29	17	11	11	15	8	8	18
Be	2	2	2	2	3	4	4	4	5	5	4	4	3	2	2	3	b.d.l.	2	3
V	141	116	140	69	147	219	155	165	218	206	205	212	132	91	89	107	67	65	141
Cr	124	231	178	36	122	136	109	144	127	142	124	145	74	79	60	57	54	51	111
Co	32	32	26	11	28	25	18	19	17	29	15	13	32	45	43	44	55	66	31
Ni	49	83	64	b.d.l.	51	54	36	37	37	82	22	29	71	34	34	36	38	44	48
Cu	62	44	27	15	12	56	34	32	11	56	22	46	31	17	34	22	19	15	31
Zn	126	64	b.d.l.	b.d.l.	94	36	b.d.l.	35	74	91	47	b.d.l.	93	61	61	55	46	41	66
Ga	20	21	16	18	29	30	26	42	39	35	33	29	21	13	13	13	13	9	23
Ge	2	3	2	b.d.l.	5	5	4	6	5	4	4	b.d.l.	2	2	2	2	2	b.d.l.	3
As	8	43	36	9	b.d.l.	8	17	9	8	29	16	22	15	9	19	6	10	b.d.l.	16
Rb	97	108	128	127	176	250	230	359	338	253	252	277	158	72	79	81	66	56	175
Sr	64	43	54	125	121	50	40	63	46	28	85	97	76	81	62	57	60	52	67
Y	35	51	39	9	48	57	46	66	48	41	58	61	38	29	28	31	27	25	41
Zr	270	325	242	205	213	138	150	264	180	162	153	203	252	377	266	297	293	306	239
Nb	17	23	15	12	21	22	20	30	25	22	22	22	18	11	11	11	11	9	18
Sn	3	3	2	1	3	5	5	7	7	5	5	3	3	1	2	2	1	1	3
Sb	1	1	1	1	1	2	2	2	2	2	2	1	1	0	0	0	0	0	1
Cs	4	7	8	6	8	11	10	17	16	10	14	12	9	4	4	5	4	3	8
Ba	543	360	397	588	691	895	634	938	1070	1020	975	1120	454	214	235	290	207	139	598
Hf	7	9	7	6	6	4	4	8	5	5	5	6	7	10	7	8	8	8	7
Ta	2	2	2	1	2	2	2	3	2	2	2	2	3	3	4	4	4	5	3
W	116	145	111	64	37	17	31	43	14	10	33	31	208	424	503	498	659	792	208

(continued)

Table 3 (continued)

	HOR 10	HOR 15	HOR 21	HOR 27	HOR 9	HOR 11	HOR 14	HOR 16	HOR 17	HOR 24	HOR 18	HOR 20	HOR 64	QGAT1	QGAT2	QGAT3	QGAT4	QGAT5	AVERAGE
Tl	0	0	b.d.l.	b.d.l.	0	0	0	0	1	0	1	0	1	0	0	1	0	0	0
Pb	19	7	b.d.l.	b.d.l.	b.d.l.	b.d.l.	b.d.l.	b.d.l.	b.d.l.	5	b.d.l.	5	8	6	6	6	16	11	9
Bi	1	b.d.l.	b.d.l.	b.d.l.	b.d.l.	b.d.l.	b.d.l.	b.d.l.	0	0	0	b.d.l.	0	b.d.l.	b.d.l.	0	0	0	0
Th	9	12	9	15	17	16	15	26	23	15	18	19	16	12	11	11	11	9	14
U	3	3	3	4	3	5	5	7	4	3	5	4	4	3	3	4	3	3	4

b.d.l below detection limit

Table 4 Rare earth elements (expressed in ppm) of La Horqueta Formation

	HOR 10	HOR 15	HOR 21	HOR 27	HOR 9	HOR 11	HOR 14	HOR 16	HOR 17	HOR 24	HOR 18	HOR 20	HOR 64	QGAT1	QGAT2	QGAT3	QGAT4	QGAT5	Average
La	33.55	43.45	34.57	48.46	61.74	76.61	57.84	84.84	74.83	45.56	66.25	71.46	47.10	32.34	33.05	32.99	32.07	25.99	50.15
Ce	69.85	90.50	72.61	99.48	124.30	143.01	115.77	167.84	151.60	93.20	134.80	145.83	96.10	69.05	67.60	68.73	67.76	54.79	101.82
Pr	8.09	10.27	8.27	11.17	13.72	15.59	12.73	18.34	16.95	10.37	15.34	16.26	11.10	7.68	7.71	7.68	7.72	6.09	11.39
Nd	33.52	42.51	34.43	45.28	53.66	63.24	50.38	72.11	66.46	40.64	63.43	65.45	42.10	29.00	29.51	29.39	28.55	23.17	45.16
Sm	7.12	8.95	7.07	8.10	9.96	12.05	9.57	13.56	10.79	8.02	14.59	12.53	8.40	6.27	6.27	6.31	5.74	5.03	8.91
Eu	1.61	2.01	1.70	1.71	2.02	2.63	1.92	2.79	2.37	1.28	2.69	2.70	1.58	1.24	1.19	1.19	1.16	0.99	1.82
Gd	6.84	9.30	7.20	5.02	9.28	11.42	8.77	12.40	9.33	7.43	13.43	11.66	7.24	5.54	5.09	5.48	5.16	4.38	8.05
Tb	1.13	1.63	1.19	0.46	1.50	1.74	1.40	2.00	1.42	1.29	2.01	1.84	1.21	0.96	0.87	0.94	0.85	0.77	1.29
Dy	6.21	9.23	6.77	1.83	8.21	9.61	7.86	11.17	8.04	7.36	10.45	10.22	6.93	5.57	5.04	5.14	4.85	4.29	7.15
Ho	1.22	1.80	1.35	0.28	1.67	1.93	1.57	2.22	1.64	1.47	1.95	2.08	1.44	1.13	1.03	1.04	1.03	0.86	1.43
Er	3.69	5.26	4.17	0.70	5.21	5.86	4.82	6.85	5.11	4.67	5.80	6.32	4.32	3.21	3.03	3.04	2.90	2.56	4.31
Tm	0.54	0.75	0.59	0.08	0.77	0.82	0.70	1.01	0.77	0.71	0.83	0.91	0.65	0.49	0.46	0.48	0.43	0.39	0.63
Yb	3.41	4.33	3.78	0.48	4.68	4.99	4.29	6.17	4.82	4.36	4.98	5.73	4.05	3.07	2.87	2.94	2.85	2.47	3.90
Lu	0.51	0.59	0.53	0.06	0.70	0.72	0.63	0.92	0.71	0.67	0.69	0.83	0.57	0.45	0.42	0.41	0.42	0.37	0.57
Σ	177.28	230.57	184.24	223.11	297.40	350.20	278.26	402.22	354.85	227.02	337.23	353.82	232.80	165.99	164.14	165.75	161.50	132.15	246.59

Table 5 Selected ratios of La Horqueta Formation

	HOR 10	HOR 15	HOR 21	HOR 27	HOR 9	HOR 11	HOR 14	HOR 16	HOR 17	HOR 24	HOR 18	HOR 20	HOR 64	QGAT1	QGAT2	QGAT3	QGAT4	QGAT5	Average
K/Rb	0.02	0.02	0.01	0.02	0.02	0.02	0.02	0.02	0.02	0.02	0.02	0.02	0.02	0.02	0.02	0.02	0.03	0.02	0.02
Rb/Sr	1.53	2.52	2.36	1.02	1.46	5.04	5.70	5.69	7.35	9.13	3.42	2.85	2.08	0.89	1.27	1.42	1.09	1.08	3.10
Ba/Rb	5.59	3.33	3.10	4.62	3.92	3.58	2.76	2.62	3.17	4.03	3.34	4.04	2.87	2.98	3.00	3.57	3.14	2.47	3.45
Ba/Sr	8.53	8.39	7.31	4.72	5.70	18.01	15.73	14.87	23.27	36.81	11.41	11.53	5.97	2.65	3.81	5.05	3.43	2.66	10.55
Th/U	3.24	3.49	3.46	3.99	5.66	3.25	3.02	3.84	5.34	4.48	3.55	4.98	4.24	3.61	3.45	3.06	3.68	3.42	3.88
Th/Sc	0.51	0.73	0.52	2.07	0.87	0.62	0.76	1.22	0.78	0.50	0.68	0.65	0.94	1.06	1.00	0.74	1.36	1.12	0.90
Zr/Sc	15.03	20.34	14.22	29.31	10.63	5.52	7.48	12.58	6.20	5.22	5.90	7.00	14.82	34.28	24.18	19.81	36.64	38.19	17.07
Zr/Hf	37.80	36.34	35.04	36.98	35.18	32.01	33.59	34.05	33.08	33.29	33.89	34.12	35.00	39.49	36.75	39.33	38.80	39.58	35.79
Zr/Nb	15.93	14.17	15.85	16.60	9.97	6.25	7.46	8.76	7.25	7.34	6.93	9.43	13.77	34.98	24.55	27.58	27.50	33.37	15.98
Zr/Y	7.84	6.38	6.23	22.66	4.39	2.42	3.27	4.01	3.78	3.99	2.64	3.33	6.72	13.11	9.54	9.57	10.69	12.41	7.39
Ti/Zr	23.52	19.29	27.72	21.50	27.84	46.39	34.61	21.82	36.30	44.25	41.05	34.30	19.63	10.94	14.83	13.13	12.74	9.06	25.50
Ti/Nb	374.69	273.24	439.51	357.04	277.71	290.06	258.04	191.11	263.17	324.76	284.27	323.57	270.27	382.65	364.17	362.23	350.46	302.48	316.08
Cr/Zr	0.46	0.71	0.73	0.17	0.57	0.99	0.73	0.55	0.71	0.88	0.81	0.71	0.29	0.21	0.23	0.19	0.18	0.17	0.52
Cr/V	0.88	2.00	1.27	0.52	0.83	0.62	0.70	0.87	0.58	0.69	0.60	0.68	0.56	0.87	0.67	0.53	0.80	0.78	0.80
Cr/Ni	2.54	2.78	2.78		2.37	2.53	3.01	3.86	3.45	1.74	3.92	4.94	1.04	2.29	1.78	1.61	1.43	1.15	2.54
Y/Ni	0.71	0.61	0.61		0.94	1.06	1.27	1.76	1.29	0.50	1.84	2.09	0.53	0.84	0.83	0.87	0.73	0.56	1.00
Sc/Cr	0.15	0.07	0.10	0.20	0.16	0.18	0.18	0.15	0.23	0.22	0.21	0.20	0.23	0.14	0.18	0.26	0.15	0.16	0.18
V/Ni	2.90	1.39	2.19		2.86	4.06	4.30	4.41	5.92	2.52	6.50	7.25	1.86	2.64	2.64	3.01	1.78	1.47	3.39
Ni/Co	1.53	2.58	2.45		1.86	2.19	2.00	2.00	2.13	2.85	2.09	2.20	2.22	0.77	0.78	0.81	0.69	0.68	1.75
La/Th	3.64	3.70	3.93	3.34	3.54	4.93	3.82	3.30	3.30	2.95	3.72	3.81	2.94	2.78	3.01	2.97	2.95	2.89	3.42
La/Sc	1.86	2.72	2.03	6.92	3.09	3.06	2.89	4.04	2.58	1.47	2.55	2.46	2.77	2.94	3.00	2.20	4.01	3.25	2.99
La/Y	0.97	0.85	0.89	5.35	1.28	1.34	1.26	1.29	1.58	1.12	1.14	1.17	1.26	1.12	1.19	1.06	1.17	1.06	1.39

(continued)

Table 5 (continued)

	HOR 10	HOR 15	HOR 21	HOR 27	HOR 9	HOR 11	HOR 14	HOR 16	HOR 17	HOR 24	HOR 18	HOR 20	HOR 64	QGAT1	QGAT2	QGAT3	QGAT4	QGAT5	Average
La/Yb	9.83	10.04	9.15	100.88	13.19	15.36	13.47	13.75	15.52	10.44	13.30	12.46	11.63	10.53	11.50	11.22	11.25	10.52	16.89
La_N/Yb_N	6.64	6.78	6.18	68.17	8.92	10.38	9.10	9.29	10.49	7.05	8.99	8.42	7.86	7.12	7.77	7.58	7.60	7.11	11.41
La_N/Sm_N	2.97	3.05	3.08	3.76	3.90	4.00	3.80	3.94	4.37	3.58	2.86	3.59	3.53	3.25	3.32	3.29	3.52	3.25	3.50
Gd_N/Yb_N	1.62	1.74	1.55	8.47	1.61	1.86	1.66	1.63	1.57	1.38	2.18	1.65	1.45	1.46	1.43	1.51	1.47	1.44	1.98
Eu_N/Eu^*	0.70	0.67	0.73	0.82	0.64	0.68	0.64	0.66	0.72	0.51	0.59	0.68	0.62	0.64	0.65	0.62	0.65	0.65	0.66
Sm/Nd	0.21	0.21	0.21	0.18	0.19	0.19	0.19	0.19	0.16	0.20	0.23	0.19	0.20	022	0.21	0.21	0.20	0.22	0.20

Table 6 Sm–Nd data of La Horqueta Formation

Sample	Sm (ppm)	Nd (ppm)	$^{147}Sm/^{144}Nd$	$^{143}Nd/^{144}Nd$	$\varepsilon Nd_{(0)}$	$\varepsilon Nd_{(t)}$	T^1_{DM} (Ga)	T^2_{DM} (Ga)	$f_{(Sm/Nd)}$
Hor 9	6.39	34.56	0.1119	0.512070	−11.07	−6.53	1.45	1.66	−0.43
Hor 10	5.52	26.07	0.1281	0.512192	−8.69	−5.01	1.50	1.55	−0.35
Hor 21	5.07	24.84	0.1233	0.512224	−8.07	−4.14	1.37	1.49	−0.37
Hor 50	4.75	23.35	0.123 1	0.512329	−6.02	−2.07	1.20	1.33	−0.37
Hor 53	7.41	36.77	0.1219	0.512247	−7.63	−3.62	1.31	1.45	0.38
Hor 66	5.72	27.23	0.1269	0.512253	−7.50	−3.76	1.38	1.46	−0.35
Hor 67	5.64	27.07	0.1259	0.512367	−5.28	−1.49	1.17	1.28	−0.36

T^1_{DM} = DePaolo et al. (1981); T^2_{DM} = DePaolo et al. (1991)

(model ages) were calculated based on the depleted mantle model (DePaolo 1981) and on the three-stage model (DePaolo et al. 1991), as indicated in Table 6.

3 Whole-Rock Geochemistry

Data are presented in Tables 2, 3, 4, and 5. The use of whole-rock geochemistry to described provenance composition has been proven to be useful in the context of the pre-Carboniferous clastic units of the San Rafael block of the Cuyania terrane, as demonstrated by Cingolani et al. (2003b), Manassero et al. (2009), and Abre et al. (2011). Therefore, despite the remobilization that could have occurred due to low-grade metamorphism, it is expected that geochemical proxies using trace and REE of La Horqueta Formation would still reflect source compositions.

In the description of the geochemical proxies that follows, the sample HOR27 is treated separately due to their unique characteristics with respect to the whole dataset. A comparison to Río Seco de los Castaños Formation is introduced, since both units of the San Rafael block show similarities.

La Horqueta Formation shows SiO_2 concentrations ranging from 44.69 to 83.17%, Al_2O_3 is between 7.32 and 24.48%, Fe_2O_3 ranges from 3.33 to 10.89%, CaO is present in low concentrations (0.59% on average), Na_2O contents ranges from 0.08 to 2.16%, whereas K_2O is between 1.23 and 6.84% (Table 2). Sample HOR27 has SiO_2, Al_2O_3, and K_2O in the range of variation of the unit, but has lower Fe_2O_3 (2.43%), and higher CaO and Na_2O contents (2.63 and 2.68%, respectively). Some of the quartz veins cutting La Horqueta Formation were also analyzed for Ag and Au with negative results.

Weathering: The Chemical Index of Alteration (CIA; Nesbitt and Young 1982) is used to evaluate the extent of primary material transformation caused by weathering. The index is calculated using mole fractions as follows: CIA = {Al_2O_3/(Al_2O_3 + CaO* + Na_2O + K_2O)} × 100, where CaO* refers to the calcium associated with silicate minerals.

For La Horqueta Formation, values range from 61 to 77, indicating intermediate weathering conditions. The exception is sample HOR27 that has a CIA value of 55, typical of unweathered crystalline rocks of granodioritic composition (Fig. 3a; Table 2; Nesbitt and Young 1989). In the ACNK diagram the samples display a general weathering trend, that starts parallel to the A-CN boundary and to the weathering path of the upper continental crust (UCC), but shows deviation toward the K apex for samples with the highest CIA, indicating K_2O enrichment comparing to UCC value (McLennan et al. 2006; Table 2). Such behavior is in accordance to XRD mineralogical and petrographical data. Comparing to data from Río Seco de los Castaños (Manassero et al. 2009) it is evident that both units have the same range of CIA variation, and similar weathering trends, although La Horqueta Formation show higher K_2O enrichment (Fig. 3a).

Weathering effects could also be detected analyzing Th/U ratios and their variation regarding Th concentrations (McLennan et al. 1993), although attempts performed on Ordovician clastic units of the San Rafael block have led to uncertain results (Abre 2007; Abre et al. this volume). Compared to UCC averages of Th (10.7 ppm) and U (2.8 ppm) according to McLennan et al. (2006), most of the samples from La Horqueta Formation (including HOR27) are enriched in Th concentrations (14 ppm on average), and show similar to enriched U concentrations (4 ppm is the average of the unit as well as the U content of HOR27). Nonetheless, the Th/U ratios are in general around 3.5–4, which is typical for unrecycled samples derived from the UCC, although a few samples have higher Th/U ratios (maximum value 6.02) indicating weathering (Fig. 3b). Sample HOR27 has a Th/U ratio of 3.99, therefore clustering along with all samples. The Río Seco de los Castaños Formation shows a narrower spread of data, since values lower than the UCC are not present (Fig. 3b).

Recycling: Resistant heavy minerals tend to be concentrated during reworking, and this effect could be deciphered by analyzing the content of elements typically carried on such heavy minerals. The Zr/Sc ratios of La Horqueta Formation range between 5.2 and 36.6, indicating that the detrital components were not recycled (Fig. 3c). Sample HOR27, with a Zr/Sc ratio of 29.31 shows the same tendency. Noteworthy are those samples with Zr/Sc ratios lower than the UCC average (14; McLennan et al. 2006), which is a response of Sc concentrations above and Zr lower than UCC average (13.6 ppm and 190 ppm, respectively; see Table 3), indicating a derivation from a source less evolved than the average UCC. The same was deduced for Río Seco de los Castaños Formation, since its narrower range of values indicate a depleted source composition and recycling was even less important comparing to La Horqueta Formation (Fig. 3c; Manassero et al. 2009).

Source composition: The average composition of the source rocks (s) could be determined through REE patterns, the character of the Eu anomaly and the content of certain trace elements which tends to be either concentrated in silicic (such as La and Th) or mafic (Sc, Cr, Co) rocks (Taylor and McLennan 1985).

The Th/Sc ratios of La Horqueta Formation range from 0.50 to 1.36 (average 0.90). Those samples with Th/Sc ratios around average UCC (0.79; McLennan et al. 2006) are explained as derived from a felsic source with a composition similar

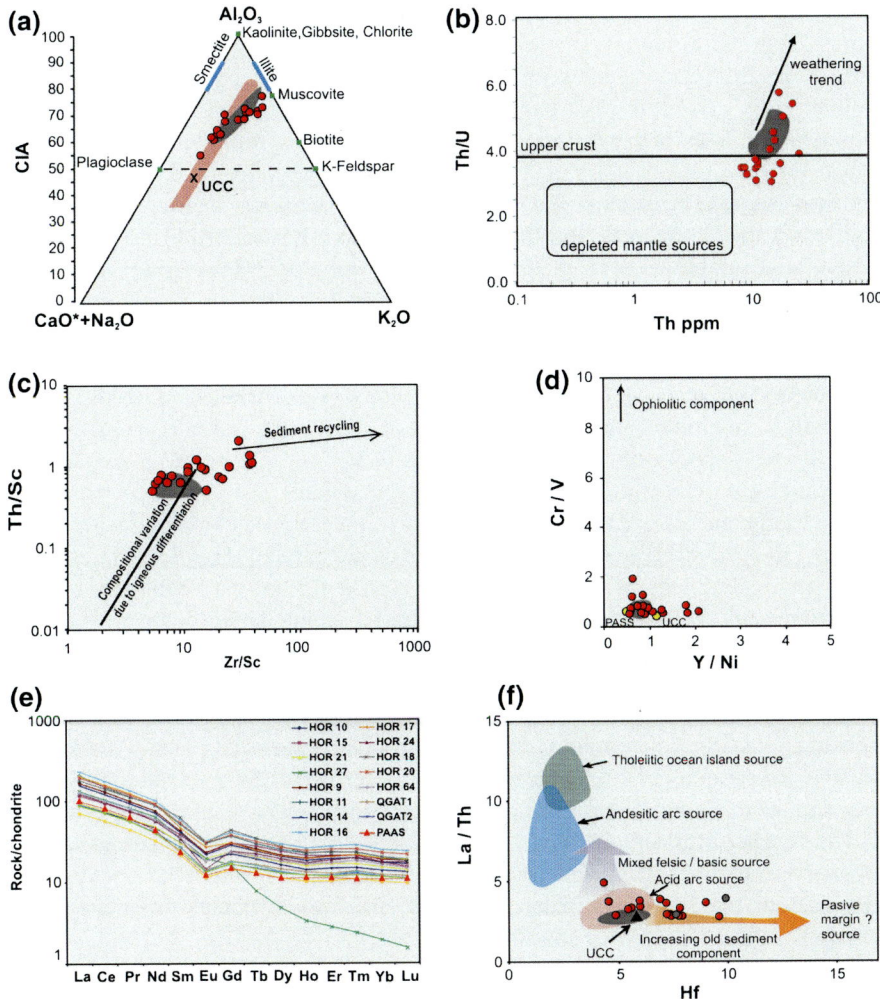

Fig. 3 a In the A-CN-K display La Horqueta Formation plot along a *vertical array* parallel to the expected weathering trend (field of *vertical lines*) for average *upper* crustal rocks; UCC values according to Taylor and McLennan (1985). b Th/U versus Th based on McLennan et al. (1993). c Th/Sc versus Zr/Sc display (McLennan et al. 2003): the ratios are typical of sedimentary rocks derived from unrecycled *upper* continental crust with a minor input of more depleted source composition. d The Y/Ni and Cr/V ratios are used to discriminate the input of a mafic source (McLennan et al. 1993). UCC values according to McLennan et al. (2006), while PAAS is following Taylor and McLennan (1985). e Chondrite normalized REE patterns; *PAAS* post-Archean Australian shales pattern (Nance and Taylor 1976) is draw for comparison. $Eu_N/Eu^* = Eu_N/(0.67 Sm_N + 0.33 Tb_N)$. f La/Th versus Hf after McLennan et al. (1980). For comparison range of variation of geochemical proxies of Río Seco de los Castaños Formation are shown as *gray* areas

to average UCC, with low Th/Sc ratios could have been derived from a depleted source (Fig. 3c, where it is clear the similarity to Río Seco de los Castaños Formation). La/Th ratios between 2.78 and 4.93 support felsic source rocks composition (Fig. 3f), as it was deduced for Río Seco de los Castaños Formation, although the latter display a narrower range of values ruling out any recycling (Manassero et al. 2009). Cr/V and Y/Ni ratios are between average values for UCC and Post-Archean Australian Shales (PAAS), as it is shown in Fig. 3d (see again similarities to Río Seco de los Castaños Formation; Manassero et al. 2009). The exception is sample HOR15 that display a Cr/V ratio of 2.0 due to Cr enrichment compared to UCC average of 83 ppm (McLennan et al. 2006), which along with Zr enrichment and scarce effects of recycling could indicate the presence of Cr-rich resistant heavy minerals such as first-cycle spinel that had been found in several sedimentary sequences of the San Rafael block (e.g., Abre et al. 2009, 2011). Although an ophiolitic source can be neglected, most of the samples show contents of Sc (up to 31 ppm), Cr (up to 231 ppm), and V (up to 219 ppm) above average UCC, indicating the influence of a less-evolved source.

The REE contents of La Horqueta Formation are enriched compared with PAAS, although the chondrite normalized REE patterns are parallel (Fig. 3e). The negative Eu anomaly (Eu_N/Eu^* of 0.66 on average) typical for detrital rocks derived from UCC is present. Noteworthy is the REE pattern of sample HOR27, which is parallel to PAAS regarding Light-REE, but shows extremely low concentrations of Heavy-REE; additionally, the Eu anomaly is the less negative of all samples analyzed (0.82; Table 5). Río Seco de los Castaños Formation also shows REE patterns parallel to PAAS with certain enrichment particularly regarding Heavy-REE and a negative Eu anomaly of 0.68 on average (Manassero et al. 2009), resulting therefore very similar to La Horqueta Formation, suggesting similar source composition.

The geochemical composition of sample HOR27 is not easily explained, since in summary, it shows CIA values typical for unweathered granodioritic rocks but a REE pattern that does not match such igneous compositions neither any other; therefore, laboratory errors cannot be rejected.

4 Isotope Geochemistry

Sm–Nd: Seven samples from La Horqueta Formation were analyzed using the Sm–Nd system and data are presented on Table 6. ε_{Nd} (t = 420 Ma) values range from −1.49 to −6.53, the $f_{Sm/Nd}$ are between −0.35 and −0.43, while the T_{DM}^1 (average crustal residence age calculated following DePaolo 1981) range from 1.17 to 1.50 Ga and T_{DM}^2 (calculated following DePaolo et al. 1991) ranges from 1.28 to 1.66 Ga. These data indicate that the average Nd isotopic signatures of the source rocks are rather a mix of both, an old upper crust and an arc component, and opposite to Río Seco de los Castaños Formation, fractionation is absent (Fig. 4a).

The εNd values are within the range of variation of data from the Cerro La Ventana Formation (Fig. 3b), which is part of the basement of the Cuyania terrane

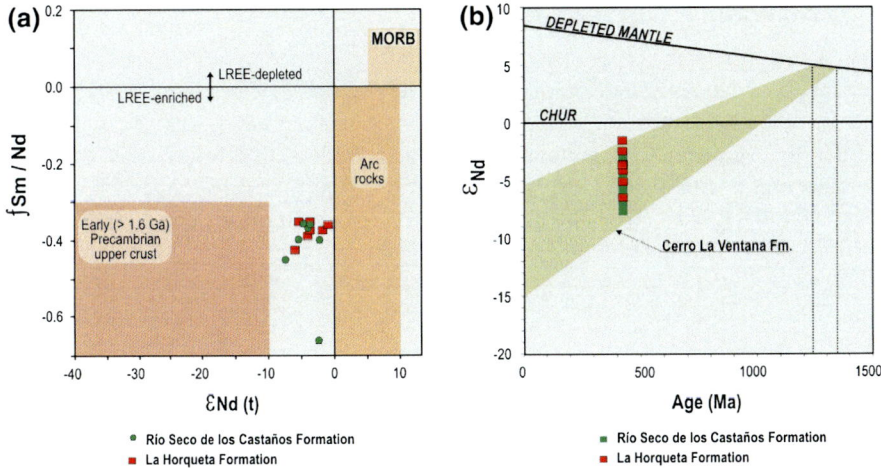

Fig. 4 **a** $f_{Sm/Nd}$ versus $\varepsilon_{Nd(t)}$ and **b** ε_{Nd} versus age. The range of Cerro La Ventana Formation Nd data (Cingolani et al. 2005b) and of Río Seco de los Castaños Formation are drawn for comparison. *CHUR* Chondritic Uniform Reservoir

outcropping within the San Rafael block (data from Cingolani et al. 2005b; Cingolani et al. this volume); The T_{DM} ages are comparable to those from Mesoproterozoic basement rocks of the Cuyania terrane studied by Kay et al. (1996) and summarized in Cingolani et al. (this volume) consistent with derivation from the nearest Grenvillian-age crustal source such as Cerro La Ventana Formation, exposed in the Ponón Trehué area. Furthermore, the Nd signature is similar to that of Río Seco de los Castaños Formation (Manassero et al. 2009), as well as to the Ordovician Pavón and Ponón Trehué Formations (Cingolani et al. 2003b; Abre et al. 2011). Ordovician to Silurian clastic sequences studied from the Precordillera *s.st.* (as part of the Cuyania terrane) also display the same range of εNd and T_{DM} values (Gleason et al. 2007; Abre et al. 2012).

Rb-Sr: Seven metapelites and six samples of micaschists were analyzed by Tickyj et al. (2001) and Tickyj et al. (this volume). The recalculated age obtained using an Isoplot/Ex Model 3 (Ludwig 2008) is 372.8 ± 8.1 Ma, initial $^{87}Sr/^{86}Sr$: 0.7164 ± 0.0012 and MSWD 8.4. The Rb-Sr data indicate that the low-grade metamorphism and folding events of La Horqueta Formation are Devonian in age. Furthermore, the age obtained agree with previous K-Ar ages reported by Linares and González (1990). The same methodology applied to Río Seco de los Castaños Formation indicate a very low-grade metamorphic age of 336 ± 23 Ma (Lower Carboniferous, Cingolani and Varela 2008). This is younger than the Rb-Sr metamorphic age of La Horqueta Formation, although both are probably linked to the final Chanic tectonic phase that occurred as a response to the accretion of Chilenia terrane at western proto-Andean Gondwana margin (Ramos et al. 1984).

5 Provenance Discussion

Geochemical analyses and particularly the Th/Sc and La/Th ratios, REE patterns and Eu anomalies indicate a derivation from a felsic source with a composition similar to average UCC, although Th/Sc and Zr/Sc ratios lower than the UCC average, along with Sc, Cr, and V concentrations suggest a provenance from source rocks slightly less evolved than the average upper continental crust. Similar conclusions were found for Río Seco de los Castaños Formation (Manassero et al. 2009) as well as for the Ordovician sequences of the San Rafael Block (Cingolani et al. 2003b; Abre et al. 2011). The agreement observed when comparing with the

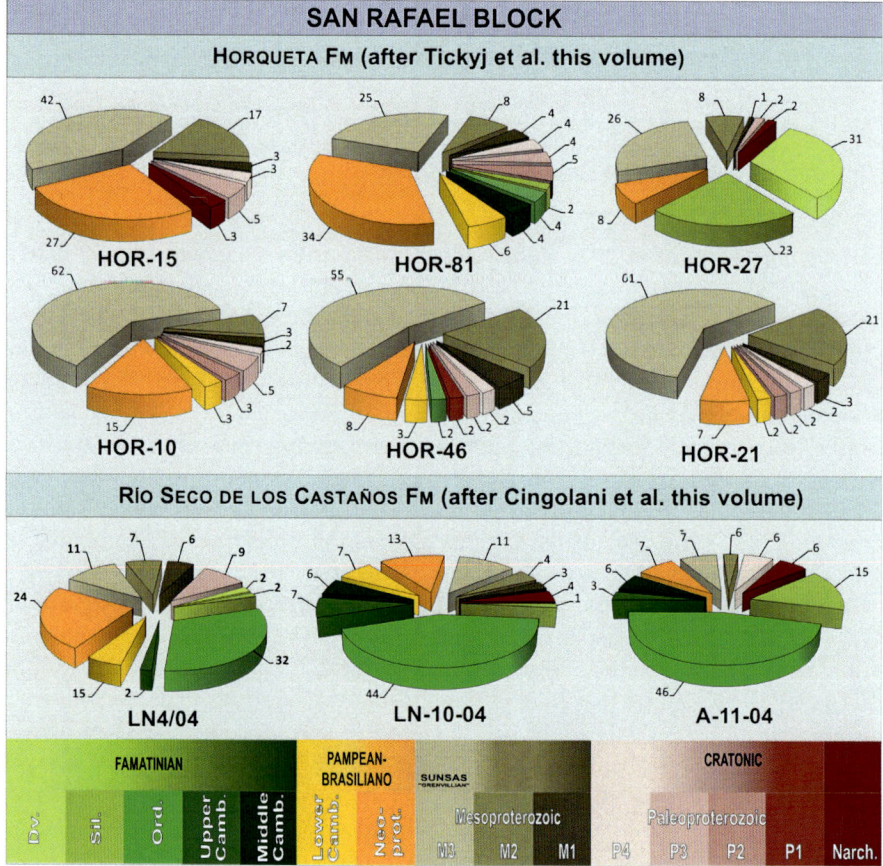

Fig. 5 Comparative U-Pb ages of the detrital zircons represented in percentage "pie diagrams" for La Horqueta and Río Seco de los Castaños Formations. Different colors are from recognized South American orogenic cycles: Archean to Paleoproterozoic; Mesoproterozoic, Brasiliano (Neoproterozoic–Early Cambrian), and Famatinian (Middle Cambrian–Devonian)

Sm–Nd signature of the Mesoproterozoic basement (Cerro La Ventana Formation) give further provenance constraints.

Zircon age patterns for La Horqueta Formation indicate four main populations, which in order of abundance correspond to the Mesoproterozoic (Grenvillian cycle), Neoproterozoic (Pampean–Brasiliano cycle), Paleoproterozoic and Upper Cambrian–Devonian (Famatinian cycle). A main derivation from the Mesoproterozoic basement of the San Rafael Block and Pampia terrane is supported, as well as a detrital input from the Río de la Plata craton and the Famatinian belt. Sample HOR27 shows however a different pattern, with a dominance of Famatinian grains, followed in abundance by the Mesoproterozoic population, the Neoproterozoic, the Paleoproterozoic and showing a few Neoarchean detrital zircon grains; it also comprises the younger detrital zircons found within the unit (ca. 0.4 Ga; Cingolani et al. 2008; Tickyj et al. this volume).

These age patterns are rather different comparing with Río Seco de los Castaños Formation which shows a dominance of Famatinian and Pampean–Brasiliano detrital zircons and lower amounts of Mesoproterozoic grains (Fig. 5). Such differences in age patterns indicate that the source rocks providing detritus to both basins were not the same.

6 Conclusions

(a) CIA values of La Horqueta Formation indicate intermediate weathering conditions, and samples with the highest CIA are enriched in K_2O comparing to UCC; some Th/U ratios support this. Zr/Sc ratios point to mainly unrecycled detritus.

(b) Th/Sc, La/Th, and Th/U ratios, REE patterns, and negative Eu anomalies are typical for detrital rocks derived from unrecycled UCC. However, Sc, Cr, and V concentrations along with low Th/Sc ratios suggest a provenance from source rocks slightly less evolved than the average upper continental crust. Sources compositions are similar to that of Río Seco de los Castaños Formation.

(c) The εNd values are within the range of variation of data from the Mesoproterozoic Cerro La Ventana Formation, which is part of the basement of the Cuyania terrane outcropping within the San Rafael Block. These isotopic data are also similar to that of the Río Seco de los Castaños Formation.

(d) Detrital zircon age patterns indicate a provenance from Mesoproterozoic (Grenvillian), Pampean–Brasiliano, and Famatinian cycles, in order of abundance.

(e) Comparison with Río Seco de los Castaños Formation indicate similar source composition based on geochemical proxies but the age of such rocks are different, according to detrital zircon age patterns. The main difference is that the Río Seco de los Castaños Formation contains larger proportion of Ordovician zircon grains while the La Horqueta Formation contains few Ordovician zircon ones and much more Grenville-aged zircons.

Acknowledgements This research was partially financed by CONICET (Grants PIPs 647; 199). We are grateful to Dr. Hugo Tickyj for field work assistance and to Dr. Héctor Ostera for several discussions and comments during revision of the manuscript.

References

Abre P (2007) Provenance of Ordovician to Silurian clastic rocks of the Argentinean Precordillera and its geotectonic implications. Ph.D. Thesis. University of Johannesburg, South Africa. UJ free web access

Abre P, Cingolani C, Zimmermann U, Cairncross B (2009) Detrital chromian spinels from Upper Ordovician deposits in the Precordillera terrane, Argentine: a mafic crust input. J S Am Earth Sci Spec Issue Mafic Ultramafic Complexes S Am Caribb 28:407–418

Abre P, Cingolani C, Zimmermann U, Cairncross B, Chemale Jr F (2011) Provenance of Ordovician clastic sequences of the San Rafael Block (Central Argentina), with emphasis on the Ponón Trehué Formation. Gondwana Res 19(1):275–290

Abre P, Cingolani C, Cairncross B, Chemale Jr F (2012) Siliciclastic Ordovician to Silurian units of the Argentine Precordillera: constraints on provenance and tectonic setting in the Proto-Andean margin of Gondwana. J S Am Earth Sci 40:1–22

Abre P, Cingolani CA, Manassero MJ (this volume). The Pavón Formation as the Upper Ordovician unit developed in a turbidite sand-rich ramp. San Rafael Block, Mendoza, Argentina. In: Cingolani C (ed) Pre-Carboniferous evolution of the San Rafael Block, Argentina. Implications in the SW Gondwana margin Springer, Berlin

Cingolani CA, Basei MAS, Llambias EJ, Varela R, Chemale Jr F, Siga Jr O, Abre P (2003a) The Rodeo Bordalesa Tonalite, San Rafael Block (Argentina): Geochemical and isotopic age constraints. 10° Congreso Geológico Chileno, Concepción, Octubre 2003, p 10 (Versión CD Rom)

Cingolani C, Manassero M, Abre P (2003b) Composition, provenance and tectonic setting of Ordovician siliciclastic rocks in the San Rafael Block: Southern extension of the Precordillera crustal fragment, Argentina. J S Am Earth Sci Spec Issue Pacific Gondwana Margin 16:91–106

Cingolani C, Varela R, Abre P (2005a) Geocronología Rb-Sr del Stock de Agua de la Chilena: Magmatismo Pérmico del Bloque de San Rafael, Mendoza. XVI Congreso Geológico Argentino, La Plata Actas en CD

Cingolani CA, Llambías EJ, Basei MAS, Varela R, Chemale Jr F, Abre P (2005b) Grenvillian and Famatinian-age igneous events in the San Rafael Block, Mendoza Province, Argentina: geochemical and isotopic constraints. In: Gondwana 12 Conference, Abstracts, p 102

Cingolani CA, Varela, R (2008) The Rb-Sr low-grade metamorphism age of the Paleozoic Río Seco de los Castaños Formation, San Rafael Block, Mendoza, Argentina. VI South American Symposium on Isotope Geology, Actas, p 4. Bariloche

Cingolani CA, Tickyj H, Chemale Jr F (2008) Procedencia sedimentaria de la Formación La Horqueta, Bloque de San Rafael (Argentina): primeras edades U-Pb en circones detríticos. XVII Congreso Geológico Argentino, San Salvador de Jujuy 3:998–999

Cingolani CA, Uriz NJ, Abre P, Manassero MJ, Basei MAS (this volume) Silurian-Devonian land-sea interaction within the San Rafael Block, Argentina: Provenance of the Río Seco de los Castaños Formation. In: Cingolani C (ed) Pre-Carboniferous evolution of the San Rafael Block, Argentina. Implications in the SW Gondwana margin Springer, Berlin

Criado Roqué P (1972) Bloque de San Rafael. In: Leanza AF (ed) Geología Regional Argentina. Academia Nacional de Ciencias, Córdoba, pp 283–295

Criado Roqué P, Ibáñez G (1979) Provincia geológica Sanrafaelino-Pampeana. In: Turner JC (ed) Segundo Simposio de Geología Regional Argentina, vol I. Academia Nacional de Ciencias, Córdoba, pp 837–869

Cuerda AJ, Cingolani CA (1998) El Ordovícico de la región del Cerro Bola en el Bloque de San Rafael, Mendoza: sus faunas graptolíticas. Ameghiniana 35(4):427–448

DePaolo DJ, Wasserburg GJ (1976) Nd isotopic variations and petrogenetic models. Geophys Res Lett 3:249–252

DePaolo DJ (1981) Neodymium isotopes in the Colorado front range and crust-mantle evolution in the Proterozoic. Nature 291:193–196

DePaolo DJ, Linn AM, Schubert G (1991) The continental crustal age distribution, methods of determining mantle separation ages from Sm-Nd isotopic data and application to the southwestern United States. J Geophys Res 96:2071–2088

Dessanti RN (1956) Descripción Geológica de la Hoja 27c-Cerro Diamante (Provincia de Mendoza). Dirección Nacional de Minería, Boletín 85:79. Buenos Aires

Gleason JD, Finney SC, Peralta SH, Gehrels GE, Marsaglia KM (2007) Zircon and whole-rock Nd–Pb isotopic provenance of Middle and Upper Ordovician siliciclastic rocks, Argentine Precordillera. Sedimentology 54:107–136

González Díaz EF (1981) Nuevos argumentos a favor del desdoblamiento de la denominada "Serie de La Horqueta" del Bloque de San Rafael, provincia de Mendoza. Congreso Geológico Argentino, No 7, Actas 3:241–256. San Luis, Argentina

Kay SM, Orrell S, Abruzzi JM (1996) Zircon and whole rock Nd–Pb isotopic evidence for a Grenville age and a Laurentian origin for the basement of the Precordillera in Argentina. J Geol 104:637–648

Linares E, González RR (1990) Catálogo de edades radimétricas de la República Argentina, años 1957–1987. Serie B (Didáctica y Complementaria) 19. Asociación Geológica Argentina, Buenos Aires, p 630

Ludwig KR (2008) User's Manual for Isoplot 3.6. A geochronological toolkit for Microsoft Excel. In: Berkeley Geochronology Center, Special Publication No 4. Berkeley, USA, p 77

Manassero M, Cingolani C, Abre P (2009) A Silurian-Devonian marine platform-deltaic system in the San Rafael block, argentine Precordillera-Cuyania terrane: lithofacies and provenance. In: Königshof P (ed) Devonian change: case studies in palaeogeography and palaeoecology. The Geological Society, London, Special Publications, vol 314, pp 215–240

McLennan SM, Nance WB, Taylor SR (1980) Rare earth element-thorium correlations in sedimentary rocks, and the composition of the continental crust. Geochim Cosmochim Acta 44:1833–1839

McLennan SM, Hemming S, McDaniel DK, Hanson GN (1993) Geochemical approaches to sedimentation, provenance, and tectonics. In: Johnsson MJ, Basu A (eds) Processes controlling the composition of clastic sediments: Geological Society of America, Special Paper, vol 284, pp 21–40

McLennan SM, Bock B, Hemming SR, Hurowitz JA, Lev SM, McDaniel DK (2003) The roles of provenance and sedimentary processes in the geochemistry of sedimentary rocks. In: Lentz DR (ed) Geochemistry of sediments and sedimentary rocks: evolutionary considerations to minerals deposit-forming environments. GeoText, vol 4. Geological Association of Canada, pp 7–38

McLennan SM, Taylor SR, Hemming SR (2006) Composition, differentiation, and evolution of continental crust: constraints from sedimentary rocks and heat flow. In: Brown M, Rushmer T (eds) Evolution and differentiation of the continental crust. Cambridge, p 377

Nance WB, Taylor SR (1976) Rare earth element patterns and crustal evolution. Australian post-Archean sedimentary rocks. Geochimica et Cosmochimica Acta 40:1539–1551

Nesbitt HW, Young GM (1982) Early Proterozoic climates and plate motions inferred from major element chemistry of lutites. Nature 199:715–717

Nesbitt HW, Young GM (1989) Formation and diagenesis of weathering profiles. J Geol 97:129–147

Patchett PJ, Ruiz J (1987) Nd isotopic ages of crust formation and metamorphism in the Precambrian of eastern and southern Mexico. Contrib Miner Petrol 96:523–528

Ramos V, Jordan TE, Allmendinger RW, Kay SM, Cortés JM, Palma MA (1984) Chilenia: un terreno alóctono en la evolución paleozoica de los Andes Centrales. 9° Congreso Geológico Argentino (Bariloche). Actas 2:84–106. Buenos Aires

Taylor SR, McLennan SM (1985) The continental crust. Its Composition and Evolution, Blackwell, London 312 pp

Tickyj H, Cingolani CA, Varela R, Chemale Jr F (2001) Rb-Sr ages from La Horqueta Formation, San Rafael Block, Argentina. III South American Symposium on Isotope Geology. Extended Abstracts, pp 628–631. Pucón. Chile

Tickyj H, Cingolani CA, Varela R, Chemale Jr F (this volume) Low-grade metamorphic conditions and isotopic age constraints of the La Horqueta pre-Carboniferous sequence, Argentinian San Rafael Block. In: Cingolani C (ed) Pre-Carboniferous evolution of the San Rafael Block, Argentina. Implications in the SW Gondwana margin. Springer, Berlin

Silurian-Devonian Land–Sea Interaction within the San Rafael Block, Argentina: Provenance of the Río Seco de los Castaños Formation

Carlos A. Cingolani, Norberto Javier Uriz, Paulina Abre, Marcelo J. Manassero and Miguel A.S. Basei

Abstract The Río Seco de los Castaños Formation (RSC) is one of the 'pre-Carboniferous units' outcropping within the San Rafael Block assigned to Upper Silurian–Lower Devonian age. We review the provenance data obtained by petrography and geochemical-isotope analyses as well as the U–Pb detrital zircon ages. Comparison with La Horqueta Formation is also discussed. The main components of this marine fine-grained siliciclastic platform are sandstones and mudstones. The conglomerates are restricted to channel fill deposits developed mainly at the Lomitas Negras location. A low anchizone for the RSC was indicated by illite crystallinity index. From the geochemical proxies described above (Manassero et al. in Devonian Change: Case studies in Palaeogeography and Palaeoecology.

Electronic supplementary material The online version of this chapter (doi:10.1007/978-3-319-50153-6_10) contains supplementary material, which is available to authorized users.

C.A. Cingolani (✉) · M.J. Manassero
Universidad Nacional de La Plata and Centro de Investigaciones Geológicas,
Diag. 113 n. 275, CP1904 La Plata, Argentina
e-mail: carloscingolani@yahoo.com

M.J. Manassero
e-mail: mj.manassero@gmail.com

C.A. Cingolani · N.J. Uriz
División Geología, Museo de La Plata, UNLP, Paseo del Bosque s/n,
B1900FWA La Plata, Argentina
e-mail: norjuz@gmail.com

P. Abre
Centro Universitario Regional Este, Universidad de la República,
Ruta 8 Km 282, Treinta y Tres, Uruguay
e-mail: paulinabre@yahoo.com.ar

M.A.S. Basei
Centro de Pesquisas Geocronologicas (CPGeo), Instituto de Geociencias,
Universidade de São Paulo, São Paulo, Brazil
e-mail: baseimas@usp.br

© Springer International Publishing AG 2017
C.A. Cingolani (ed.), *Pre-Carboniferous Evolution of the San Rafael Block, Argentina*, Springer Earth System Sciences,
DOI 10.1007/978-3-319-50153-6_10

Geological Society, 2009) a provenance from an unrecycled crust with an average composition similar to depleted compared with average Upper Continental Crust is suggested. T_{DM} ages are within the range of the Mesoproterozoic basement and Palaeozoic supracrustal rocks of the Precordillera-Cuyania terrane. ε_{Nd} values of the RSC are similar to those from sedimentary rocks from the Lower Palaeozoic carbonate-siliciclastic platform of the San Rafael Block. These data suggest an Early Carboniferous (Mississippian) low-metamorphic (anchizone) event for the unit. It is correlated with the 'Chanic' tectonic phase that affected the Precordillera-Cuyania terrane and also linked to the collision of the Chilenia terrane in the western pre-Andean Gondwana margin. As final remarks we can comment that the studied RSC samples show dominant source derivation from Famatinian (Late Cambrian-Devonian) and Pampean-Brasiliano (Neoproterozoic-Early Cambrian) cycles. Detritus derived from the Mesoproterozoic basement are scarce. U–Pb data constrain the maximum sedimentation age of the RSC to the Silurian–Early Devonian.

Keywords Silurian–Devonian · Provenance analysis · Río Seco de los Castaños unit · Chanic phase · Cuyania terrane

1 Introduction

The Río Seco de los Castaños Formation (González Díaz 1972, 1981) is one of the 'pre-Carboniferous units' outcropping within the San Rafael block (Fig. 1). This sequence was first part of the La Horqueta metasedimentary unit (Dessanti 1956), but it was redefined based on its sedimentary characteristics by González Díaz (1981) and was assigned to the Devonian by Di Persia (1972). Contributions by González Díaz (1972), Nuñez (1976) and Criado Roqué and Ibañez (1979) described other sedimentary features of this foreland marine sequence. Rubinstein (1997) found acritarchs and other microfossils assigned to the Upper Silurian age near the 144 Road (km 702) outcrops. Poiré et al. (1998, 2002) recognized some trace fossil associations that helped to interpret different sub-environments of deposition within a wide siliciclastic marine platform. More recently Pazos et al. (2015) record the presence of relevant ichnogenus along the Atuel River outcrops. Manassero et al. (2009) presented a sedimentary description and stratigraphy, geochemical and provenance facies analysis of this unit. Rapid deposition and storm action on the platform are suggested by the presence of hummocks and swaleys facies. Furthermore, plant debris (Morel et al. this volume) indicates that the continental source was not far away.

It is well known that the Upper Silurian–Lower Devonian is a time of great changes not only of ecosystems but of climates as well, caused probably by complex interactions between the fast-developing terrestrial biosphere, marine

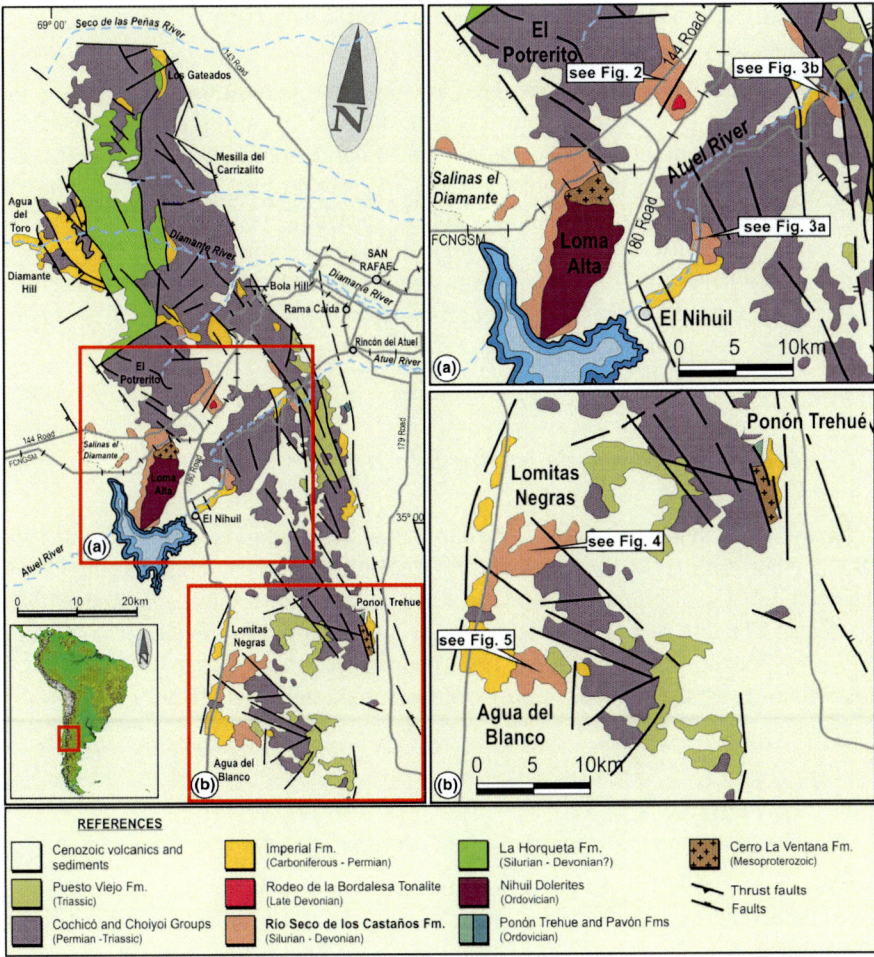

Fig. 1 Geological sketch map of the San Rafael Block showing different outcrops of the Río Seco de los Castaños Fm (RSC)

ecosystems and the atmosphere. Within this framework the Río Seco de los Castaños Formation (RSC) was deposited within a basin influenced by both, land and sea environments.

Based on these records, the main focus of the present paper is to review the provenance data obtained by petrography and geochemical-isotope analyses as well as to describe the recently acquired U–Pb detrital zircon ages. The data comparison with La Horqueta Formation is also discussed here.

2 Geological Aspects and Recognized Outcrops

Neither the base nor the top of the RSC are exposed. At the Loma Alta section this unit is separated by an unconformity or tectonic contact from the Mesoproterozoic mafic rocks (basement) and the Ordovician dolerite rocks. In other regions it is separated by unconformity from the Carboniferous-Lower Permian (El Imperial Formation) a fossiliferous marine-glacial/continental sedimentary unit locally forming deeply incised channels. The great angular unconformity is clearly showed at the Atuel River creek. The outcrops are rather isolated since they have been dismembered by Mesozoic and Cenozoic tectonism, according to Cuerda and Cingolani (1998) and Cingolani et al. (2003a) they are located at (Fig. 1)

2.1 Road 144-Rodeo de la Bordalesa

Trace fossils such as the *Nereites-Mermia* facies were mentioned (Poiré et al. 1998, 2002). Microfossils were described by Rubinstein (1997), although they were assigned to "La Horqueta Formation" (Fig. 2). In this region, the Rodeo de la Bordalesa Tonalite intruded the RSC. It has a magmatic arc geochemical signature and a crystallization age of 401 ± 4 Ma (Lower Devonian; Cingolani et al. 2003b, this volume), which also constrain the depositional age of the RSC.

2.2 Atuel River Creek

González Díaz (1972, 1981) described the type-section of the RSC in this region. Two main outcrops are recorded, one located about 12 km to the NE of the El Nihuil town (Fig. 3a) and the other near the Valle Grande dam (Fig. 3b), where the Seco de los Castaños River becomes an affluent of the Atuel River (González Díaz 1972). In the first locality the Formation comprises more than 600–700 m of tabular, green sandstones and mudstones with sharp contacts. It shows regional folding and dippings between 50° and 72° to the SE or NE. Above RSC inclined strata lay Upper Paleozoic horizontally bedded sedimentary rocks, displaying, therefore, a remarkable angular unconformity. In the Atuel Creek area fragments of primitive vascular plants are described and assigned to the Lower Devonian and marine microfossils such as prasinophytes, spores and acritarchs were found by D. Pöthe de Baldis (cf. Morel et al. 2006; this volume) in the RSC, indicating shallow water conditions near the coastline.

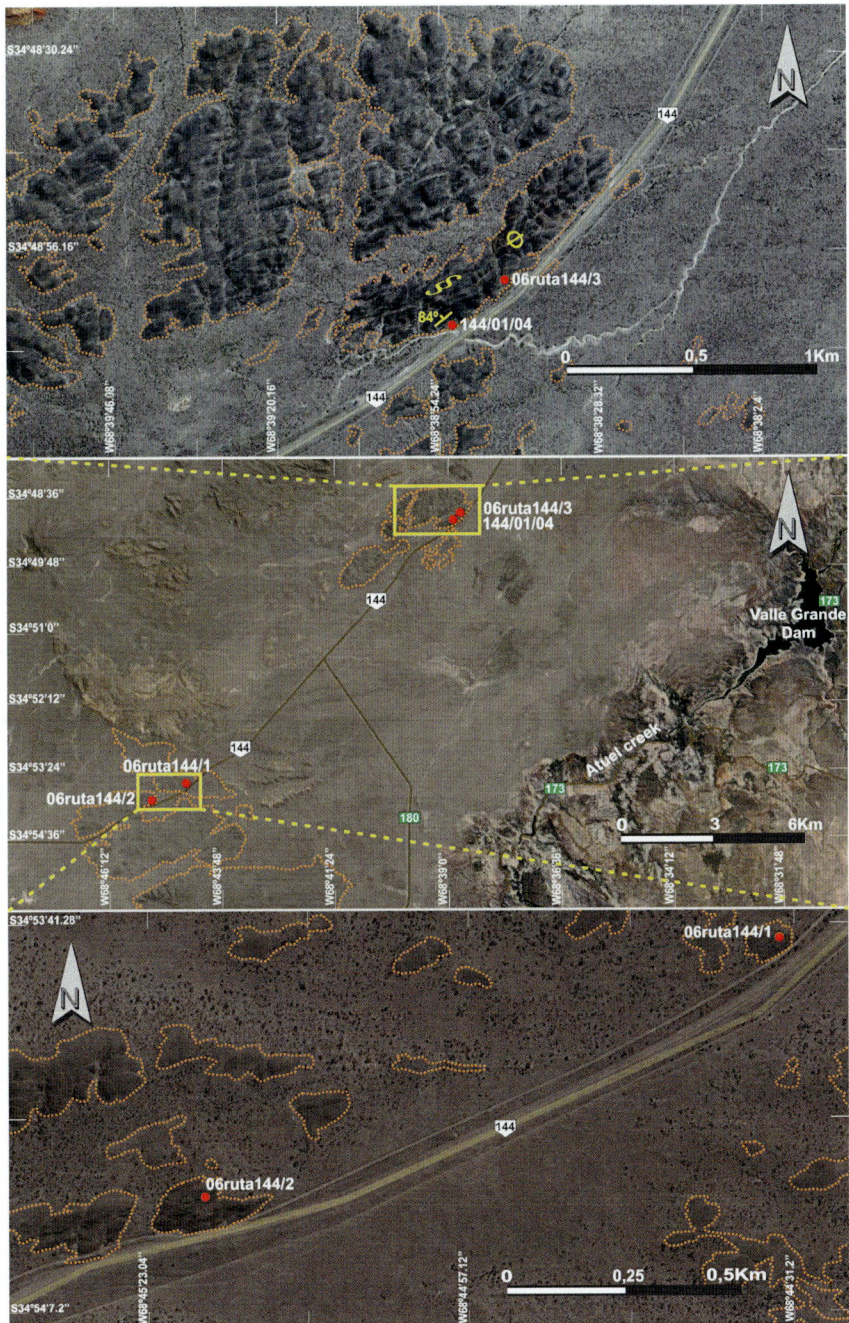

Fig. 2 Images showing the outcrops of the Río Seco de los Castaños Fm close to the 144 Road and position of the studied samples (*red dots*). The presence of microfossils and ichnofossils are denoted within the *upper image*

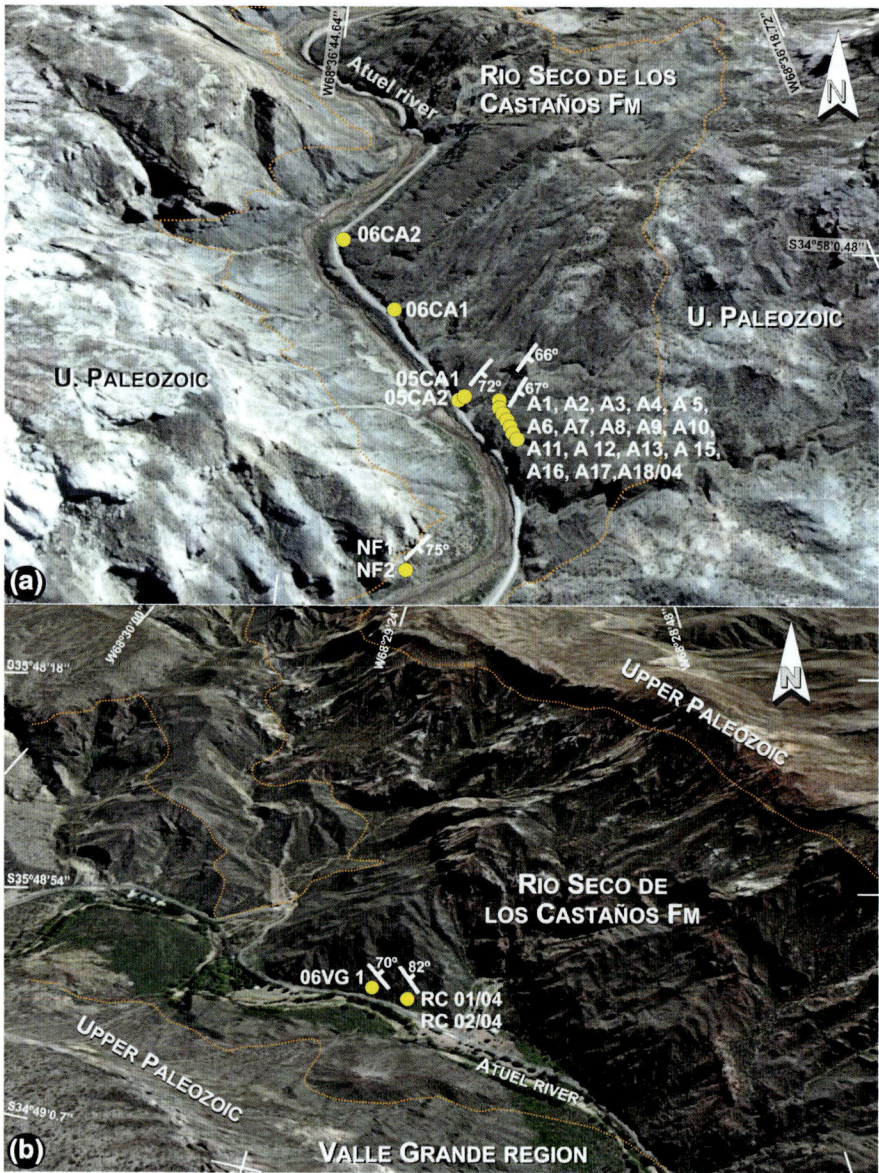

Fig. 3 Google-satellite images of **a** the Atuel River section and **b** Valle Grande region. The unconformity with the Upper Paleozoic succession is shown, as well as sampled locations (*yellow dots*)

2.3 Nihuil Area

The RSC (Fig. 1) is developed close to the Mesoproterozoic basement and Ordovician MORB-type dolerite rocks called 'El Nihuil mafic body' at the Loma Alta region (Cingolani et al. 2000; this volume).

2.4 Lomitas Negras and Agua del Blanco Areas

This region comprises the southernmost outcrops (Figs. 4 and 5). A Devonian coral known as '*Pleurodyctium*' was mentioned at Agua del Blanco, while conglomerates with limestone clasts bearing Ordovician fossils are described from the Lomitas Negras region (Di Persia 1972). Both successions are clearly folded and show substratal structures and wave ripples at the base of the sandy beds. After Di Persia (1972) the Lomitas Negras succession reaches a thickness of 2550 m.

3 Sedimentological Analysis

The main components of this marine fine-grained siliciclastic platform are sandstones and mudstones (Manassero et al. 2009). The conglomerates are restricted to channel fill deposits located mainly at the Lomitas Negras section. Main lithotypes recognized in the RSC platform are,

Mudstones: Comprise 50–90% of thin-beds, greenish in colour, usually with lamination and slight bioturbation commonly in repetitive sequences. The dark tonality and the scarcity of organic activity suggest anoxic conditions in low energy environments.

Heterolithics: Comprises thin-bedded sandstones and intercalated mudstones, with good lateral continuity and tabular-planar beds of few centimetres thick and grey to green colours. It is a very common facies, that exhibit sharp contacts and in many cases wave and current ripple structures, and also climbing ripples (Fig. 6). Normal grading and bioturbation are the dominant internal structures. Represents a well-oxygenated environment interpreted as a proximal or shallow marine platform, with dominance of a sub-tidal environment. The trace fossils are developed over a soft substrate with moderate energy.

Laminated siltstones: These rocks comprise bedded siltstones that range in thickness from several tens of centimetres to 1 m. They are intercalated with fine-grained sandstones with sharp contacts. Some coarser grained beds show small-scale ripple cross-lamination.

Sandstones: Comprise fine to medium-grained, grey and green, medium-bedded (10–15 cm thick) sandstones. They not only show massive and sharp contacts but also current and wave ripple marks (wave index 12–20) suggesting seawater-depths

1. 05LN1	6. 05LN6	11. 05LN11	16. 05LN16	21. 05LN21	26. 05LN26	31. LN4/04	36. LN09/04
2. 05LN2	7. 05LN7	12. 05LN12	17. 05LN17	22. 05LN22	27. 05LN27	32. LN5/04	37. LN10/04
3. 05LN3	8. 05LN8	13. 05LN13	18. 05LN18	23. 05LN23	28. LN1/04	33. LN6/04	38. LN11/04
4. 05LN4	9. 05LN9	14. 05LN14	19. 05LN19	24. 05LN24	29. LN2/04	34. LN7/04	39. LN12/04
5. 05LN5	10. 05LN10	15. 05LN15	20. 05LN20	25. 05LN25	30. LN3/04	35. LN8/04	40. LN13/04

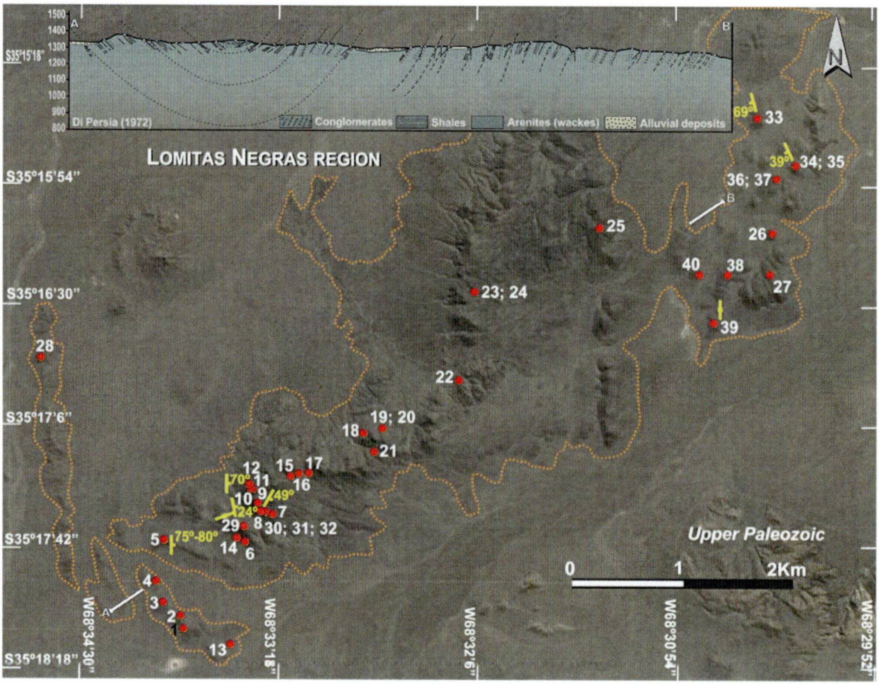

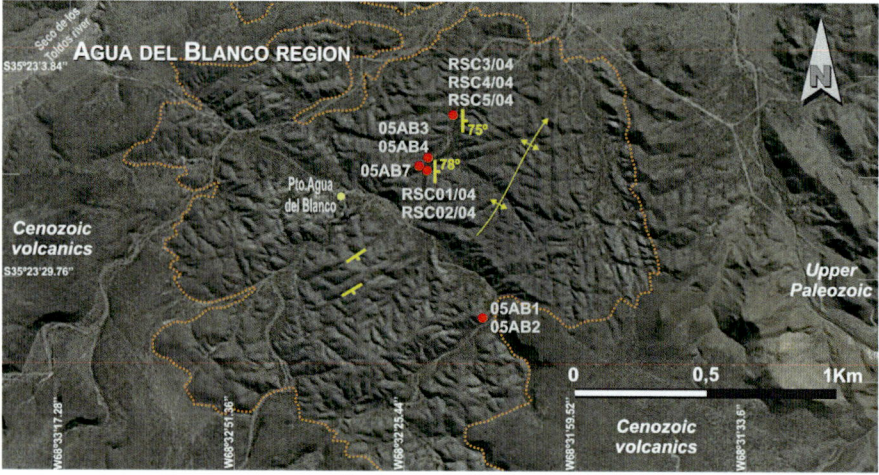

Fig. 4 Image showing the samples (*red dots*) along the Lomitas Negras section. At the upper part of the figure it is reproduced the SE-NW synclinal stratigraphic profile of RSC after Di Persia (1972)

Fig. 5 In this image the folded structure and sampling (*red dots*) of the southernmost outcrops of the RSC (Agua del Blanco region) are shown. The contact with the Upper Paleozoic units is also shown

Fig. 6 Tabular facies of sandstones and mudstones showing the change from traction to suspension processes within the platform

of 20 m. Deformational structures such as contorted beds and dish structures are present and scarce flute marks can develop to the base of the beds.

Rapid deposition and storm action on the platform is recognized by the presence of hummocks and swales in this facies. Furthermore, plant debris (Morel et al. this volume) indicates that the continental source was not far away. The erosive base of some beds implies a high sedimentation rate and the dominance of thin-beds with fine sediments suggests the action of low density gravity flows in the platform. Within the last described facies a *charcoal bed* (10–15 cm thick) that might be a marker horizon, was also found (Fig. 7a). It is composed of a mixture of silty-quartz, illite-kaolinite clays and amorphous organic matter with a TOC (total organic carbon) of 1%. Its presence is restricted to the section of Atuel creek. Recently Pazos et al. (2015) record the ichnogenera *Dictyodora* Weiss, which constitutes one of the most diverse, documented outside Europe and North America. The ichnospecies recognised include *D. scotica* and *D. tenuis*—and a new ichnospecies, *D. atuelica*. The succession studied by Pazos et al. (2015) contains abundant microbial mats (wrinkle marks) as either extended surfaces or patches. Wave-dominated deltas have facies sequences that coarsen upwards from shelf mud through silty-sand to wave and storm influenced sands, capped with lagoon or strand-plains where these peat beds can develop to the top of each cycle. This seems to be the case for the Atuel section (Manassero et al. 2009) where several

Fig. 7 a The RSC outcrop show the intercalated charcoal bed at road 173 (km 12) at the Atuel River section; up sequence to the NE and **b** typical vertical and differentially weathered strata at Aisol creek outcrop (West located at the left side)

prograding sequences with intense wave action have been described. Deformation by 'Chanic' tectonic phase is evident (Fig. 7a, b).

Conglomerates: Both, clast- and matrix-supported conglomerates wtih erosive bases are usually restricted to 2–3 m wide and 1 m deep channels. This facies is only present at the Lomitas Negras section (Fig. 4), developing lenticular and laterally discontinuous beds. They are poorly sorted and the matrix is medium to coarse sand. Clasts range from 2 to 10 cm long and show chaotic disposition without stratification; they are mainly composed of wackes, marls, limestones, siltstones, phyllites, quartz and feldspars. Some limestone clasts bear Ordovician fossils (Nuñez 1976; Criado Roqué and Ibañez 1979).

To the top, the channels could reach several metres wide and two or three metres thick. The conglomerates tend to have a sub-vertical position, due to the regional folding of the sequence (Fig. 4). As they are harder than the associated fine-to-medium-grained sedimentary rocks, they result into a strong geomorphologic control. The thickness of sandstones and mudstones associated to this facies suggests high energy, a relatively instability of the coastline and close continental source areas bearing plant remains (Fig. 8).

Fig. 8 Plant remains from Lomitas Negras section. Distance between *white dots* is 1 mm

4 Petrography and Diagenesis Studies

25 thin section samples were analyzed under the microscope. The minerals recorded by point counting are quartz (monocrystalline, polycrystalline and metamorphic), K-feldspar (microcline), plagioclase, opaque minerals, hematite and sedimentary or metamorphic rock fragments (Fig. 9). The presence of detrital biotite and scarce muscovite suggest short transport and reworking of sediments (Manassero et al. 2009). Many of the medium-grained sandstones (2–1.5Φ) are wackes (more than 15% matrix) and are composed of subangular quartz, with normal and wavy extinction, feldspars and fragments of polycrystalline quartz. Samples from the Lomitas Negras section show higher proportions of polycrystalline quartz.

The rocks are classified as feldspathic-wackes and quartz-wackes. In the Q-F-L diagrams, the sandstones of the Río Seco de los Castaños Formation show a cluster of data in both the recycled orogen and continental block fields (Fig. 10). Feldspars and biotite are widespread altered to chlorite, giving the typical greenish colours to the rocks. In the Lomitas Negras section the abundance of polycrystalline quartz displace the data to the recycled field. Although the data is showing some dispersion, we assume an uplifted igneous-metamorphic basement or recycled orogen as source areas.

The clay minerals fraction was studied using XRD, and it shows a dominance of illite (40–60%), kaolinite (25–40%) and chlorite which ranges from 10 to 20%, although it can go up to 35% when interlayered with smectite (Manassero et al. 2009). Muscovite and interstratified chlorite/smectite are very scarce. The less than 2 μm fraction contains as well very small amounts of quartz and plagioclases. A low anchizone for the RSC was indicated by illite crystallinity index.

5 Geochemistry and Isotopic data

All the samples ($n = 14$) analyzed by geochemical methods from the RSC are claystones, except for one siltstone and sandstone (Manassero et al. 2009). CIA values are between 61 and 78. In the A-CN-K diagram the samples follow a general weathering trend which is broadly parallel to the A-CN join, regarding the average upper continental crust (UCC) composition, although K-metasomatism of some samples is evident. The RSC is moderately to highly weathered. Compared with Post-Archaean Australian Shales (PAAS) most of the samples are depleted in Th and U, although some samples are enriched in both. The Th/U ratios range from below to above the PAAS value of 4.7 but above the upper continental crust average indicating weathering processes, in accordance to CIA. The Zr/Sc and Th/Sc ratios indicate that recycling was not important for the RSC Formation, and an input from a source geochemically less evolved than average UCC. The Cr/V and Y/Ni ratios are 0.79 ± 0.33 and 0.86 ± 0.3 (average) respectively, indicating source rock(s) more mafic than the average UCC.

Fig. 9 Photomicrographs of medium-fine quartz-feldspathic wackes. RSC4: sample from Atuel River Creek. LN 3 and 4: samples from Lomitas Negras. On the left with crossed nicols. High matrix content and subangular character of minerals is shown. *Qz* quartz; *Pl* plagiocase; *Msc* muscovite; *Lt* lithics; *A* amphibole; *Fe* Fe oxides. Scale bar 200 μm

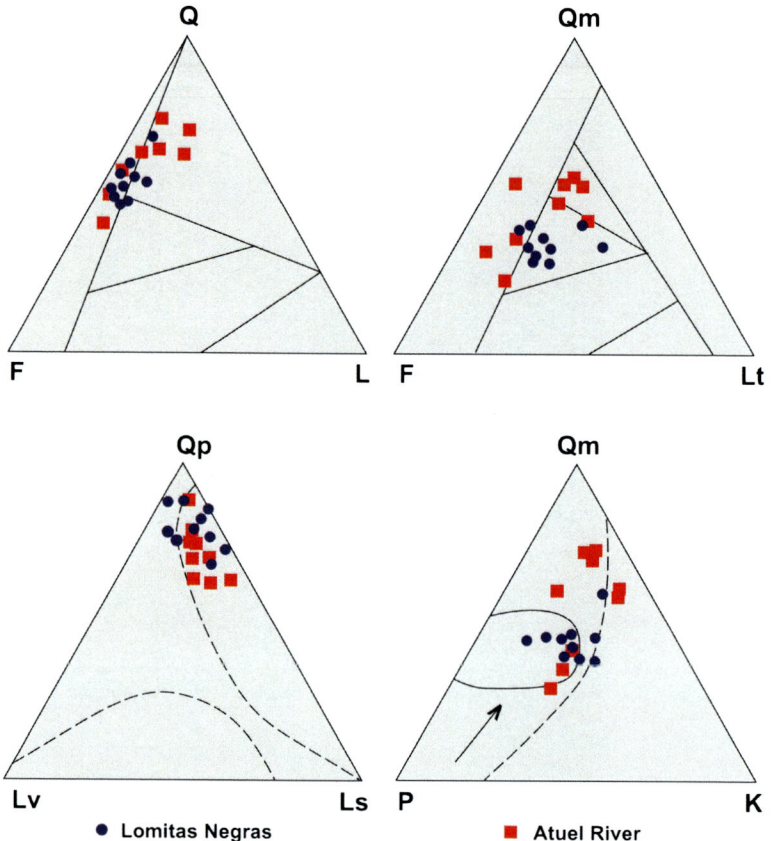

Fig. 10 Ternary diagrams after Dickinson and Suczek (1979) and Dickinson et al. (1979, 1983) plotting sandstone samples from the Atuel River Creek and Lomitas Negras sections. *F* feldspars; *FK* K-feldspars; *L* lithoclasts; *P* plagioclases; *Q* quartz (including polycrystalline quartz); *Qm* monocrystalline quartz; *Qp* polycrystalline quartz; *Ls* sedimentary lithoclasts; *Lv* volcanic lithoclasts (modified from Manassero et al. 2009)

The chondrite normalized REE diagram for the RSC is broadly similar to the PAAS pattern, showing a moderately enriched light rare earth elements pattern, a negative Eu-anomaly and a rather flat heavy rare earth elements distribution. Eu-anomaly of 0.81 along with Eu concentrations that double that from the PAAS supports the influence of a depleted source.

From the geochemical proxies described above (Manassero et al. 2009) a provenance from an unrecycled crust with an average composition similar to depleted compared with average UCC is suggested.

5.1 Sm–Nd Data

As were presented in Manassero et al. (2009) the RSC samples ($n = 7$) shows ε_{Nd} (t) values (where $t = 420$ Ma is the proxy age of sedimentation) ranging from -2.5 to -7.7 (average -4.5 ± 1.7). ε_{Nd} values are between those typical for the upper continental crust or older crust and those typical for a juvenile component (Fig. 11). Samples with the less negative ε_{Nd} (t) display the lowest Th/Sc ratios, indicating that the more juvenile the source the more depleted its geochemical signature. The $f_{Sm/Nd}$ against ε_{Nd} (t) diagram shows a data cluster between fields of arc-rocks and old crust. $f_{Sm/Nd}$ values out of the range of variation of the upper crust (-0.4 to -0.5) could be indicating Sm–Nd fractionations due to secondary processes.

T_{DM} ages are within the range of the Mesoproterozoic basement and Palaeozoic supracrustal rocks of the Precordillera-Cuyania terrane. ε_{Nd} values of the RSC are similar to those from sedimentary rocks from the Lower Palaeozoic carbonate-siliciclastic platform of the San Rafael Block, which show ε_{Nd} (t) between -0.4 and -4.9 (Cingolani et al. 2003a) and they are also in the range of variation of ε_{Nd} values of the Mesoproterozoic basement of the San Rafael Block (the Cerro La Ventana Formation; Cingolani et al. 2005) recalculated at 420 Ma. Although some $f_{Sm/Nd}$ values are below or above average values for the upper crust, all samples but one has $f_{Sm/Nd}$ values in the range of variation of the Cerro La Ventana Formation (Cingolani et al. 2005; Cingolani et al. this volume).

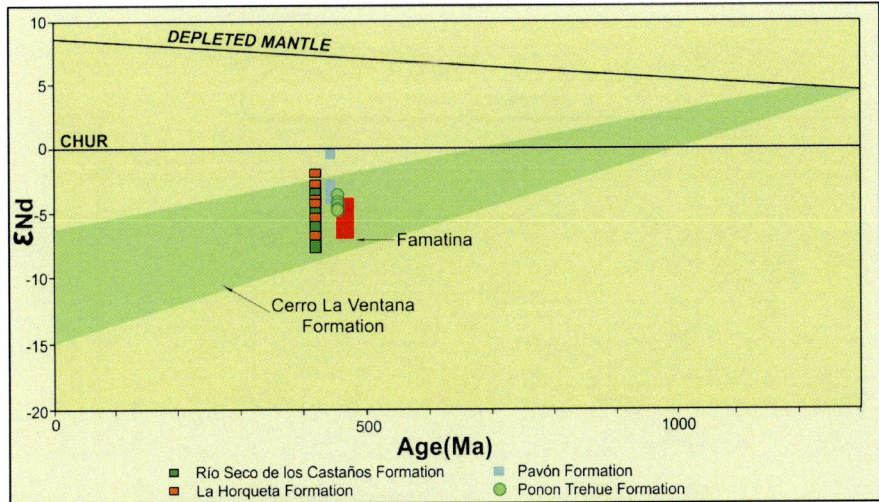

Fig. 11 Sm–Nd data of 7 whole rock samples from RSC that plot within the εNd range of Mesoproterozoic basement. For comparison the Pavón and Ponón Trehué Fms and Famatina arc (Ordovician) are also plotted (Abre et al. 2011, 2012)

5.2 Rb–Sr Whole Rock Data

Determination of Rb and Sr contents was performed by XRF, and the isotopic composition on natural Sr by mass spectrometry. The sample preparation, chemical attacks and Sr concentration with cation exchange resin were carried out in the clean laboratory of the Centro de Investigaciones Geológicas (CIG, University of La Plata, Argentina) and mass spectrometry was performed in the Centro de Pesquisas Geocronológicas (CPGeo), São Paulo, Brazil (Cingolani and Varela 2008). The results were plotted on isochron diagram, using the Isoplot model after Ludwig (2001).

Eight fine-grained samples were selected for analysis by Rb–Sr systematic; their location (all from the Atuel river type-section) is shown in Fig. 3. The Rb content varies between 165 and 312 ppm, while the Sr concentration ranges from 29 to 88 ppm. The $^{87}Rb/^{86}Sr$ ratios are between 7.5 and 24.4 in agreement with the relative high concentration of Rb and low contents of Sr. The expansion in the isochronic diagram (Fig. 12; Table 1) is acceptable for metasedimentary rocks. We interpret that during the low-metamorphic event Rb and Sr underwent isotopic homogenization, and therefore whole rock alignment is present (MSWD = 7.4). The high value of the initial isotopic ratio $^{87}Sr/^{86}Sr$ (0.7243) suggests a provenance for the sedimentary detritus from an evolved continental crust source. If we reject the sample A-02-04 that is out of the main alignment, the obtained Rb–Sr age is 336 ± 23 Ma.

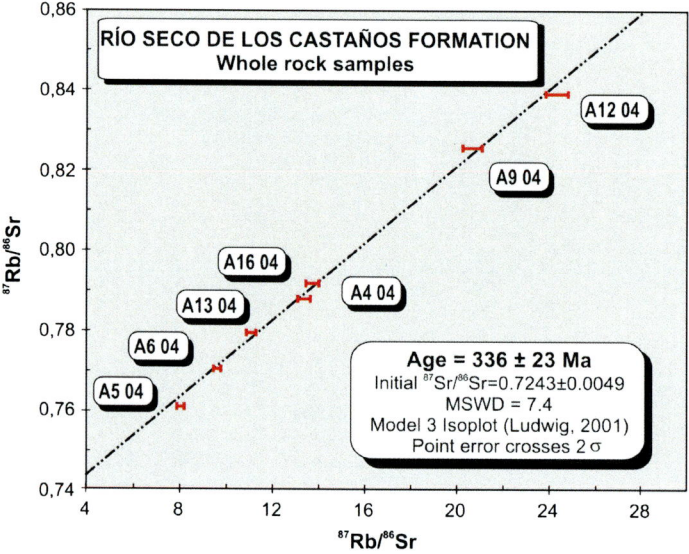

Fig. 12 Rb–Sr isochrone diagram for the RSC whole rock samples

Table 1 Rb–Sr systematic data

Lab. No. (1)	Field No.	Rb (ppm) (2)	Sr (ppm) (2)	$^{87}Rb/^{86}Sr$	Error	$^{87}Sr/^{86}Sr$ (3)	Error
CIG 1350	A2/04	228.6	88.2	7.54	0.15	0.754476	0.000018
CIG 1351	A4/04	207.0	43.8	13.79	0.28	0.791519	0.000012
CIG 1352	A5/04	164.6	59.5	8.05	0.16	0.761021	0.000089
CIG 1353	A6/04	167.6	50.6	9.65	0.19	0.770341	0.000024
CIG 1354	A9/04	311.8	44.0	20.75	0.42	0.825325	0.000016
CIG 1355	A12/04	244.7	29.4	24.40	0.49	0.838778	0.000016
CIG 1356	A13/04	219.0	57.4	11.12	0.22	0.779242	0.000018
CIG 1357	A16/04	199.0	43.3	13.41	0.27	0.787614	0.000019

(1) CIG: La Plata; 1350 was eliminated, (2) FRX: CPGeo. São Paulo, and (3) mass spectrometer: CPGeo. São Paulo

These data suggest an Early Carboniferous (Mississipian) low-metamorphic (anchizone) event for the unit. Recalculating the data using all samples and following Isoplot 3.5 model 3 (Ludwig 2012), the obtained age is 346 ± 30 Ma, with enhanced error by using the eight samples. Another 'pre-Carboniferous' siliciclastic unit of the San Rafael Block, called La Horqueta Formation, which shows Rb-Sr whole rock ages (Tickyj et al. 2001) of 371 ± 61 and 379 ± 15 (Late Devonian), with an initial $^{87}Sr/^{86}Sr$ ratio of 0.7150, a considerable difference with respect to the RSC unit is evident, supporting the interpretation that both received detritus from different sources.

It is important to note that in the upper section of the Carboniferous-Lower Permian unconformable El Imperial unit, Rocha Campos et al. (2011) obtained U–Pb SHRIMP zircon ages of 297 ± 5 Ma (Carboniferous-Permian boundary) from tuff levels.

The Rb–Sr whole rock isotopic data of the metapelites record the very low-grade metamorphism during the Early Carboniferous. It is in correlation with the 'Chanic' tectonic phase that affected the Precordillera-Cuyania terrane (Ramos et al. 1986) linked to the collision of the Chilenia terrane in the western pre-Andean Gondwana margin. Rb–Sr data also help to constrain the depositional age and the discussion about the source areas of the RSC detritus.

5.3 U–Pb Detrital Zircon Age Data

U–Pb analyses of detrital zircons have been intensively used as an important tool to study the sedimentary provenance and the age of source(s) of detritus (Cingolani et al. 2014). Three samples from the RSC were analyzed by the U–Pb zircon systematic using LA-ICP-MS equipment, at the Centro de Pesquisas Geocronológicas, University of São Paulo (A-11-04; LN-10-04) and at Isotope

Laboratory of University of Río Grande do Sul, Porto Alegre, Brazil (LN-4-04). The obtained results are as follows:

Sample A-11-04 (Atuel river section, 34°57′47″S–68°36′40″W; Fig. 13; Table 3 in Supplementary Material): It is characterized by a main mode (61%) of detrital zircons of *Ordovician to Early Silurian* ages (433–480 Ma). In order of abundance, secondary records show the following age groups: *Mesoproterozoic*, with 1043–1392 Ma (10.5%), *Cambrian* with ages of 490 and 520 Ma (8.9%), *Neoproterozoic* with 548 and 731 Ma (7.5%), *Paleoproterozoic* with ages between 1686 and 1888 Ma (5.9%) and *Neoarchean* with ages between 2582 and 2628 Ma (5.9%).

Sample LN-10-04 (Lomitas Negras section, 35°15′52″S-68°30′19″W; Fig. 13; Table 4 in Supplementary Material): Shows also a dominance of *Ordovician* zircon grains (46.5%), with 435 and 486 Ma. A second group comprises *Cambrian* zircons (18.3%) with 493 and 537 Ma. In less proportion we found *Mesoproterozoic* grains (15.5%) with ages ranging from 1024 to 1352 Ma and *Neoproterozoic* zircons (12.7%) with ages from 543 to 956 Ma. Finally, we found two minor groups, one of *Neoarchean* zircons (4.2%) with ages of 2619–2686 Ma and another of *Paleoproterozoic* zircons (2.8%) with data of 1402–1530 Ma.

Sample LN-4-04 (Lomitas Negras section, 35°17′32.00″S–68°33′26.00″W; Fig. 13; Table 5 in Supplementary Material): *Ordovician* zircon grains (27%) are also dominant in this sample, with ages between 444 and 483 Ma. As a second group, we found Neoproterozoic grains (24%) with 545 and 997 Ma and Mesoproterozoic zircon grains (M3, M2, and M1; 25%) with ages ranging from 1019 to 1567 Ma. Upper *Cambrian* aged zircons show 9% and those of *Lower Cambrian* are 2%. Finally, two minor groups of zircons are present, one of *Paleoproterozoic* (8%) ages ranging from 1770 to 2425 Ma and another *Silurian–Devonian* (4%). In this sample most of the detrital zircons are coming from Ordovician, Neoproterozoic and Mesoproterozoic source ages in the same proportion.

As final remarks we can comment that the studied RSC samples show (Fig. 14) dominant derivation from Famatinian (Late Cambrian-Devonian) and Pampean-Brasiliano (Neoproterozoic-Early Cambrian) cycles. Detritus derived from the Mesoproterozoic basement are scarce. U–Pb data constrain the maximum sedimentation age of the RSC to the Silurian-Early Devonian.

5.4 Comparison with La Horqueta Formation

A comparison between RSC and La Horqueta Formation (Tickyj et al., this volume; Abre et al., this volume) is shown on Table 2. The RSC present low anchizone metamorphic grade dated by Rb–Sr as Lower Carboniferous; fossil (plants, acritarchs, ichnogenera) are preserved in different outcrops; a provenance from

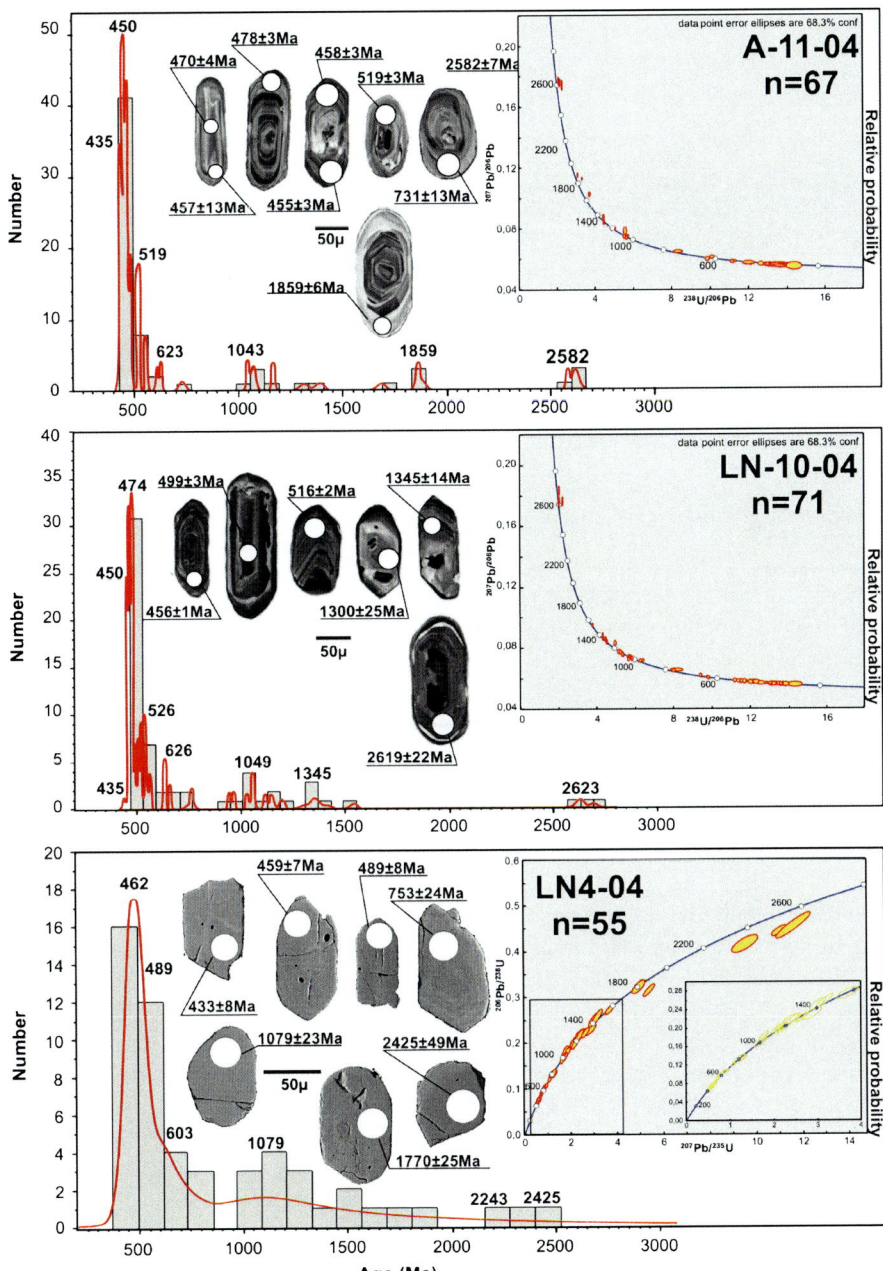

Fig. 13 Frequency histograms and probability curves of detrital zircon ages from RSC samples obtained by LA-ICP-MS. On each sample, the number of analyzed grains and obtained pattern ages was represented. On the right Tera-Wasserburg or Concordia diagrams for each sample is shown

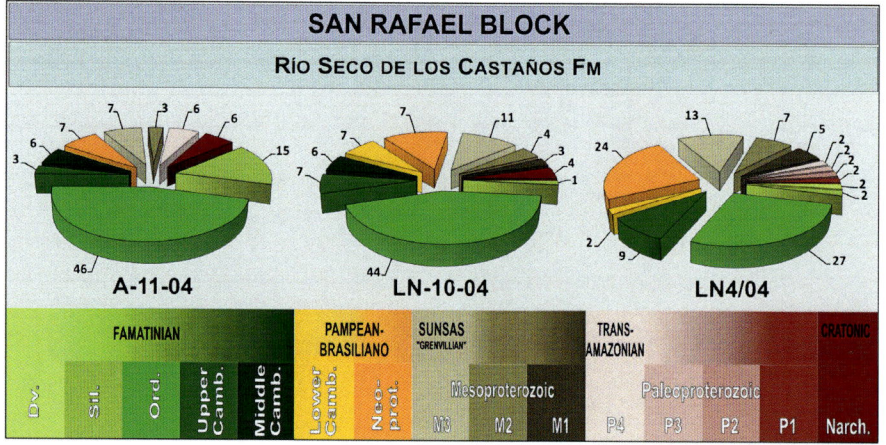

Fig. 14 Percentage of U–Pb detrital zircon ages of each sample represented in 'pie-diagrams'. Colours distinguished the main South American tectonic cycles

Famatinian rocks is accounted by U–Pb zircon patterns; it comprises conglomerate facies and a tonalite intrusive (401 Ma). On the other way, the La Horqueta Formation underwent a low grade of metamorphism, the fossil record is absent, Rb–Sr ages are older, as well as the main source rocks (determined by U–Pb detrital zircon ages), and it is intruded by a younger felsic plutonic body.

5.5 Lu–Hf Systematic

It is known that zircon preserves the initial $^{176}Hf/^{177}Hf$ ratio of the original magma, providing record of the Hf composition of their source environment at the time of crystallization. This ratio can be used to determine Hf model ages. Thus, the Hf isotopic composition of zircons can be utilized as a petrological tracer of a host rocks origin.

17 zircon grains (Table 6) were selected from the sample A-11-04 for Lu–Hf analysis by LA-ICP-MS at the Centro de Pesquisas Geocronológicas, University of São Paulo, Brazil. The $\varepsilon Nd(t)$ values of Ordovician and Neoproterozoic zircons range between −14.78 and −4.20 which reveal zircon derived from recycling old crust. Only one sample (Mesoproterozoic age zircon) show positive value 5.40 that linked with juvenile crust (Fig. 15).

Table 2 Comparison between RSC and La Horqueta Formation

Units	P–T conditions	Tectonic vergence	Fossil record	Rb–Sr (whole rock)	Detrital U–Pb zircon age cycles	Sedimentary environments	Intrusives	Sm–Nd data	Lithofacies	Ore deposits
Rio Seco de los Castaños Fm	Low anchizone	East vergence folded	Ichnofossils: Nereites-Mermia Dictyodora Microfossils (acritarchs) Plants: Lycophythes	336 ± 23 Ma Lower Carboniferous	1. Famatinian 2. Pampean-Brasiliano 3. Mesoproterozoic	Marine platform-siliciclastic deltaic system Different facies	Devonian Rodeo Bordalesa Tonalite (401 ± 3 Ma)	εNd ($t = 420$ Ma) -2.5 to -7.7	Several from heterolithic to conglomerate	
La Horqueta Fm	Low-grade Hydrothermal veins	East vergence folded cleavage	No records	371 ± 61 Ma; 379 ± 15 Ma Upper Devonian	1. Mesoproterozoic 2. Pampean-Brasiliano	Turbiditic marine siliciclastic	Permian Agua de la Chilena Granodiorite	εNd ($t = 420$ Ma) -1.49 to -6.53	Heterolithic	Sulphides La Picaza, Rodeo and others

Table 6 Lu–Hf systematic data of zircons from sample A-11-04 (Laboratory CPGeo SPL 846)

Sample SPL 846-A-11-04

Grains/spot	^{176}Hf/^{177}Hf	±2 se	^{176}Lu/^{177}Hf	±2 se	U-Pb Age (T1) Ma	ε Hf (0)	^{176}Hf/^{177}Hf (T1)	ε Hf (T1)	^{176}Hf/^{177}Hf DM (T$_{U-Pb}$)	T DM (Ma)	^{176}Hf/^{177}Hf DM (T$_{DM}$)	ε Hf (TDM)
1.1	0.282275	0.000036	0.001125	0.000021	1073	−17.56	0.282253	5.40	0.282446	1513	0.282122	10.68
3.1	0.282192	0.000050	0.002810	0.000059	447	−20.51	0.282168	−11.51	0.282902	2101	0.281685	8.56
5.1	0.282378	0.000032	0.000795	0.000009	453	−13.93	0.282371	−4.20	0.282898	1645	0.282024	10.21
25.1	0.282273	0.000026	0.001578	0.000033	456	−17.66	0.282259	−8.11	0.282896	1894	0.281840	9.31
37.1	0.282254	0.000170	0.003872	0.000055	450	−18.31	0.282222	−9.57	0.282900	1981	0.281774	9.00
46.1	0.282253	0.000027	0.001586	0.000033	454	−18.34	0.282240	−8.83	0.282897	1938	0.281807	9.15
46.2	0.282273	0.000029	0.001654	0.000021	459	−17.66	0.282258	−8.07	0.282894	1893	0.281840	9.31
11.1	0.282195	0.000041	0.002395	0.000045	462	−20.40	0.282174	−10.98	0.282891	2079	0.281701	8.64
18.1	0.282152	0.000036	0.002067	0.000053	460	−21.93	0.282134	−12.44	0.282893	2169	0.281634	8.31
58.1	0.282260	0.000028	0.001081	0.000017	460	−18.09	0.282251	−8.30	0.282893	1909	0.281828	9.26
19.1	0.282091	0.000043	0.003368	0.000052	521	−24.08	0.282058	−13.78	0.282849	2300	0.281536	7.84
7.1	0.282183	0.000031	0.001846	0.000061	519	−20.84	0.282165	−10.06	0.282850	2065	0.281712	8.69
21.1	0.282166	0.000028	0.002229	0.000043	453	−21.40	0.282147	−12.10	0.282898	2145	0.281652	8.40
22.1	0.282318	0.000028	0.000665	0.000022	465	−16.10	0.282312	−6.20	0.282896	1775	0.281928	9.70
36.1	0.282218	0.000035	0.000872	0.000011	450	−19.60	0.282211	−9.90	0.282900	2005	0.281757	8.90
50.1	0.282199	0.000030	0.001405	0.000017	454	−20.30	0.282187	−10.70	0.282897	2056	0.281718	8.70
53.1	0.282029	0.000032	0.002130	0.000064	459	−26.30	0.282010	−16.90	0.282894	2444	0.281428	7.30

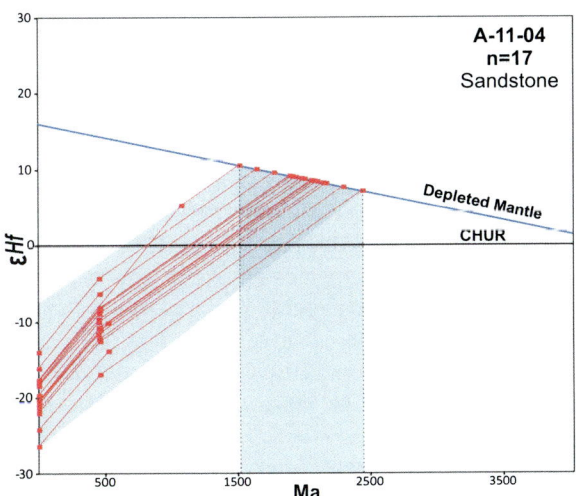

Fig. 15 εHf diagram obtained for A-11-04 sample

6 Discussion and Interpretation

As we concluded in Manassero et al. (2009) the relatively scarce diversity of sub-environments, dominance of fine-to-medium detrital grain sizes, lack of tractive sedimentary structures, and the important thickness of the beds associated with gravity flow processes are typical of a distal (below wave base) to proximal, silty-siliciclastic, marine platform-deltaic system. In this case, the sedimentary input was continuous, due to the absence of internal discontinuities. The dominant processes acting on this palaeo-environment were wave and storm action, prevailing the settling of fine material over the tractive processes. The presence of primitive vascular plant debris in the Atuel and Lomitas Negras sections (Morel et al. this volume) suggests closely related vegetated areas. The hydraulic regimes were moderate and the sea level changes in this sequence have generated very few sedimentary unconformities, but widespread lateral bed continuity.

Similar siliciclastic environments (and probably equivalent from a stratigraphical point of view), are interpreted as overfeed sedimentary foreland systems with great thickness (high sedimentary rates) and low textural maturity, e.g. the Villavicencio and Punta Negra Formations both from the Cuyania terrane. They have been described by other authors (González Bonorino 1975; Edwards et al. 2001, 2009; Peralta 2005, 2013; Cingolani et al. 2013). However, the channelled conglomerates and organic matter-rich beds lithofacies (charcoal) present in the RSC, allow us to distinguish this unit from other similar environments found within the Cuyania terrane.

The main detrital zircon age populations found in the RSC, indicate that two main sources are responsible for the vast majority of observed ages. The main peak corresponds to Ordovician ages that could have been derived from the Famatinian orogenic belt which is developed in the Pampia terrane (which major magmatic event is Ordovician). A second group of sources is characterized by Mesoproterozoic ages between 1000 and 1150 Ma that may confirm partial derivation from the easternmost igneous-metamorphic complex (Cerro La Ventana Formation).

The Sm–Nd signature of the Río Seco de los Castaños Formation agree well with the Mesoproterozoic basement and the carbonate-siliciclastic platform (same range of variation of the εNd (t) and T_{DM} ages), supporting both provenances (Fig. 11). The continental source areas (Cerro La Ventana Formation and the Ordovician sedimentary units) were located not far away towards the east within the San Rafael block (Fig. 1). The detrital material was westwards funnelled (conglomerate channels) from these positive areas into the outer platform areas also laterally associated with a prograding deltaic system along coastal sectors. The basin was deepening towards the west (open sea). Short transport is deduced from petrographical and sedimentological features. The limestone conglomerate-clasts support a provenance from rocks that belong to an Ordovician carbonate-siliciclastic platform, which is also located to the east.

Acknowledgements Field and laboratory work were financially supported by CONICET (grants PIPs 0647, 199), ANPCyT (grant PICT 07027) and University of La Plata (Projects 11/573, 11/704). We thank to colleagues Daniel Poiré, Eduardo Morel, Peter Königshof, Pablo Pazos, and Eduardo Llambías for field work assistance and suggestions. Thanks to Prof. Koji Kawashita for helping us in U–Pb laboratory work and data interpretation. Finally we acknowledge to Genaro Arena and his family as field guide experts for Lomitas Negras and Agua del Blanco regions.

References

Abre P, Cingolani CA, Zimmermann U, Cairncross B, Chemale Jr F (2011) Provenance of Ordovician clastic sequences of the San Rafael Block (Central Argentina), with emphasis on the Ponón Trehué Formation. Gondwana Res 19(1):275–290

Abre P, Cingolani CA, Cairncross B, Chemale Jr F (2012) Siliciclastic Ordovician to Silurian units of the Argentine Precordillera: constraints on Provenance and tectonic setting in the Proto-Andean margin of Gondwana. J South Am Earth Sci 40:1–22

Abre P, Cingolani CA, Chemale Jr F, Uriz NJ (this volume) La Horqueta Formation: geochemistry, isotopic data and provenance analysis. In: Cingolani C (ed) Pre-Carboniferous evolution of the San Rafael Block, Argentina. Implications in the SW Gondwana margin. Springer

Cingolani CA, Varela R (2008) The Rb–Sr low metamorphic age of the Río Seco de los Castaños Formation, San Rafael Block, Argentina. VI South American Symposium on Isotope Geology, Actas. San Carlos de Bariloche, Argentina, pp 1–4

Cingolani CA, Llambías EJ, Ortiz LR (2000) Magmatismo básico pre-Carbónico del Nihuil, Bloque de San Rafael, Provincia de Mendoza, Argentina. IX Congreso Geológico Chileno, Puerto Varas 2:717–721

Cingolani CA, Manassero MJ, Abre P (2003a) Composition, provenance, and tectonic setting of Ordovician siliciclastic rocks in the San Rafael block: southern extension of the Precordillera crustal fragment, Argentina. J South Am Earth Sci 16:91–106

Cingolani CA, Basei MAS, Llambías EJ, Varela R, Chemale Jr F, Siga O Jr, Abre P (2003b) The Rodeo Bordalesa Tonalite, San Rafael Block (Argentina): Geochemical and isotopic age constraints. 10° Congreso Geológico Chileno (CD ROM). Concepción. Octubre 2003. 10 p. Versión CD Rom

Cingolani CA, Llambías EJ, Basei MAS, Varela R, Chemale Jr F, Abre P (2005) Grenvillian and Famatinian-age igneous events in the San Rafael Block, Mendoza Province, Argentina: geochemical and isotopic constrains: In: Pankhurst RJ, Veiga GD (eds) Gondwana 12. Academia Nacional de Ciencias, Mendoza, p 103

Cingolani CA, Manassero MJ, Basei MAS, Uriz NJ (2013). Provenance of the Villavicencio Fm (Lower Devonian) in the southern sector of the Precordillera, Mendoza, Argentina: new sedimentary and geochronological data. 7°Congreso Uruguayo de Geología y 1° Simposio de Minería y Desarrollo del Cono Sur, Actas 191–196, Montevideo, Uruguay

Cingolani CA, Manassero MJ, Uriz NJ, Basei MAS (2014) Provenance insights of the Silurian-Devonian Río Seco de los Castaños unit, San Rafael Block, Mendoza: U-Pb zircon ages. XIV Congreso Geológico Argentino, Acta CD-ROM. Resumen: Tectónica Preandina, S21–10. Córdoba

Cingolani CA, Basei MAS, Varela R, Llambías EJ, Chemale Jr F, Abre P, Uriz NJ (this volume) The Mesoproterozoic Basement at the San Rafael Block, Mendoza Province (Argentina): geochemical and isotopic age constraints. In: Cingolani CA (ed.) The pre-carboniferous evolution of the San Rafael Block, Argentina. Implications in the SW Gondwana margin. Springer

Criado Roqué P, Ibañez G (1979) Provincia geológica Sanrafaelino-Pampeana. In: Turner JC (ed) Segundo Simposio de Geología Regional Argentina. Academia Nacional de Ciencias, Córdoba, I:837–869

Cuerda AJ, Cingolani CA (1998) El Ordovícico de la región del Cerro Bola en el Bloque de San Rafael, Mendoza: sus faunas graptolíticas. Ameghiniana 35(4):427–448

Dessanti RN (1956) Descripción geológica de la Hoja 27c-cerro Diamante (Provincia de Mendoza). Dirección Nacional de Geología y Minería. Boletín 85, 79 p. Buenos Aires

Dickinson WR, Suczek ChA (1979) Plate tectonics and sandstone compositions. Am Assoc of Petrol Geologists Bull 63(12):2164–2182

Dickinson WR, Helmold KP, Stein JA (1979) Mesozoic lithic sandstones in central Oregon. J Sediment Geol 49(2):0501–0516

Dickinson WR, Beard LS, Brakenridge GR, Erjavec JL, Ferguson RC, Inman KF, Knepp RA, Lindberg FA, Ryberg PT (1983) Provenance of North American Phanerozoic sandstones in relation to tectonic setting. Geol Soc Am Bull 94:222–235

Di Persia J (1972) Breve nota sobre la edad de la denominada Serie de la Horqueta-Zona Sierra Pintada, Departamento de San Rafael, Provincia de Mendoza. IV Jornadas Geológicas Argentinas 3:29–41

Edwards D, Morel E, Poiré DG, Cingolani CA (2001) Land plants in the Devonian Villavicencio Formation, Mendoza Province, Argentina. Rev Palaeobot Palynol 116:1–18. Elsevier

Edwards D, Poiré DG, Morel E, Cingolani CA (2009) Plant assemblages from SW Gondwana: further evidence for high—latitude vegetation in the Devonian of Argentina. In: Bassett MG (ed). Early Palaeozoic Peri—Gondwana Terranes: New Insights from Tectonics and Biogeography, vol 325. Geological Society, London, Special Publications, pp 233–255

González Bonorino G (1975) Sedimentología de la Formación Punta Negra y algunas consideraciones sobre la geología regional de la Precordillera de San Juan y Mendoza. Revista de la Asociación Geológica Argentina 30:223–246

González Díaz E (1972) Descripción geológica de la Hoja 27d San Rafael, Mendoza. Servicio Minero-Geológico, Buenos Aires, Boletín 132, 127 pp

González Díaz E (1981) Nuevos argumentos a favor del desdoblamiento de la denominada Serie de La Horqueta del Bloque de San Rafael, Provincia de Mendoza. 8° Congreso Geológico Argentino, Actas 3:241–256, Buenos Aires

Ludwig, KR (2001) SQUID 1.02, A User Manual, A Geochronological Toolkit for Microsoft Excel. Berkeley: Berkeley Geochronology Center Special Publication

Ludwig KR (2012) A geochronological toolkit for Microsoft Excel, version 3.76. Berkeley Geochronology Center, Special Publication N5, 75 pp. Berkeley

Manassero MJ, Cingolani CA, Abre P (2009) A Silurian-Devonian marine platform-deltaic system in the San Rafael Block, Argentine Precordillera-Cuyania terrane: lithofacies and provenance. In Königshof P (ed) Devonian Change: Case studies in Palaeogeography and Palaeoecology, vol 314. Geological Society, Special Publication, pp 215–240, London

Morel EM, Cingolani CA, Ganuza DG, Uriz NJ (2006) El registro de Lycophytas primitivas en la Formación Río Seco de los Castaños, Bloque de San Rafael, Mendoza. 9° Congreso Argentino de Paleontología y Bioestratigrafía, Córdoba. Abstract

Morel EM, Cingolani CA, Ganuza D, Uriz NJ, Bodnar J (this volume) Primitive vascular plants and microfossils from the Río Seco de los Castaños Formation, San Rafael Block, Mendoza Province, Argentina. In: Cingolani CA (ed) The pre-Carboniferous evolution of the San Rafael Block, Argentina. Implications in the SW Gondwana margin. Springer

Nuñez E (1976) Descripción geológica de la Hoja 28c El Nihuil, provincia de Mendoza. Subsecretaría de Minería, Servicio Nacional Minero Geológico, (unpublished report)

Pazos P, Heredia A, Fernández DE, Gutiérrez C, Comerio M (2015) The ichnogenus Dictyodora from late Silurian deposits of central-western Argentina: Ichnotaxonomy, ethology and ichnostratigrapical perspectives from Gondwana. Palaeogeogr Palaeoclimatol Palaeoecol 439:27–37. doi:10.1016/j.palaeo.2015.02.008

Peralta SH (2005) Formación Los Sombreros: un evento diastrófico extensional del Devónico (inferior?-medio?) en la Precordillera Argentina. XVI Congreso Geológico Argentino 4:322–326, La Plata

Peralta SH (2013) Devónico de la sierra de la Invernada, Precordillera de San Juan, Argentina: Revisión estratigráfica e implicancias paleogeográficas. Revista de la Asociación Geológica Argentina 70(2):202–215

Poiré DG, Cingolani CA, Morel EM (1998) Trazas fósiles de la Formación La Horqueta (Silúrico), Bloque de San Rafael, Mendoza, Argentina. III Reunión Argentina de Ichnología y I Reunión de Icnología del Mercosur. Resúmenes, Mar del Plata, Argentina

Poiré DG, Cingolani CA, Morel EM (2002) Características sedimentológicas de la Formación Río Seco de los Castaños en el perfil de Agua del Blanco: pre-Carbonífero del Bloque de San Rafael, Mendoza. 15° Congreso Geológico Argentino, Actas 1:129–133, Calafate

Ramos VA, Jordan TE, Allmendinger RW, Mpodozis C, Kay SM, Cortéz JM, Palma MA (1986) Paleozoic terranes of the central Argentine Chilean Andes. Tectonics 5:8555–8880

Rocha-Campos AC, Basei MA, Nutman AP, Kleiman LE, Varela R, Llambías E, Canile FM, da Rosa O. de CR (2011) 30 million years of Permian volcanism recorded in the Choiyoi igneous province (W Argentina) and their source for younger ash fall deposits in the Paraná Basin: SHRIMP U–Pb zircon geochronology evidence. Gondwana Res 19:509–523

Rubinstein C (1997) Primer registro de palinomorfos silúricos en la Formación La Horqueta, Bloque de San Rafael, Provincia de Mendoza. Argentina. Ameghiniana 34(2):163–167

Tickyj H, Cingolani CA, Varela R, Chemale Jr F (2001) Rb-Sr ages from La Horqueta Formation, San Rafael Block, Argentina. III South American Symposium on Isotope Geology. Pucón, Chile, 4p

Tickyj H, Cingolani CA, Varela R, Chemale Jr F (this volume) Low-grade metamorphic conditions and isotopic age constraints of the La Horqueta pre-Carboniferous sequence, Argentinian San Rafael Block. In: Cingolani C (ed) Pre-Carboniferous evolution of the San Rafael Block, Argentina. Implications in the SW Gondwana margin. Springer

Primitive Vascular Plants and Microfossils from the Río Seco de los Castaños Formation, San Rafael Block, Mendoza Province, Argentina

Eduardo M. Morel, Carlos A. Cingolani, Daniel Ganuza, Norberto Javier Uriz and Josefina Bodnar

Abstract In this contribution we describe fossil plant remains from Río Seco de los Castaños Formation, at San Rafael Block, Mendoza Province, Argentina. The fossil plants comprise non-forked and forked axes without or with delicate lateral expansions, which are assigned to *Bowerophylloides cf. mendozaensis* and *Hostinella* sp. We refer them to primitive land plants and discuss about their systematic affiliation. Furthermore, we mention the presence of a diverse acritarch assemblage present in the same lithostratigraphic unit. On the basis of the taxonomical information and stratigraphic correlation, we could infer that Río Seco de

E.M. Morel (✉) · D. Ganuza · J. Bodnar
División Paleobotánica, Museo de La Plata, UNLP, Paseo del Bosque s/n, B1900FWA La Plata, Argentina
e-mail: emorel@fcnym.unlp.edu.ar

D. Ganuza
e-mail: dganuza@fcnym.unlp.edu.ar

J. Bodnar
e-mail: jbodnar@fcnym.unlp.edu.ar

E.M. Morel
Comisión de Investigaciones Científicas de la Provincia de Buenos Aires (CIC), La Plata, Argentina

C.A. Cingolani
Universidad Nacional de La Plata and Centro de Investigaciones Geológicas, Diag. 113 n. 275, CP1904 La Plata, Argentina
e-mail: carloscingolani@yahoo.com

C.A. Cingolani · N.J. Uriz
División Geología, Museo de La Plata, UNLP, Paseo del Bosque s/n, B1900FWA La Plata, Argentina
e-mail: norjuz@gmail.com

C.A. Cingolani · J. Bodnar
Consejo Nacional de Investigaciones Científicas y Técnicas (CONICET), La Plata, Argentina

los Castaños Formation has an Early Devonian age. The taphonomical conditions of this fossil association would indicate that the plants were transported some distance from their presumed coastal and riverbank habitats. Finally, studying the amount and the percentage of kaolinite within charcoal levels, warm to cool temperate paleoclimatic conditions were deduced.

Keywords Vascular plants · Microfossils · Lower Devonian · San Rafael block · Mendoza

1 Introduction

The main focus of this chapter is the study of the first record of primitive vascular plants from Río Seco de los Castaños Formation (RSC) that is one of the "pre-Carboniferous" units of the San Rafael block (González Díaz 1972, 1981). Based on the stratigraphical and paleontological evidence, the age of this formation is considered between the Late Silurian and Early Devonian. The fossil remains described here were found at Atuel River Creek, where the type section of RSC is located about 12 km NE of El Nihuil town (Fig. 1).

At this outcrop the unit comprises near 600 m of marine siliciclastic sedimentary rocks, mainly green sandstones and grey mudstones (Fig. 2). The fossiliferous stratum with fossil plants is placed 60 cm above a charcoal bed. This charcoal is interbedded with a massive sandstone body, and it is composed of a mixture of silty-quartz, illite-kaolinite clays and amorphous organic matter (Morel et al. 2006; Manassero et al. 2009). In order to study the origin of the charcoal, three kinds of analyses were done: Total Organic Carbon (in ACME Lab, Canada), isotopic deviation of ^{13}C (=δ ^{13}C) in Activation Laboratories LTD, Ontario, Canada and pyrolysis quantification (in GeoLab Sur S.A., Buenos Aires, Argentina).

The sedimentary rocks which contain the fossil plants were interpreted as deposits of suspension and fall out from low density turbidity currents in a distal platform (Manassero et al. 2009). The paleoenvironmental conditions are shallow marine water near the coastline as can be inferred from the taphonomical attributes of the fossil plants. A similar siliciclastic environment has been recognized in the Upper Silurian-Lower Devonian Villavicencio Formation of southern Precordillera (Edwards et al. 2001). However, the RSC unit has two distinctive sedimentological characteristics: conglomerate channels and organic-matter-rich beds.

The plant debris comprises non-forked and forked axes without or with delicate lateral expansions. Sporangia were not preserved in any sample. Although these plants were previously assigned to lycophytes (Morel et al. 2006, 2007), here we discuss their possible filiations and systematic relationships.

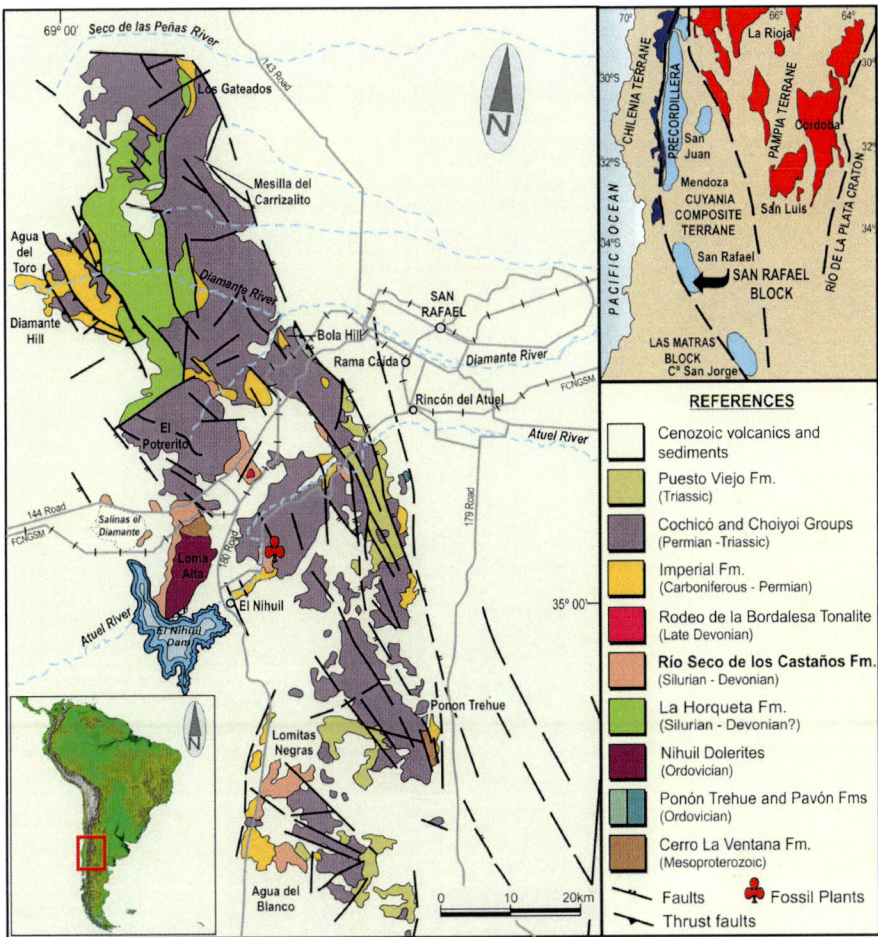

Fig. 1 Geological sketch map of the San Rafael BlockSan Rafael Block, showing the outcrops of the Río Seco de los Castaños Fm. The *red symbol* indicates the fossil plants locality

2 Paleontological Background

The paleontological contributions about RSC are scarce. The first reference corresponds to the coral *Pleurodyctium* sp. by Di Persia (1972) found in Agua del Blanco region, but lacks an original description. Pöthe de Baldis (1999) in an unpublished report has mentioned a palynological association, in the type section of the RSC unit at Atuel creek, and in the same sequence where the plant debris was found. Those palynomorphs have a very high-grade of thermal alteration (<4 of Staplin scale) making the generic assignment relatively difficult. Acritarchs, algae prasinophytes, and probably plant spores were recognized, which are listed below and reproduced in Fig. 3.

Fig. 2 Stratigraphic column of the San Rafael Block

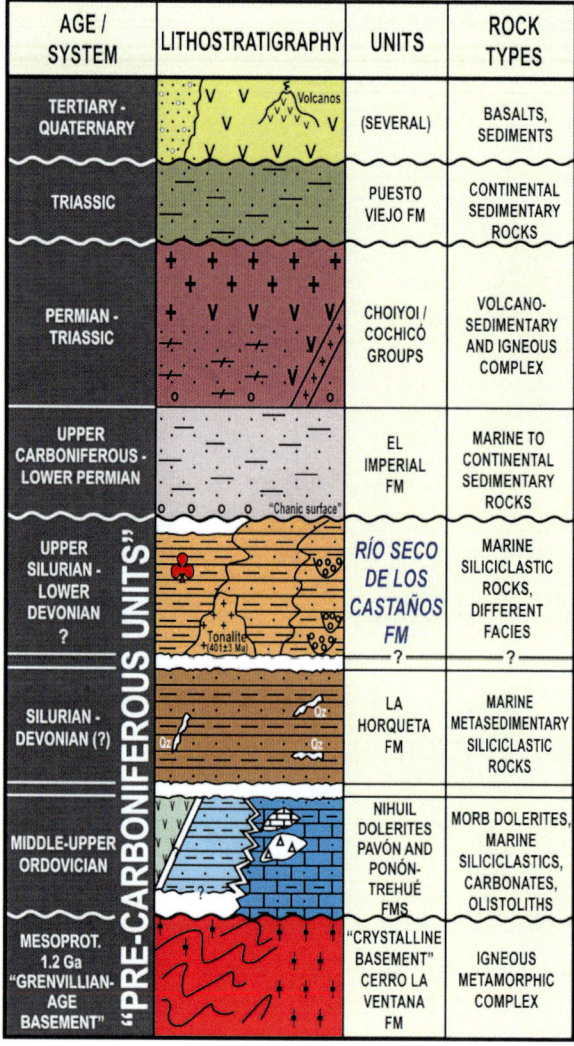

Acritarcha: *Ammonidium alloiteaui* (Deunff) Deunff, *Ammonidium* cf. *Ammonidium hydraferum* (Stockmans and Williére) Pöthe de Baldis, *Lophosphaeridium* sp., *Micrhystridium* sp. aff. *Micrhystridium stellatum* Deflandre, *Protoleiosphaeridium* sp., *Veryhachium trispinosum* (Eisenack) Stockmans and Williére.
Prasynophyceae: *Veliferites* sp. cf. *Veliferites jachalensis* Pöthe de Baldis, *Cymatiosphera* sp.
Plant spores: *Streelispora*?

Although the palynomorph elements are not diverse, the presence of acritarchs would indicate a shallow marine paleoenvironment, near the coast as it is suggested

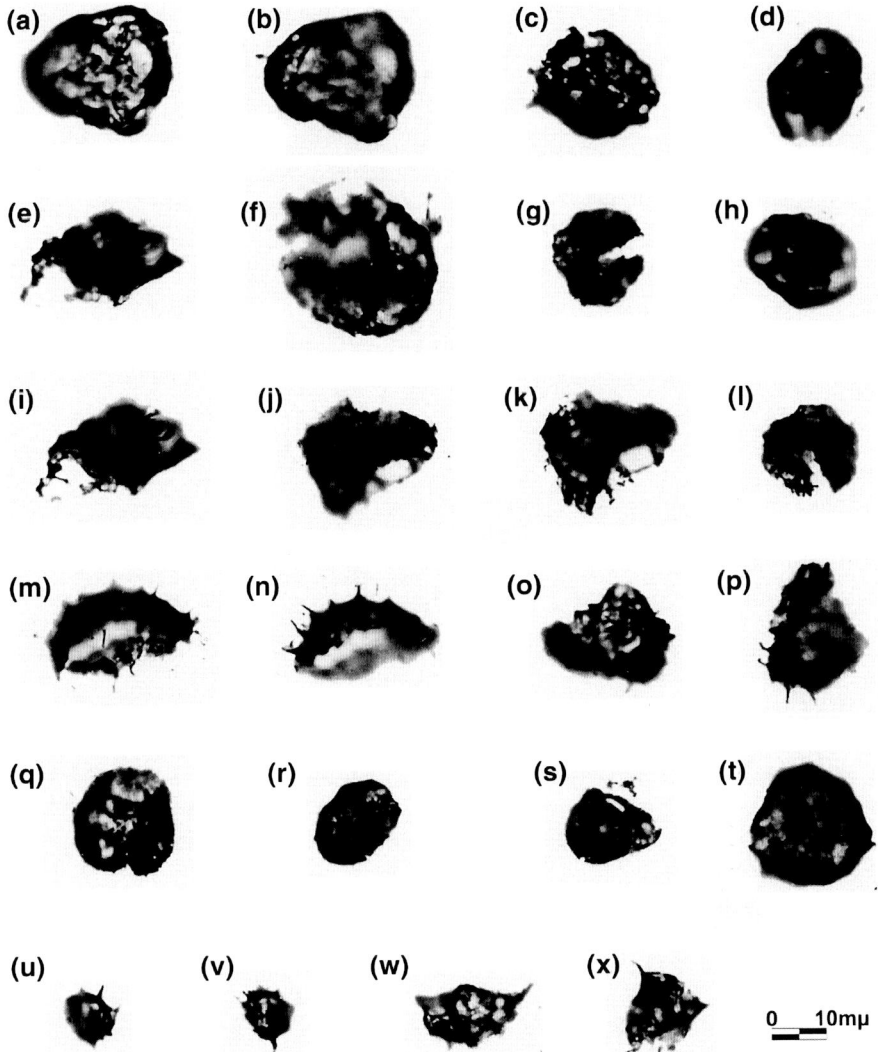

Fig. 3 Microfossil assemblage after Pöthe de Baldis (1999). **a** and **b** *Streelispora* sp.; **c, d** and **h** Spores?; **e** and **i** *Veliferites jachalensis*; **f** *Cymatiosphaera* sp.; **g** and **l** Spores?; **j** and **k** *Duvernaysphaera*? sp.; **m** and **n** *Ammonidium* sp. cf. *Ammonidium hydraferum*; **o** and **p** *Ammonidium* sp. cf *Ammonidium alloiteaui*; **q** *Lophosphaeridium* sp.; **r** and **s** *Protoleiosphaeridium* sp., **t** *Micrystridium* sp.; **u** and **v** *Micrystridium* sp. cf. *Micrystridium stellatum*; **w** and **x** *Veryhachium trispinosum*

by the presence of some spores. The presence of *Veliferites* sp. cf. *Veliferites jachalensis* Pöthe de Baldis, would imply a Late Silurian age for this association, since this taxon is known from the Silurian (Ludlow) of the Argentine Precordillera (Los Espejos Formation).

It is important to mention that the palynoflora studied by Rubinstein (1997) from outcrops located near the 144 road at km 702 and assigned to 'La Horqueta Formation' by this author, were afterwards considered as part of the RSC sensu González Díaz (1981) based on sedimentary rock attributes and the presence of the ichnofacies of *Nereites-Mermia* (Cuerda and Cingolani 1998, Poiré et al. 2002 and discussion in Manassero et al. 2009). Rubinstein (1997) described the following Upper Silurian palynoflora:

Ammonidium ludloviense (Lister) Dorning (1981)
Dactylofusa cabotti (Cramer) Fensome et al. (1990)
Dactylofusa striatifera (Cramer) Fensome et al. (1990)
Evittia denticulata denticulata (Cramer) Le Hérissé (1989)
Fimbriaglomerella divisa Loeblich Jr. and Drugg (1968)
Leiofusa estrecha Cramer (1964)
Tylotopalla sp. cf *T. pyramidalis* (Lister) Dorning (1981)
Baltisphaeridium spp.
Leiofusa sp.
Leiosphaeridia spp.
Retisphaeridium sp.
Veryhachium? sp.

This association shares some genera with the assemblage mentioned by Pöthe de Baldis (1999), such as *Ammonidium* and *Veryhachium*.

Poiré et al. (2002) have recognized different ichnogenera in the Agua del Blanco region, as follows, *Arenicolites, Bergaueria, Cochlichnus, Cruziana, Gordia, Mammlichnus, Palaeophycus, Phycodes, Rusophycus* and *Teichichnus*. This ichnofacies represents a well oxygenated environment and it is interpreted as a proximal to shallow marine platform, with dominance of subtidal environment. The trace fossils are developed in soft substrates of moderate energy (Manassero et al. 2009). On the other hand, Pazos et al. (2013) documented the ichnogenera on the 144 road (km 702) locality that contains dominantly *Nereites irregularis* Schafhäutl, *Helminthopsis* and less abundant arthropod trackways and grazing traces. Microbial mats are very abundant but clearly not related to *Nereites*. At the Atuel creek locality *Nereites* is also the most abundant ichnofossil but it contains more than one ichnospecies: *Nereites irregularis* and one beautifully preserved specimen of *Nereites cambrensis* Murchinson, the type ichnospecies of the ichnogenus. Other specimens are assignable to the questionable *Nereites delpeyi* Borrouilh, a dubious valid ichnospecies that probably is a junior synonym of *Nereites missouriensis* Weller. Other well preserved trace fossil is *Dictyodora* in upper relief expressions. This ichnogenus (Poiré et al. 1998; Pazos et al. 2013, 2015) contains several ichnospecies, with an apparently ichnostratigraphic value. At the Atuel creek RSC outcrop the ichnogenera *Dictyodora* Weiss is recorded. The ichnospecies recognised include *Dictyodora scotica* and *Dictyodora tenuis*, and a new ichnospecies, *Dictyodora atuelica*. The succession studied by Pazos et al. (2015) contains abundant microbial mats (wrinkle marks), either as extended surfaces or patches.

3 Paleobotany

Division Tracheophyta

Incertae Sedis
Genus **Bowerophylloides** Edwards et al. (2001)
Bowerophylloides* cf. *mendozaensis Edwards ct al. (2001) (Fig. 4a–f)

Description Impression of delicate herbaceous stem fragments, with dichotomizing sterile axes covered by crowded narrow enations, in spiral arrangement. The bases of enations are fusiform when axes are defoliated. The fossil plants present stems with enations and unbranched axes with irregularly shaped patches. Description is based on 7 fragments, all of which bear enations showing varying degree of fragmentation. No anatomical details have been preserved. Branching is

Fig. 4 a–f *Bowerophylloides* cf. *mendozaensis* Edwards et al. (2001). **a** LPPB 13814b. **b** LPPB 13816b. **c** LPPB 13817a. **d** LPPB 13829a. **e** LPPB 13818a. **f** LPPB 13830. **g–h** *Hostinella* sp. **g** LPPB 13831. **h** LPPB 13832. *Scale bar* 0.5 cm

isotomic-dichotomous (in one specimen: LPPB 13814). The largest specimen reaches 10.7 mm. Stem width (excluding enations) ranges from 0.5 to 1.3 mm. Enations in profile attached to the sides of the stem show swollen decurrent bases extending into linear structures that are straight or distally slightly curved. They are inserted at 45–50° angles to the stem. The regular spacing between these enations suggests a regular phyllotaxis, which can be defined as spiral. The bases of enations are subcircular to fusiform in outline, have a central prominence, and range from 0.1 to 0.3 mm wide and 0.2 to 0.3 mm long. The shape of the free distal part of the enations, are probably linear to slightly spatulate (LPPB 13816). Enations show no evidence for a central trace.

Comparisons This material is assigned to *Bowerophylloides mendozaensis* Edwards et al. (2001), because of morphological similarities to material from Villavicencio Formation (Mendoza province; *cf.* Edwards et al. 2001). The specimens described above have sterile axes covered by narrow, elongate projections (enations), directed towards the apex, truncated in the apex, with a maximum length of 0.31 cm and a maximum width of 0.03 cm near the base. Every dimension of these samples agrees with the original material described by Edwards et al. (2001) (Table 1).

This type material of *Bowerophylloides mendozaensis* Edwards et al. (2001) had been originally assigned to *Baragwanathia* (Cuerda et al. 1987). Such identification was rejected because (1) there is little direct morphological similarity with Australian *Baragwanathia* with its elongate flexuous leaves (*e.g.* Lang and Cookson 1935), and (2) there is no anatomical evidence of microphylls.

In RSC samples, it is also impossible to confirm the vascular status of the enations, viz. whether or not they are microphylls, with obvious consequences for unequivocal identification of these fossils as the leafy shoots of lycophytes

Table 1 Comparative table of the RSC plant specimens and type material of *Bowerophylloides*

LPPB	13816a	13818a	13817a	13814b	13822a	13819	13829	13830	*Bowerophylloides* (Edwards et al. 2001)
Axis length above	9.3	7.5	6.9	5.3	7.1	5.0	10.7	9.95	$(3 = n) \times 2.6 + 2.7$ below and above
Axis width	0.5	0.5	1.0	0.9	0.6	0.35	1.26	0.8	1.35 – 1.9 $(4 = n) \times 0.6$ above dichotomy
Enation length	3.1	1.0	0.9	–	0.9	1.5	2.63	4.6–4.4	<2.0 But probably greater
E. basal width	0.3	0.1	0.16	–	0.5	0.3			0.19–0.32
E. basal length	–	–	0.2	0.3			0.57	0.42–0.55	0.27–1.8
"Lamina" width							0.14	0.17–0.22	0.14–0.30
Insertion angle	45–50°	40–45	45–50°		50°	45–50°	40–47°	40–45°	20–35° distally— c50° proximaly

Therefore, the assignation to *Bowerophylloides*, a genus belonging to primitive *incertae sedis* vascular plants, is supported by the following features: (1) *Bowerophylloides* has longer and wider axes but dimensions of the enations are comparable, *i.e.* they fall within the range of the *Bowerophylloides* material, (2) in neither of them is possible to determine a phyllotaxy, but regularity of spacing of emergences in profile, together with the shape of enation bases in face view, suggests a spiral phyllotaxy, (3) angles of insertion of RSC material match with those of the *Bowerophylloides* type specimen, and (4) the material certainly fits the generic diagnosis of *Bowerophylloides*, although spatulate leaves would be needed to allow precise specific identification.

Genus **Hostinella** Barrande ex Stur 1882
***Hostinella* sp.** (Figure 4g, h)

Description Fragments of axes with dichotomous isotomous branching, usually with little change in diameter of the daughter branches. The axes are branched only once. Some specimens show evidence of central strand (LPPB 13832). The fragments reach 28.1 mm of length, and a width of 0.6–1 mm before dichotomy and 0.6–0.8 mm above dichotomy.

Studied material: LPPB 13831, 13832.

Comparisons Isotomously branching axes are traditionally referred to genus *Hostinella*. This taxon had been used for a variety of naked axes with dichotomous or pseudomonopodial branching with occasionally, bud-like protuberances in the upper angle of the dichotomy (Gensel and Andrews 1984). This genus has not great significance as biological entities, thus the RSC samples cannot be assigned to any group of vascular plants.

4 Discussion

Taphonomical considerations
The plant assemblage presents a low diversity and high fragmentation of the specimens. This can be explained by taphonomical conditions, since the plant debris were preserved in heterolithic siltstone/sandstone strata, inferred as fair weathered deposits in a subtidal coastal environment on a shallow shelf. The plants would have been transported some distance from their presumed coastal and riverbank habitats, for which we have no direct information (Poiré and Morel 1996; Edwards et al. 2009).

Systematic considerations

Despite that *Bowerophylloides mendozaensis* was considered as an enigmatic "leafy" shoot taxon (Edwards et al. 2001), some characters present in the genus confirm that it belongs to a primitive vascular plant. Particularly the enations show evidence of a regular phyllotaxy which can be defined as spiral, and they have a swollen basis that remains attached to the axis. These characteristics could link this taxon to the lycophyte lineage (*cf.* Kenrick and Crane 1997).

Correlations

The low diversity and high fragmentation of this plant assemblage makes difficult the correlation with other coeval paleofloras. The occurrence of *Bowerophylloides*, present in Lower Devonian Villavicencio Formation, would suggest a similar age for RSC fossil assemblage. The microflora of Villacencio Formation (Rubinstein 1993) shows abundant spores and only three species of acritarchs, *Veryachium* cf. *downiei* Stockmans and Willière, *Veryachium* cf. *lairdi* Deunff, and *Micrhystridium* sp. These genera are also present in RSC Formation (see Paleontological background). Even though the biochrons of these acritarch taxa are extended, their presence is another element for correlation.

Charcoal level

The Total Organic Carbon analysis gave a value of 1.08%, which represent an important proportion. The isotopic deviation of ^{13}C (= δ ^{13}C) analysis showed a value of −26.3‰. Such a deviation is an average amount of the land vegetation, since the land plants are classified in two main groups: C3 and C4, according to their metabolic photosynthetic mechanism. The 85% of the vascular plants are of C3 type, and show very low values of δ ^{13}C, between −22 and −30‰. The plants that generated the charcoal level here analysed are comprised into the last mentioned interval of δ ^{13}C values. Otherwise, the values of C4 plants range between −10 and −14‰, much higher that the amounts obtained in this work. C4 plants correspond to the 15% of land plant and mainly comprise tropical herbs. The pyrolysis quantification analysis is shown in Table 2.

Table 2 Pyrolysis quantification data of the charcoal sample 05CA2

Lab record	Sample	TOC	S1	S2	S3	T °C	S1/TOC	S3/TOC	S1/S1 + S2
			mg/g					O index	Productivity index
LC-07-005	05CA2	0.37	0.01	0	0.29	274 °C	3	78	1

These data indicate severe maturity for charcoal material. The charcoal amount and the record of kaolinite in this bed could indicate, following the paleogeographic reconstruction of Scotese et al. (1999), that the paleoclimatic conditions were warm to cool temperate.

Acknowledgements Financial support for this chapter was provided by grant Projects 647 and 199 CONICET (Argentina). The authors want to express their thanks to Dianne Edwards (Cardiff, UK) for numerous suggestions and comments in the early draft of the paper. We also thank to Marcelo Manassero, Paulina Abre and María Cecilia Amenábar for kind comments.

References

Cramer FH (1964) Some acritarchs of the San Pedro formation (Gedinniano) of the Cantabrian Mountains in Spain. Bull Soc Belge Géol 73:33–38

Cuerda AJ, Cingolani CA (1998) El Ordovícico de la región del Cerro Bola en el Bloque de San Rafael, Mendoza: sus faunas graptolíticas. Ameghiniana 35:427–448

Cuerda AJ, Cingolani C, Arrondo O, Morel E, Ganuza D (1987) Primer registro de plantas vasculares en la Formación Villavicencio, Precordillera de Mendoza, Argentina. Cuarto Congreso Latinoamericano de Paleontología. Actas I:179–183 (Santa Cruz de la Sierra Bolivia)

Di Persia (1972) Breve nota sobre la edad de la denominada Serie de la Horqueta-Zona Sierra Pintada, Departamento de San Rafael, Provincia de Mendoza. 4° Jornadas Geológicas Argentinas 3:29–41

Dorning KJ (1981) Silurian acritarchs from the type Wenlock and Ludlow of Shropshire, England. Rev Palaeobot Palynol 34:175–203

Edwards D, Morel E, Poiré DG, Cingolani CA (2001) Land plants in the Devonian Villavicencio formation, Mendoza Province, Argentina. Rev Palaeobotany Palynol 116:1–18

Edwards D, Poiré DG, Morel E, Cingolani CA (2009) Plant assemblages from SW Gondwana: further evidence for high—latitude vegetation in the Devonian of Argentina. Bassett MG (ed) Early Palaeozoic Peri—Gondwana terranes: new insights from tectonics and biogeography, vol 325. Geological Society, London (Special Publications), pp 233–255

Fensome RA, Williams GL, Barss MS, Freeman MJ, Hill JM (1990) Acritarchs and fossil prasinophytes: an index to genera, species and intraspecific taxa. American Association of Stratigraphic Palynologists, Contribution Series N° 25, 771 pp

Gensel PG, Andrews HN (1984) Plant life in the Devonian. Praeger Publishers, New York, 380 pp

González Díaz EF (1972) Descripción geológica de la Hoja 27d, San Rafael, Mendoza. Servicio Minero—Geológico, Boletín, vol 132. 127 pp (Buenos Aires)

González Díaz EF (1981) Nuevos argumentos a favor del desdoblamiento de la denominada Serie de La Horqueta del Bloque de San Rafael, Provincia de Mendoza. 8° Congreso Geológico Argentino. Actas 3:241–256 (Buenos Aires)

Kenrick P, Crane PR (1997) The origin and early diversification of land plants, a cladistic study. Smithsonian Institution Press, Washington, 441 pp

Lang WH, Cookson IC (1935) On a flora, including vascular land plants, associated with Monograptus, in rocks of Silurian age, from Victoria, Australia. Phil Trans R Soc London 224B:421–449

Le Hérissé A (1989) Acritarches et kystes d'algues prasinophycées du Silurien de Gotland, Suede. Palaeontographia Italica 76:57–302

Loeblich AR Jr., Drugg WS (1968) New acritarchs from the Early Devonian (late Gedinnian) Haragan Formation of Oklahoma, USA. Tulane Stud Geol 6:129–137

Manassero MJ, Cingolani CA, Abre P (2009) A Silurian—Devonian marine platform—deltaic system in the San Rafael Block, Argentine Precordillera—Cuyania terrane: lithofacies and provenance. In: Königshof P (ed) Devonian change: case studies in palaeogeography and palaeoecology, vol 314. The Geological Society, London (Special Publications), pp 215–240

Morel EM, Cingolani CA, Ganuza DG, Uriz NJ (2006) El registro de Lycophytas primitivas en la Formación Río Seco de los Castaños, Bloque de San Rafael, Mendoza. 9° Congreso Argentino de Paleontología y Bioestratigrafía. Resúmenes: 47 (Córdoba)

Morel EM, Cingolani CA, Ganuza DG, Uriz NJ (2007) Primitive lycophytes record at the Río Seco de los Castaños Formation, San Rafael Block, Mendoza province, Argentina. Field Meeting of the IGCP 499, Devonian land-sea interaction: evolution of ecosystems and climate. pp 106–108 (San Juan)

Pazos P, Heredia A, Cingolani CA (2013) Nereites ichnofacies in the Río Seco de los Castaños Formation, Mendoza, Argentina: age, facies and trace-fossil content. II Reunión Argentina de Icnología. Abstracts (Santa Rosa)

Pazos P, Heredia A, Fernández DE, Gutiérrez C, Comerio M (2015) The ichnogenus Dictyodora from late Silurian deposits of central-western Argentina: ichnotaxonomy, ethology and ichnostratigrapical perspectives from Gondwana. Palaeogeogr Palaeoclimatol Palaeoecol 439:27–37

Poiré DG, Morel E (1996). Procesos sedimentarios vinculados a la depositación de niveles con plantas en secuencias silúrico-devónicas de la Precordillera. Argentina. Actas VI Reunión Argentina de Sedimentología, pp 205–210

Poiré DG, Cingolani C, Morel E (1998) Trazas fósiles de la Formation Horqueta (Silúrico), Bloque de San Rafael, Mendoza, Argentina. Tercera Reunión de Icnología y Primera Reunión de Icnología del Mercosur. Resúmenes: 24 (Mar del Plata)

Poiré DG, Cingolani CA, Morel E (2002) Características sedimentológicas de la Formación Río Seco de los Castaños en el perfil de Agua del Blanco: Pre-Carbonífero del Bloque de San Rafael, Mendoza. XV Congreso Geológico Argentino I:129–133 (Calafate)

Pöthe de Baldis ED (1999) Informe palinológico del Proyecto: Investigaciones en las unidades del Precámbrico y Paleozoico inferior del Bloque de San Rafael. Instituto de Geología (INGEO), Universidad Nacional de San Juan, Mendoza (5p, unpublished)

Rubinstein CV (1993) Primer registro de miosporas y acritarcos del Devónico inferior en el "Grupo Villavicencio", Precordillera de Mendoza, Argentina. Ameghiniana 30:219–220

Rubinstein CV (1997) Primer registro de palinomorfos silúricos en la Formación la Horqueta, Bloque de San Rafael, Provincia de Mendoza Argentina. Ameghiniana 34:163–167

Scotese CR, Boucot AJ, McKerrow WS (1999) Gondwanan palaeogeography and palaeoclimatology. J Afr Earth Sc 28:99–114

The Rodeo de la Bordalesa Tonalite Dykes as a Lower Devonian Magmatic Event: Geochemical and Isotopic Age Constraints

Carlos A. Cingolani, Eduardo Jorge Llambías, Miguel A.S. Basei, Norberto Javier Uriz, Farid Chemale Jr. and Paulina Abre

Abstract One of the 'pre-Carboniferous units' from the San Rafael Block is the sedimentary Río Seco de los Castaños Formation, which is distributed in isolated outcrops within the Block. At the Rodeo de la Bordalesa area two small intrusives in the mentioned unit were mapped, composed of tonalitic rocks, lamprophyre ('spessartite-kersantite') and aplite dykes. We present in this paper, geochemical and isotopic data from the gray tonalitic rocks with abundant mafic enclaves and late magmatic aplite veins. The country rocks are a folded sequence of feldspathic sandstones, wackes, and shales. The Rodeo de la Bordalesa tonalite dykes are characterized by high to medium potassium concentration, with metaluminous composition and I-type calc-alkaline signature. The 401 ± 4 Ma U–Pb zircon age corresponds to the emplacement time and it is confirmed by the K–Ar biotite age.

C.A. Cingolani (✉) · E.J. Llambías
Centro de Investigaciones Geológicas, Universidad Nacional de La Plata, Diag. 113 n. 275, CP1904 La Plata, Argentina
e-mail: carloscingolani@yahoo.com

E.J. Llambías
e-mail: llambias@cig.museo.unlp.edu.ar

M.A.S. Basei
Centro de Pesquisas Geocronológicas (CPGeo), Universidade de São Paulo, São Paulo, Brazil
e-mail: baseimas@usp.br

C.A. Cingolani · N.J. Uriz
División Geología, Museo de La Plata, UNLP, Paseo del Bosque s/n, B1900FWA La Plata, Argentina
e-mail: norjuz@gmail.com

F. Chemale Jr.
Programa de Pós-Graduação em Geologia, Universidade do Vale do Rio dos Sinos, 93.022-000 São Leopoldo, RS, Brazil
e-mail: faridcj@unisinos.br; faridchemale@gmail.com

P. Abre
Centro Universitario Regional Este, Universidad de la República, Ruta 8 Km 282, Treinta y Tres, Uruguay
e-mail: paulinabre@yahoo.com.ar

The Rb–Sr whole rocks and biotite age of 374 ± 4 Ma could be related to deformation during the 'Chanic' tectonic phase. Nd model ages (T_{DM}) show an interval between 1 and 1.6 Ga, indicating Mesoproterozoic age derivation, whereas the negative ε_{Nd} is typical from crustal sources. The crystallization age for the Rodeo de la Bordalesa tonalite corresponds to a Lower Devonian time and suggests that part of the Late Famatinian magmatic event is present in the San Rafael Block. The dykes are contemporaneous with the large peraluminous batholith in Pampeanas Ranges, with the transpressional shear belts during 'Achalian' event and could be correlated with the Devonian magmatism present in the southern part of the Frontal Cordillera. The geochemical and geochronological data allow us to differentiate the Rodeo de la Bordalesa tonalite from the mafic rocks exposed at the El Nihuil area.

Keywords San Rafael Block · Tonalite dykes · Geochronology · Lower Devonian · Magmatic event

1 Introduction

The San Rafael Block (SRB) lies in west-central Mendoza province, Argentine (35° S–68°30'W), and has SSE-NNW structural Cenozoic trend in the pre-Andean region. To the North and South the Cuyo and Neuquén sedimentary basins bound it, respectively. To the East the SRB passes into the Pampean plains vanishing under the modern basaltic back arc volcanism and sedimentary cover; the boundary to the West is defined by the Andean foothill (Fig. 1). Paleontological and geological

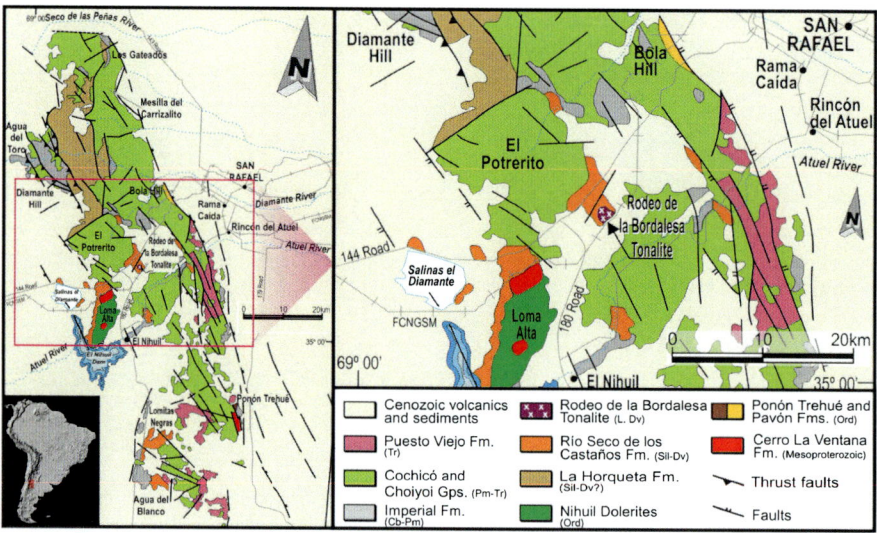

Fig. 1 Regional geological sketch map of the San Rafael Block and location of the Rodeo de la Bordalesa study region

evidence allow interpreting the SRB as a southern extension of the Cuyania terrane (Ramos 2004 and references there in). Diverse igneous-metamorphic and sedimentary units of Precambrian to Middle Paleozoic age are present and are known as 'pre-Carboniferous units' since they are located below the Upper Paleozoic regional unconformity (Dessanti 1956).

One of these units is the sedimentary Río Seco de los Castaños Formation, which is distributed in isolated outcrops within the SRB (Fig. 1). At the Rodeo de la Bordalesa area, Dessanti (1956) mapped two small intrusives composed of tonalitic rocks, lamprophyre ('spessartite-kersantite'), and aplite dykes.

We present here, geochemical and isotopic data from Rodeo de la Bordalesa tonalite intrusive rocks that contribute to characterize and constrain the emplacement of these magmatic rocks, as well as to the knowledge of the correlation of the Late Famatinian event in western Argentina. The sample locations are shown in Fig. 2.

2 Geological Background

The Rodeo de la Bordalesa tonalite was first described as intruded in the 'La Horqueta Series' by Dessanti (1956) and mentioned by Davicino and Sabalúa (1990) as tonalite dikes ('trondjhemites') emplaced in La Horqueta sequence. After González Díaz (1964, 1981), Cuerda and Cingolani (1998) and Cingolani et al. (2003) works, the area was remapped and tonalites host rocks were assigned to the Río Seco de los Castaños Formation (Manassero et al. 2009; Cingolani et al. 2011, this volume). The Río Seco de los Castaños Formation (RSC) outcrops at (Fig. 1): **a**. Road 144-Rodeo de la Bordalesa: locations, where Rubinstein (1997) found Silurian acritarchs and other microfossils, and trace fossils were mentioned by several authors (Criado Roqué and Ibáñez 1979; Poiré et al. 1998; Pazos et al. 2015). Cingolani et al. (2003) as preliminary work, constraints the isotopic age and composition of the tonalitic intrusive body; **b**. Atuel River creek: this is the type section of the sequence, near Valle Grande area (González Díaz 1964). The beds are folded and show dipping of 50°–72° to the SE or NE. **c**. El Nihuil area: comprise a sedimentary sequence close to the Mesoproterozoic basement and to the Ordovician mafic rocks called 'El Nihuil Mafic Unit' (Cingolani et al. 2003). **d**. Lomitas Negras and Agua del Blanco areas: comprise the southern outcrops of RSC, where Di Persia (1972) mentioned a coral (*Pleurodyctium*) of Devonian age and conglomerates with limestone clasts bearing Ordovician fossils.

The Rodeo de la Bordalesa intrusive rock crops out near the deactivated railroad tracks ('Ferrocarril General San Martín'), as a gray tonalitic body with abundant mafic enclaves (less than 30 cm) and comprising 10–30 cm thick late magmatic aplite veins (Fig. 2a–d). At this area the RSC is a folded sequence of feldspathic sandstones, wackes, and shales (Cingolani et al. 2003; Manassero et al. 2009).

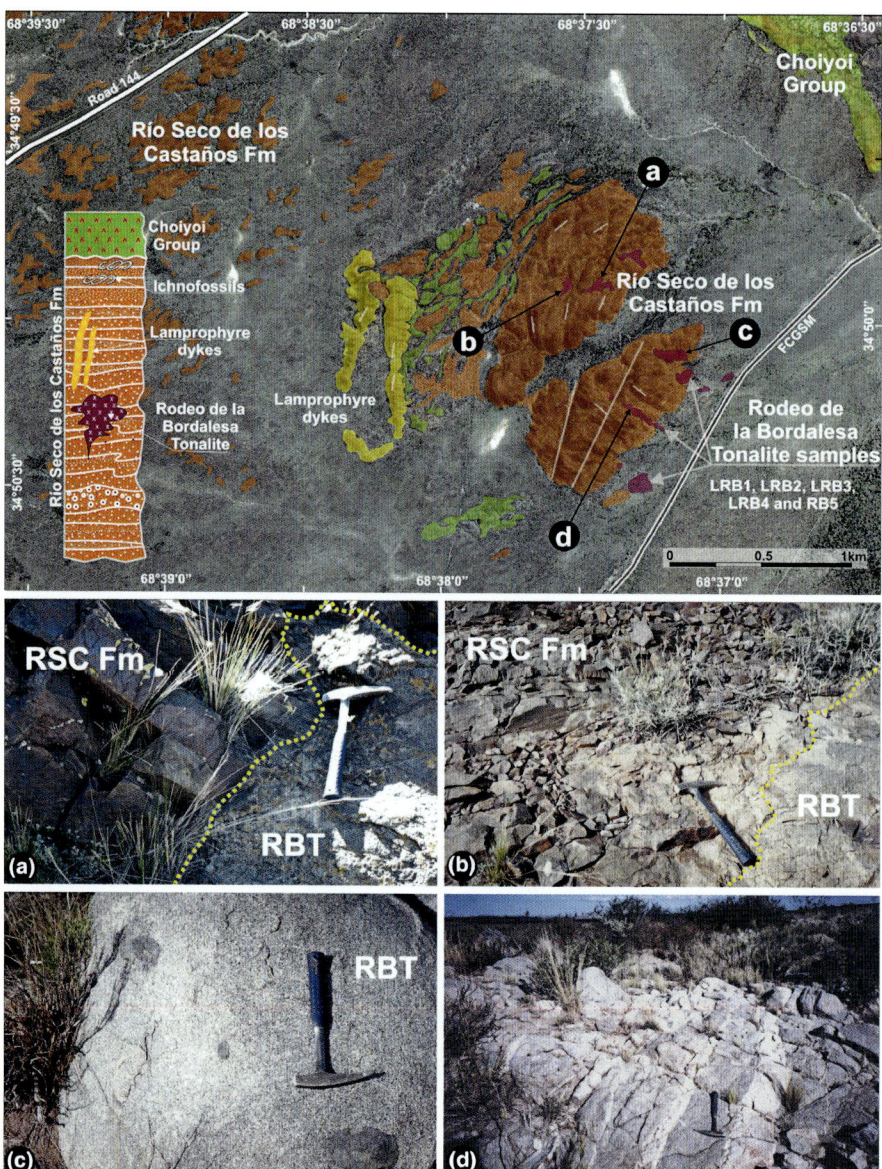

Fig. 2 Image showing the geological sketch map with stratigraphic column of Rodeo de la Bordalesa region. **a, c** Intrusive field relationship between the tonalite (RBT) and Río Seco de los Castaños Formation (RSC). **b** Inclusions in the tonalite. **d** The tonalite rocks intruded by aplite veins. FCGSM: railroad tracks

Previous geochronological data yielded biotite K–Ar ages of 475 ± 17 Ma and 452 ± 8 Ma (González 1971; González Díaz 1981) for the intrusive rocks, which are in disagreement with the intrusive character into Silurian-Lower Devonian country rocks.

3 Petrography and Geochemistry Aspects

The Rodeo de la Bordalesa tonalite consists of small laminar bodies intruded into the RSC unit; these intrusives are either parallel or crosscut the stratification, and although their composition is similar, they have different textures. The largest one (*ca.* 70 m thick) is close to the old railroad tracks and its country rocks (feldspathic sandstones, wackes and shales) develop a metamorphic contact characterized by recrystallized biotite and minor muscovite (Fig. 3). The tonalite shows a medium-grained equigranular texture and it is composed of zoned plagioclase (average An_{40}), green amphibole (sometimes with a core of clinopyroxene), biotite, and interstitial quartz. Zircon and apatite are present as accessory minerals (Fig. 4). The other body also intrudes the RSC and crops out northward of the previously described; it consists of dykes and small irregular bodies of porphyritic tonalite. Phenocrysts consist of zoned plagioclase (average An_{50}), scarce clinopyroxene surrounded by amphibole and biotite. The groundmass is composed of plagioclase, scarce biotite, and interstitial quartz.

Five samples (Table 1) were analyzed for major, trace and rare earth elements (ACTLABS, Canada). They plot in the TAS diagram adapted to plutonic rocks by Bellieni et al. (1995) into the field of tonalites (Fig. 5).

Modal composition indicates an I-type signature and in the AFM diagram (Irvine and Baragar 1971) samples show a calc-alkaline trend (Fig. 6a). They are characterized by high to medium potassium concentrations (after Peccerillo and Taylor 1976); with A/CNK index ranging from 0.90 to 0.95 they are regarded as metaluminous rocks (Fig. 6b).

The extended multielement diagram normalized to primitive mantle (Taylor and McLennan 1985) show depression of Nb and Ti and low enrichment of HFSE, typical of calc-alkaline series (Fig. 7a). The REE patterns show LREE enrichment and flat HREE behavior, also characteristic of calc-alkaline rocks (Fig. 7b).

To constrain the tectonic environment of emplacement three discrimination diagrams were applied (Fig. 8), from which it is deduced that the tonalites intruded within an active continental margin since they plot in the field of volcanic arc granitoids (Fig. 8a) from Pearce et al. (1984), while in the Whalen et al. (1987) diagram plot into the I-type field (Fig. 8b). Furthermore, Harris et al. (1986) diagram allows the discrimination of pre-collisional calc-alkaline arc-related granitoids from syn- to post-collisional intrusions and within plate intrusions. In this regard, the late- and post-collisional character of samples from the Rodeo de la Bordalesa (Fig. 8c) agree well with an emplacement within the RSC folded sedimentary rocks afterwards the 'Chanic' tectonic phase.

Fig. 3 Representative photomicrographs of the RSC hornfels at the contact with the tonalite rocks. *Pl* plagioclase; *Qz* quartz; *Fd* K-feldspar; *B* biotite. *Zr* zircon. The abundance of recrystallized biotite is evident

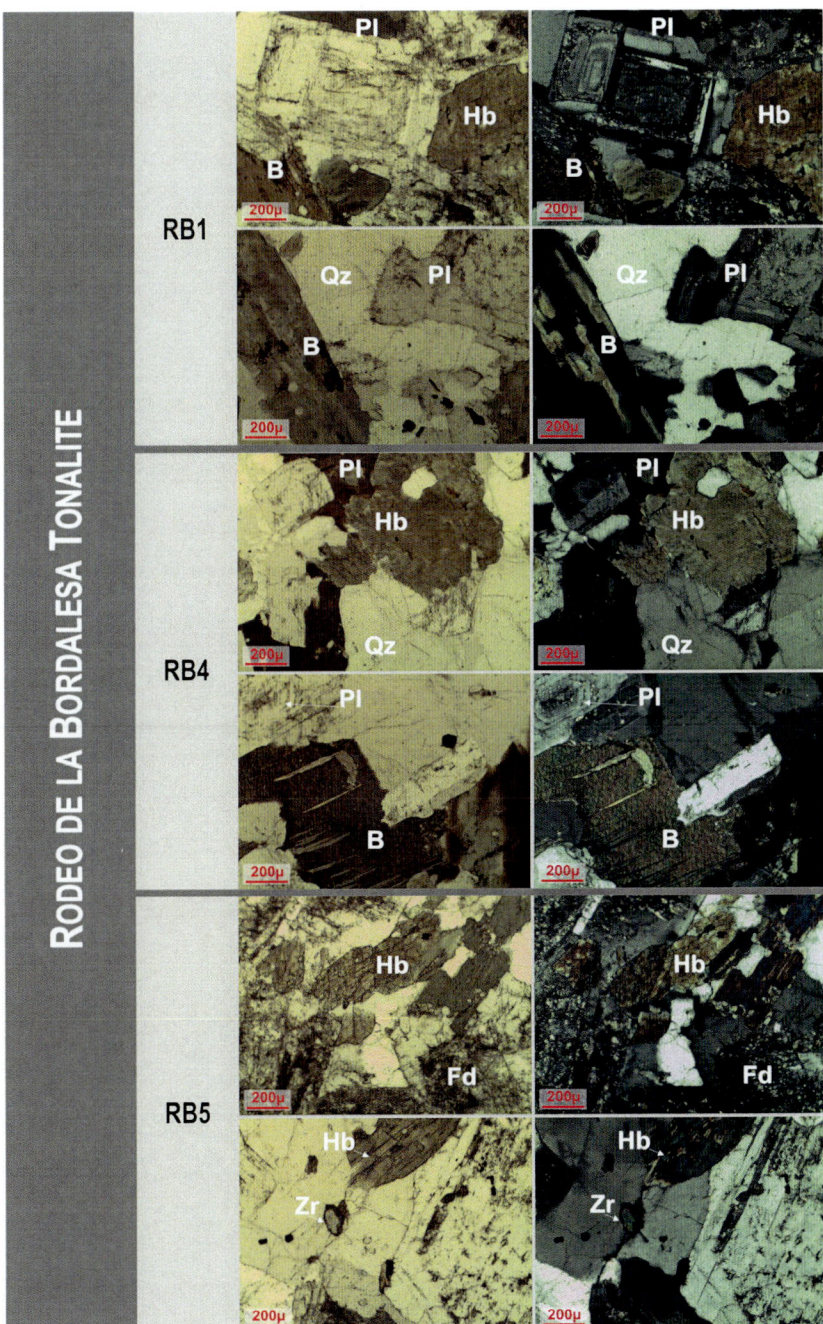

Fig. 4 Photomicrographs of the Rodeo de la Bordalesa tonalite samples showing equigranular texture. *Pl* zoned plagioclase; *Hb* hornblende; *Qz* quartz; *Fd* K-feldspar; *B* biotite, *Zr* zircon

Table 1 Geochemical data (ACTLABS, Canada) of the studied samples. Major elements in wt% and trace elements (including REE) in ppm

Sample	SiO$_2$	TiO$_2$	Al$_2$O$_3$	Fe$_2$O$_3$	MnO	MgO	CaO	Na$_2$O	K$_2$O	P$_2$O$_5$	LOI	Total	SiO$_2$an	Na$_2$Oan	KOan	TAS	FeOt
LRB1	60.08	0.67	16.48	5.91	0.1	3.74	5.16	3.15	2.55	0.17	1.85	99.86	60.16	3.15	2.55	5.71	5.32
LRB2	57.1	0.81	16.04	7.11	0.12	4.85	5.73	3.26	2	0.17	2.77	99.96	57.12	3.26	2.00	5.26	6.40
LRB3	54.9	0.71	17.2	6.7	0.11	4.32	6.35	3.12	2.18	0.18	4.65	100.4	54.68	3.11	2.17	5.28	6.03
LRB4	58.81	0.79	16.44	7.18	0.11	4.46	5.77	3.27	2.2	0.2	1.66	100.9	58.30	3.24	2.18	5.42	6.46
RB5	57.4	0.73	18.19	6.79	0.12	2.76	6.44	3.41	1.79	0.25	0.92	98.79	58.10	3.45	1.81	5.26	6.11

Sample	Sr	Cs	Rb	Ba	Th	U	Ta	Nb	Ce	Zr	Hf	Y	V	Cr	Co	Ni	Ga	Tl	Pb	Sc
LRB1	432	7.6	98	662	8.14	2.52	1.16	8.9	50.5	137	3.8	20	124	116	23	28	20	0.56	16	20
LRB2	469	4.6	68	618	5.7	1.52	0.96	7.2	41.4	116	3.3	20	172	141	33	33	18	0.44	−5	25
LRB3	449	4.7	74	589	5.53	1.72	0.86	6.5	39.2	116	3.2	19	140	73	24	22	18	0.59	13	22
LRB4	409	5.1	72	522	6.15	1.64	1.05	7.2	39.4	112	3.1	19	141	112	17	22	15	0.13	−5	21
RB5	474	1.5	48	594	2.91	0.79	1.4	7.5	33.6	120	2.9	17	92	34	17		17	0.07		15

Sample	La	Ce	Pr	Nd	Sm	Eu	Gd	Tb	Dy	Ho	Er	Tm	Yb	Lu
LRB1	25.3	50.5	5.793	23	4.76	1.141	4.03	0.62	3.4	0.67	2.1	0.321	1.89	0.291
LRB2	20	41.4	4.868	20.1	4.33	1.153	3.68	0.6	3.36	0.68	2.12	0.324	1.9	0.299
LRB3	19.1	39.2	4.599	19.2	4.25	1.141	3.45	0.56	3.29	0.66	2.08	0.301	1.88	0.301
LRB4	20.1	39.4	4.625	19.1	4.02	1.038	3.41	0.57	3.22	0.65	2.03	0.301	1.8	0.288
RB5	16.7	33.6	3.96	16.4	3.36	1.21	3.3	0.5	2.92	0.61	1.82	0.275	1.72	0.265

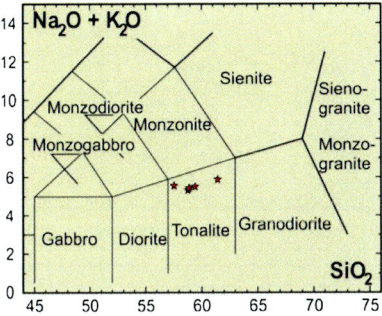

Fig. 5 TAS diagram modified by Bellieni et al. (1995)

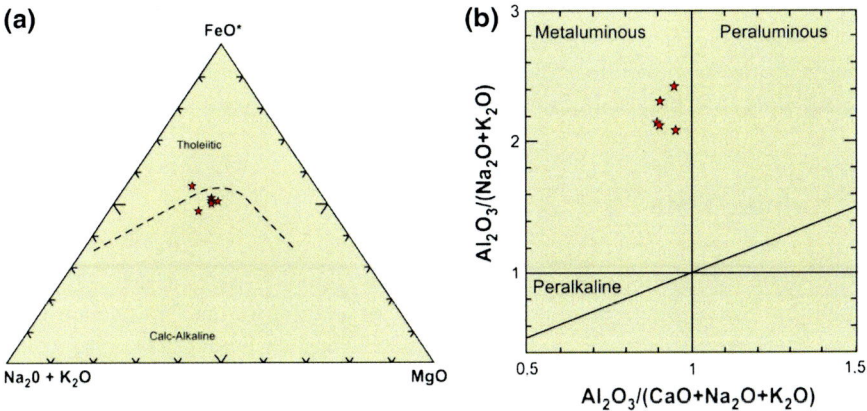

Fig. 6 a Ternary AFM diagram. b A/CNK discrimination diagram after Maniar and Piccoli (1989)

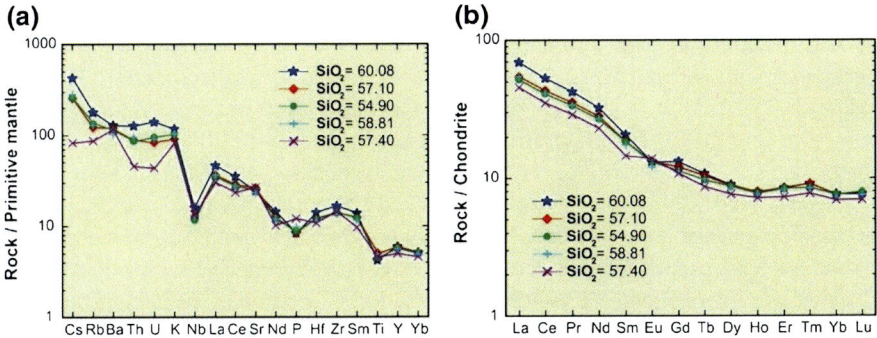

Fig. 7 a Multielements spider-diagram normalized to primitive mantle. SiO_2 content of each sample is shown. b Chondrite normalized REE diagram for the five samples

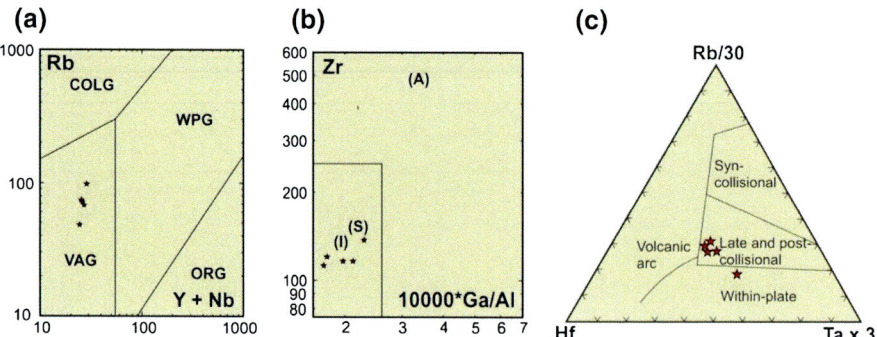

Fig. 8 Tectonic discrimination diagram of granitoids. **a** Pearce et al. (1984) diagram; COLG: collisional, WPG: within plate, VAG: volcanic arc, ORG: orogenic granitoids. **b** Whalen et al. (1987) diagram. **c** Harris et al. (1986) ternary diagram

All these characteristics allowed us to differentiate the Rodeo de la Bordalesa tonalite from the tholeiitic mafic rocks (mainly gabbros, amphibolites and porphyritic dolerites) exposed at the El Nihuil area (Cingolani et al. 2000).

4 Isotopic Data

To constrain the age of the Rodeo de la Bordalesa intrusive rocks new U–Pb, K–Ar, Rb–Sr and Sm–Nd data have been obtained, in addition to the Ordovician biotite K–Ar dates reported by González (1971) and González Díaz (1981) and Middle Devonian age (whole rock 380 ± 20 Ma) by Linares et al. (1987).

a. U–Pb (ID-TIMS): The procedure for U–Pb zircon analyses at Centro de Pesquisas Geocronológicas, Instituto de Geociencias, USP (Brazil) is as follow: After 10 kg of sample were crushed and reduced to 140–200-mesh grain-sizes the portion rich in heavy minerals was treated with bromoform ($d = 2.89$ g/cm^3) and methyl iodide ($d = 3.3$ g/cm^3), and the fraction containing the heavy minerals was processed in the Frantz separator at 1.5 A and split in several zircon-rich magnetic fractions. The final purification of each fraction was done by handpicking. The dissolution of the zircon crystals was carried out with HF and HNO$_3$ in Teflon micro bombs in which a mixed ^{205}Pb/^{235}U spike was added. A set of 15 micro bombs arranged in a metal jacket is left for three days in a stove at 200 °C. Then, the HF is evaporated and HCl (6 N) added to the micro bombs, replaced in the stove for 24 h. After the evaporation of HCl 6 N, the residue is dissolved in HCl (3 N). U and Pb are concentrated and purified by passing the solution in an anionic exchange resin column. The solution enriched in U and Pb is, after addition of phosphoric acid, evaporated until the formation of a microdrop. The sample is deposited in a rhenium filament and the isotopic composition is determined with Finnigan MAT 262 solid source mass spectrometer. After reduction of the data (PBDAT), the results (Table 2) are plotted in appropriate diagrams using the software ISOPLOT/EX (Ludwig 1999, 2001).

Table 2 U–Pb (ID-TIMS) analytical data from CPGeo, Sao Paulo, Brazil. The studied sample is LRB1

SPU	207/235#	Error (%)	206/238#	Error (%)	207/206#	Error (%)	206/204*	Pb (ppm)	U (ppm)	Weight (mg)	206/238 Age
1927	0.498825	0.81	0.064676	0.79	0.05594	0.160	882	17.1	251.5	0.0703	404
1928	0.497372	0.99	0.06454	0.97	0.05589	0.174	754	13.3	196.4	0.0672	403
1929	0.488365	0.55	0.064299	0.54	0.05509	0.109	2482	17.8	269.7	0.0727	402
1930	0.492999	1.03	0.064428	0.99	0.05550	0.245	886	12.2	183.2	0.0703	403

SPU Laboratory number

(M-5) *Magnetic fractions* Numbers in parentheses indicated the tilt used on Frantz at 1.5 amp. current

Radiogenic Pb corrected for blank and initial Pb; U corrected for blank

*Not corrected for blank or nonradiogenic Pb

Total U and Pb concentrations corrected for analytical blank

Ages given in Ma using Ludwig Isoplot/Ex program (2000), decay constants by Steiger and Jager (1977)

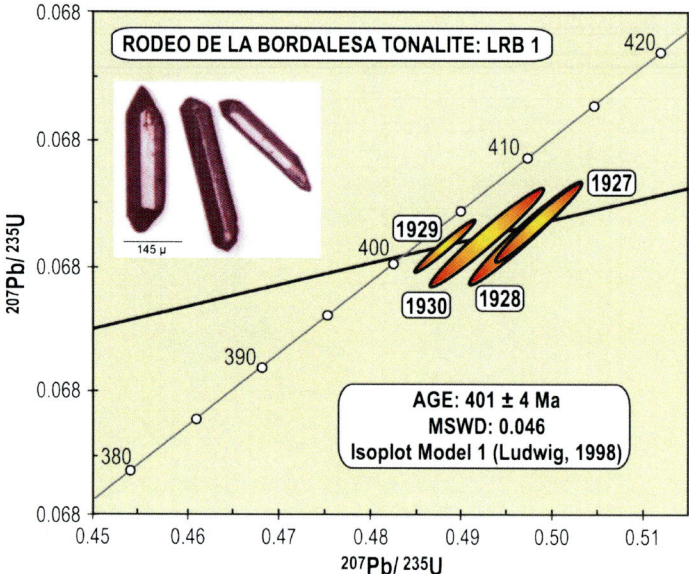

Fig. 9 U–Pb Concordia diagram. The inset show photographs under microscope of some dated zircon crystals with 300–450 μ size

Table 3 K–Ar analytical data from CPGeo, São Paulo, Brazil

Lab No SPK	Field No.	Mineral	Rock type	K (%)	Error (%)	^{40}ArRad	^{40}ArAtm (%)	Age (Ma)	Error (Ma)
7731	RB-04A21	Biotite	Tonalite	6.2488	3.0413	108.95	6.24	401.30	17.10

As we can see on the Concordia diagram (Fig. 9) the U–Pb average age obtained in four zircon fractions by ID-TIMS is 401 ± 4 Ma and that corresponds to Early Devonian (Emsian) time (IUGS International Stratigraphic Chart 2015).

b. K–Ar: Biotite fresh minerals were separated from one tonalite sample (RB-04) and dated using the K–Ar methodology at the Centro de Pesquisas Geocronológicas, Instituto de Geociencias, USP (Brazil), and the duplicate obtained data are presented in Table 3. The biotite gave an age of 401 ± 17 Ma. This value is very close and confirms the zircon U–Pb (ID-TIMS) age.

c. Rb–Sr: The Rb–Sr method was applied using five whole rock samples from the main tonalite outcrop near de railroad tracks. The biotite separate from one whole rock was also used. Rb and Sr XRF analyses as well as the mass spectrometry for Sr were carried out at the Laboratorio de Geología Isotópica, Universidade Federal do Rio Grande do Sul, Porto Alegre (Brazil). The sample preparation and extraction of natural Sr through cation exchange columns were performed at the Centro de Investigaciones Geológicas, Universidad Nacional de

Table 4 Rb–Sr analytical data

Field No.	Lab. No.	Rb (ppm)	Sr (ppm)	$^{87}Rb/^{86}Sr$	Error	$^{87}Sr/^{86}Sr$	Error
01LRB1	CIG 1248	52.9	506.6	0.3023	0.006	0.707074	0.000021
01LRB2	CIG1249	40.8	488	0.242	0.0048	0.706387	0.000019
01LRB10	CIG1251	54.7	464	0.3413	0.0068	0.70719	0.00002
01LRB16	CIG 1252	64.6	306.6	0.6101	0.0122	0.709615	0.000021
01LRB21	CIG1253	52.3	415.5	0.3644	0.0073	0.707392	0.000023
Biotite	CIG 1268	313.07	19.49	47.64	0.24	0.959464	0.000294

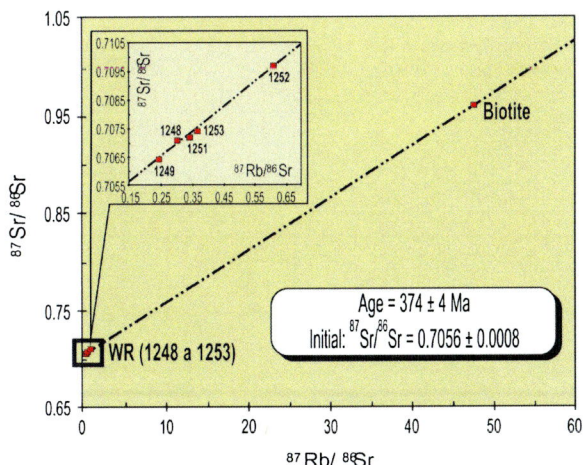

Fig. 10 Rb–Sr isochronic diagram using whole rock samples and biotite

La Plata. As it is shown in Table 4, the samples show low Rb (40–60 ppm) and high Sr contents (300–500 ppm), with a low Rb/Sr ratio (0.10–0.20). Rb–Sr whole rock diagram (Fig. 10) shows an alignment of five samples within a very low range of $^{87}Rb/^{86}Sr$ (0.24–0.61), and defines an 'age' of 600 ± 100 Ma with an IR: 0.7043. Because the error is too high we utilized a biotite separation as Rb-rich mineral. For the biotite sample the Rb/Sr ratio is 26. The age obtained with the five whole rocks and the biotite is 374 ± 4 Ma, with an IR: 0.7056 ± 0.0006 as we can see on the diagram from Fig. 10.

d. Sm–Nd: To apply the Sm–Nd method, five whole rock tonalitic samples (RB1–RB 5) were used (Table 5). The isotope dilution technique for Sm–Nd analyses (using a combined ^{149}Sm–^{150}Nd spike) as well as the mass spectrometry for Sm and Nd were carried out at the Laboratorio de Geología Isotópica, Universidade Federal do Rio Grande do Sul, Porto Alegre (Brazil). The isotopic ratios were measured using the VG 354 mass spectrometer with multiple collector system. The samples do not define an acceptable alignment. The model ages (T_{DM}) calculated according to DePaolo (1981) for the whole rock samples are in the range of 1 and 1.6 Ga. The ε_{Nd} (400 Ma) for these samples is in between −4.45 and −10.20, indicating crustal source (Fig. 11).

Table 5 Sm–Nd analytical data

Field No.	Sm	Nd	$^{147}Sm/^{144}Nd$	$^{147}Nd/^{144}Nd$	Error	ε_{Nd} (0)	Age	ε_{Nd} (t)	T_{DM} (Ga)
RB-1	4.02	18.66	0.130219	0.512406	10	−4.52	400	−0.70	1.2
RB-2	2.93	15.06	0.117629	0.512410	25	−4.45	400	0.10	1.0
RB-3	3.79	19.15	0.119617	0.512354	15	−5.55	400	−1.12	1.1
RB-4	3.88	18.92	0.124118	0.512115	240	−10.20	400	−6.04	1.6
RB-5	3.78	18.78	0.121827	0.512381	20	−5.02	400	−0.72	1.1

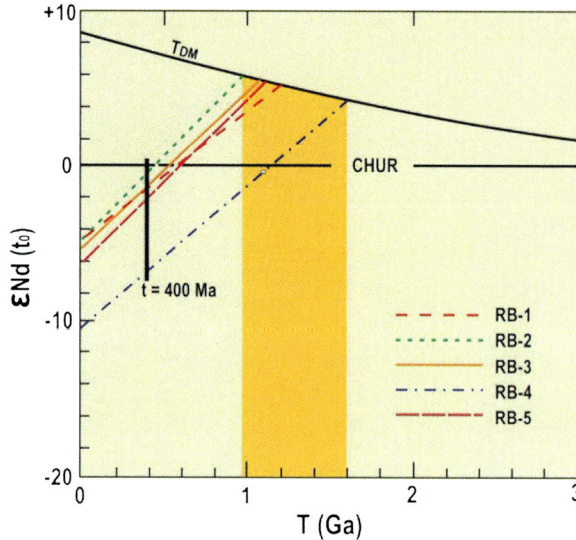

Fig. 11 The ε_{Nd} evolution diagram of samples at $t = 400$ Ma

5 Concluding Remarks

Based on the newly obtained data the following statements can be made:

– The Rodeo de la Bordalesa tonalite dykes at San Rafael Block are characterized by high to medium potassium contents, with a metaluminous character and I-type calc-alkaline signature. It forms part of a magmatism that could be related to a post-collisional tectonic event.
– We interpret the *ca.* 400 Ma U–Pb zircon age obtained within a concordia diagram, as the crystallization age which corresponds to the emplacement time. This data are confirmed by the K–Ar biotite age. The Ordovician K–Ar ages (González 1971) are not supported by our geochronological data and are also not consistent with the RSC paleontological record (Cingolani et al. this volume).

- The Rb–Sr whole rocks and biotite age of 374 ± 4 Ma, could be linked to the 'Chanic' tectonic phase, in agreement with other geochronological data (Toubes and Spikermann 1976, 1979). Cingolani and Varela (2008) presented a Rb–Sr isochronic whole rock age of 336 ± 23 Ma for the anchimetamorfic event that affected the Río Seco de los Castaños unit, implying an Early Carboniferous (Mississipian) low-grade metamorphism for the RSC. Tickyj et al. (2001), based on similar isotopic studies determined isochronic whole rock ages ranging from 371 ± 62 to 379 ± 15 Ma for the La Horqueta sequence, from which suggested an Upper Devonian low-grade metamorphism. Similar data were obtained in metasedimentary rocks from Precordillera (Cucchi 1971; Buggish et al. 1994; Ramos et al. 1998; Davis et al. 1999) that strongly suggests Upper Devonian-Lower Carboniferous age for the synmetamorphic ductile deformation in the western side of Cuyania terrane in connection with the 'Chanic' tectonic phase.
- Nd model ages (T_{DM}) show an interval between 1 and 1.6 Ga that corresponds to Mesoproterozoic age derivation and the negative ε_{Nd} is in accordance to crustal sources.
- The crystallization age for the Rodeo de la Bordalesa tonalite dykes corresponds to a Lower Devonian time (Pragian-Emsian boundary) according to IUGS time scale and suggests that part of the Late Famatinian magmatic event is present in the San Rafael Block. The tonalite rocks are contemporaneous with the large peraluminous batholith exposed in Pampean Ranges (Rapela et al. 1992; Dahlquist et al. 2014), with the transpressional shear belts during 'Achalian' event (Sims et al. 1998); it could be as well correlated with the Devonian magmatism present in Pampa de los Avestruces (Tickyj et al. 2009) in the southern part of the Frontal Cordillera and some places studied recently by Tickyj et al. (2015) near Agua Escondida Mine District in the southern sector of the SRB.
- The geochemical and geochronological data allow us to differentiate the Rodeo de la Bordalesa tonalite from the mafic rocks (mainly porphyritic dolerites with tholeiitic signature) exposed at the El Nihuil area.

Acknowledgements Field and laboratory work was financed by PIP-CONICET grants 647 and 199. We thank Ricardo Varela for Rb–Sr whole rock laboratory preparation. Diego Licitra, Leonardo Ortiz and Natalia Hernández helped us in field work, mineralogical separation, and petrographical descriptions. We acknowledge to Mario Campaña for technical assistance. To Alejandro Capelli for kind permission to access at the outcrops near the deactivated railroad tracks ('Ferrocarril General San Martin'). Domingo Solorza from Puesto Cortaderas, helped us in a field work activities.

References

Bellieni G, Visentin J, Zanettin B (1995) Use of the chemical TAS diagram (total alkali-silica) for classification of plutonic rocks: problems and suggestions. PLINIUS (Supplemento italiano all'Eur J Miner) 14:49–52

Buggish W, von Gosen W, Henjes-Kunst E, Krumm S (1994) The age of early Paleozoic deformation and metamorphism in the Argentine Precordillera-Evidence from K–Ar data. Zentralbl Geol Palaontol Teil I:275–286

Cingolani CA, Llambías EJ, Ortiz LR (2000) Magmatismo básico pre-Carbónico del Nihuil, Bloque de San Rafael, Provincia de Mendoza, Argentina. In: 9° Congreso Geológico Chileno, vol 2. Puerto Varas, pp 717–721

Cingolani CA, Basei MAS, Llambías EJ, Varela R, Chemale Jr F, Siga O Jr, Abre P (2003) The Rodeo Bordalesa Tonalite, San Rafael Block (Argentina): Geochemical and isotopic age constrains. In: 10° Congreso Geológico Chileno, Concepción, p 10 (CD Rom)

Cingolani CA, Varela R (2008) The Rb-Sr low-grade metamorphism age of the Paleozoic Río Seco de los Castaños Formation, San Rafael Block, Mendoza, Argentina. VII South American symposium on Isotope Geology, S.C. Bariloche, Argentina, p 4 (CD-ROM)

Cingolani CA, Varela R, Chemale Jr F, Uriz NJ (2011) Geocronología U–Pb de las monzodioritas de la Boca del Río, Cacheuta-Mendoza, Argentina. In: 18° Congreso Geológico Argentino, Simposio de Tectónica pre-Andina. Actas. Neuquén, p 2 (CD Rom)

Cingolani CA, Uriz NJ, Abre P, Manassero MJ, Basei MAS (this volume) Silurian-Devonian land-sea interaction within the San Rafael Block, Argentina: Provenance of the Rio Seco de los Castaños Formation. In: Cingolani C (ed) Pre-Carboniferous evolution of the San Rafael Block, Argentina. Implications in the SW Gondwana margin. Springer (Chapter 10)

Criado Roqué P, Ibáñez G (1979) Provincia geológica Sanrafaelino-Pampeana. In: Turner JCM (ed) Geología Regional Argentina. Academia Nacional de Ciencias, I, pp 745–769

Cucchi RJ (1971) Edades radimétricas y correlación de metamorfitas de la Precordillera, San Juan-Mendoza, Rep. Argentina. Rev Asoc Geol Argent 28(3):241–267

Cuerda AJ, Cingolani CA (1998) El Ordovícico de la región del cerro Bola en el Bloque de San Rafael, Mendoza: sus faunas graptolíticas. Ameghiniana 35(4):427–448. Buenos Aires

Dahlquist JA, Alasino PH, Bello C (2014) Devonian F-rich peraluminous A-type magmatism in the proto-Andean foreland (Sierras Pampeanas, Argentina): geochemical constraints and petrogenesis from the western central region of the Achala batholiths. Mineral Petrol 108:391–417

Davicino RE, Sabalúa JC (1990) El cuerpo básico de El Nihuil, Depto. San Rafael, Pcia. de Mendoza, Rep. Argentina. In: 11° Congreso Geológico Argentino, San Juan, Actas, vol 1, pp 43–47

Davis J, Roeske S, McClelland W, Snee L (1999) Closing the ocean between the Precordillera terrane and Chilenia: early Devonian ophiolite emplacement and deformation in the Southwest Precordillera. In: Ramos VA, Keppie, JD (eds) Laurentia-Gondwana connections before Pangea. Geological Society of America, Special Paper 336. Boulder, Co. USA, pp 115–138

DePaolo DJ (1981) Neodymiun isotopes in the Colorado Front Range and implications for crust formation and mantle evolution in the Proterozoic. Nature 291:193–197

Dessanti RN (1956) Descripción geológica de la Hoja 27c-cerro Diamante (Provincia de Mendoza). Dirección Nacional de Geología y Minería. Boletín, vol 85. Buenos Aires, p 79

Di Persia CA (1972) Breve nota sobre la edad de la denominada Serie de la Horqueta- Zona Sierra Pintada. Departamento de San Rafael, Provincia de Mendoza. 4ª Jornadas Geológicas Argentinas, vol 3. Mendoza, pp 29–41

González RN (1971) Edades radimétricas de algunos cuerpos eruptivos de Argentina. Rev Asoc Geol Argent 26(3):411–412

González Díaz EF (1964) Rasgos geológicos y evolución geomorfológica de la Hoja 27 d, San Rafael y su zona vecina occidental (Pcia. de Mendoza). Asoc Geol Argent Rev XIX(3):151–188

González Díaz EF (1981) Nuevos argumentos a favor del desdoblamiento de la denominada "Serie de la Horqueta" del Bloque de San Rafael, Provincia de Mendoza. In: 8° Congreso Geológico Argentino, Actas, vol 3. San Luis, pp 241–256

Harris NBW, Pearce JA, Tindle AG (1986) Geochemical characteristics of collision-zone magmatism. In: Coward MP, Rios AC (eds) Collision tectonics. Geological Society of London, Special Bulletin, vol 19, pp 67–81

Irvine TN, Baragar WRA (1971) A guide to the chemical classification of the common rocks. Can J Earth Sci 8:523–548

IUGS (2015) International Chronostratigraphic Chart. International Commission on Stratigraphy, International Union of Geological Sciences (IUGS)

Linares E, Parica C, Parica P (1987) Catálogo de edades radimétricas determinadas para la República Argentina (IV años 1979–1980 realizadas por INGEIS y sin publicar y V años 1981–1982 publicadas). Publicaciones Especiales, Asociación Geológica Argentina, Serie B (Didáctica y Complementaria), vol 15, pp 1–49

Ludwig KR (1999) Using Isoplot/Ex, version 2. A geochronological toolkit for Microsoft excel. Berkeley Geochronological Center, Special Publication 1a, 47 p

Ludwig KR (2001) Squid 1.02: a user manual. Berkeley Geochronl Cent Spec Publ 2:19

Manassero MJ, Cingolani CA, Abre P (2009) A Silurian-Devonian marine platform-deltaic system in the San Rafael Block, Argentine Precordillera-Cuyania terrane: lithofacies and provenance. In: Konigshof P (ed). Devonian change: case studies in Palaeogeography and Palaeoecology. The Geological Society, vol. 314. Special Publications, London, pp 215–240

Maniar PD, Piccoli PM (1989) Tectonic discrimination of granitoids. Geol Soc Am Bull 101:635–643

Pazos PJ, Heredia AM, Fernández DE, Gutiérrez C, Comerio M (2015) The ichnogenus *Dictyodora* from late Silurian deposits of central-western Argentina: Ichnotaxonomy, ethology and ichnostratigrapical perspectives from Gondwana. Palaeogeogr Palaeoclimatol Palaeoecol 439:27–37. doi:10.1016/j.palaeo.2015.02.008

Pearce JA, Harris NBW, Tindle AG (1984) Trace element discrimination diagrams for the tectonic interpretation of granitic rocks. J Petrol 25:956–983

Peccerillo A, Taylor SR (1976) Geochemistry of Eocene calc-alkaline volcanic rocks from the Kastamonu area, northern Turkey. Contrib Miner Petrol 58:63–81

Poiré DG, Cingolani C, Morel E (1998) Trazas fósiles de la Formación Horqueta (Silúrico), Bloque de San Rafael, Mendoza, Argentina. Tercera Reunión de Icnología y Primera Reunión de Icnología del Mercosur, Resúmenes, vol 24. Mar del Plata

Ramos VA (2004) Cuyania, an exotic block to Gondwana: review of a historical success and the present problems. Gondwana Res 7:1009–1026

Ramos VA, Dallmeyer R, Vujovich GI (1998) Time constrains on the early Paleozoic docking of the Precordillera central Argentina. In: Pankhurst RJ, Rapela CW (eds) The Proto-Andean Margin of Gondwana. Geological Society of London, Special Publication, vol 142, pp 143–158

Rapela CW, Coira B, Toselli A, Saavedra J (1992) The lower Paleozoic magmatism of south-western Gondwana and the evolution of the Famatinian orogen. Int Geol Rev 34:1081–1142

Rubinstein C (1997) Primer registro de palinomorfos silúricos en la Formación La Horqueta, Bloque de San Rafael, provincia de Mendoza, Argentina. Ameghiniana 34(2):163–167

Sims JP, Ireland TR, Camacho A, Lyons P, Pieters PE, Skirrow RG, Stuart-Smith PG (1998) U–Pb, Th–Pb and Ar–Ar geochronology from the southern Sierras Pampeanas, Argentina: implications for the Palaeozoic tectonic evolution of the western Gondwana margin, In: Pankhurst RJ, Rapela CW (eds) The Proto-Andean margin of Gondwana. Geological Society of London, Special Publication, vol 142, pp 259–282

Steiger RH, Jager E (1977) Subcommission on geochronology: convention on the use of decay constants in geochronology and cosmochronology. Contribution to the geologic time scale, AAPG, Studies in Geology 6:67–71

Taylor SR, McLennan SM (1985) The continental crust: its composition and evolution. Blackwell, Oxford, 312 p

Tickyj H, Cingolani CA, Chemale Jr F (2001) Rb–Sr ages from La Horqueta Formation, San Rafael Block (Argentina). In: III South American symposium on isotope geology, Pucón, Chile (CD Rom)

Tickyj H, Fernández MA, Chemale Jr F, Cingolani C (2009) Granodiorita Pampa de los Avestruces, Cordillera Frontal, Mendoza: un intrusivo sintectónico de edad devónica inferior. In: 14° Reunión de Tectónica. Libro de resúmenes, vol 27, Río Cuarto, Argentina

Tickyj H, Tomezzoli MA, Basei MAS, Fernández MA, Blatter JM, Rodríguez N, Gallo LC (2015) Geología de la Formación Piedra de Afilar, basamento granítico del Distrito Minero Agua Escondida, Mendoza. In: 3° Simposio sobre Petrología Ígnea y Metalogénesis Asociada. Río Negro, Argentina

Toubes RO, Spikermann JP (1976) Algunas edades K–Ar para la Sierra Pintada, provincia de Mendoza. Rev Asoc Geol Argent 31(2):118–126

Toubes RO, Spikermann JP (1979) Nuevas edades K–Ar para la Sierra Pintada, provincia de Mendoza. Rev Asoc Geol Argent 34(1):73–79

Whalen JB, Currie KL, Chappel BW (1987) A-type granites: geochemical characteristics, discrimination and petrogenesis. Contrib Miner Petrol 95:407–419

Pre-Carboniferous Tectonic Evolution of the San Rafael Block, Mendoza Province

Carlos A. Cingolani and Victor A. Ramos

Abstract The pre-Carboniferous evolution of the San Rafael Block is described in different stages. The first one is referred to the Mesoproterozoic basement derived from a complex plutonic and volcanic protolith of Cerro La Ventana Formation. The signature of this basement indicates a common origin with the present eastern part of Laurentia. The carbonate platform of Cuyania terrane has been drifted away during Early Cambrian to Early Ordovician times. The Ordovician silico-carbonate sequences of the San Rafael Block are unconformably deposited over the basement near the present eastern slope of the Cuyania terrane. Detrital zircon ages show a provenance derived from Mesoproterozoic source. The El Nihuil dolerites with a tholeiitic ocean floor signature considered the southern end of the Famatinian ophiolites were interpreted as a Late Ordovician–Early Silurian extensional event. The collision of Cuyania produced a new west polarity subduction and a magmatic arc, represented by the Devonian Rodeo de la Bordalesa tonalite and the granitoids of the "Agua Escondida Mining District". The Late Silurian–Early Devonian sequences of La Horqueta and Río Seco de los Castaños formations were deformed during the collision and accretion of the Chilenia terrane against the proto-Andean margin, and recorded an east vergent cleavage developed on the previous deformed rocks. This collision produced the strong angular unconformity between the La Horqueta/Río Seco de los Castaños Formations and the El Imperial Formation (Upper Paleozoic). The new subduction with east polarity characterized the beginning of the Gondwanian cycle. The new magmatic arc was interrupted by the intense Lower Permian deformation of the San Rafael tectonic phase.

C.A. Cingolani (✉)
División Geología, Museo de La Plata and Centro de Investigaciones Geológicas (CONICET-UNLP), Diagonal 113 n. 275, La Plata 1904, Argentina
e-mail: ccingola@cig.museo.unlp.edu.ar; carloscingolani@yahoo.com

V.A. Ramos
Instituto de Estudios Andinos Don Pablo Groeber (IDEAN), Universidad de Buenos Aires–CONICET, Ciudad Universitaria, 1428 Buenos Aires, Argentina
e-mail: andes@gl.fcen.uba.ar

Keywords Cuyania terrane · Famatinian collision · Chanic phase · Chilenia terrane · Proto-Andean margin

1 Introduction

The San Rafael Block consists of a large variety of rocks of mainly Paleozoic age, which crop out in the south-central part of Mendoza province, at 35°S–68°30′W (see previous chapters for details). These rocks were exposed as a result of uplift during an episode of broken foreland associated with a flat subduction period produced at some point of the Andean evolution during latest Miocene and Early Pliocene (Ramos et al. 2014). Geological and paleontological evidence suggest that the San Rafael Block is the southern part of the Cuyania terrane. Diverse "pre-Carboniferous units" crop out in this region, as known since the early studies of Dessanti (1945, 1956). There are composed by igneous, metamorphic, and sedimentary rocks. New insights presented in previous chapters shed light in the stratigraphy, petrography, geochemistry and isotopic data, which as a whole are the bases to constraining the pre-Carboniferous tectonic evolution of the San Rafael Block.

2 The Basement of San Rafael Block

The Mesoproterozoic basement is derived from a plutonic and volcanic protolith complex that characterizes the Cerro La Ventana Formation. The mafic to intermediate gneisses, foliated quartz diorites, diorites, and tonalites, partially grade to amphibolites and migmatites. Their geochemical signature as well as the low Sr initial ratio shows a primitive calc-alkaline signature, associated with a magmatic arc. Their characteristics are similar to the Las Matras tonalite–trondhjemites suite from La Pampa, which also have a Grenvillian age (Sato et al. 2000; Varela et al. 2011) (Fig. 1).

Whole-rock samples analyzed by the Sm–Nd method define an age of 1228 ± 63 Ma, confirming previous Rb–Sr ages (Cingolani and Varela 1999; Cingolani et al. 2005). The *ca.* 1.2 Ga model ages (T_{DM}) and the Epsilon Nd are indicative of a depleted source, less evolved than CHUR (Chondritic Uniform Reservoir), at the time of crystallization. Euhedral zircon fractions analyzed by U–Pb (ID-TIMS, *Isotope dilution by Thermo Ionization Mass Spectrometer*) and *in situ* LA-ICP-MS methods indicate an age of *ca.* 1.2 Ga, interpreted as a Mesoproterozoic zircon crystallization ages (Cingolani et al. 2005, 2012), confirm previous data mentioned by Astini et al. (1995) and Keller (1999). The El Nihuil Mafic Unit comprises meter-scale-deformed gabbroic and orthogneissic rocks of intermediate compositions bearing a foliation similar to the one found in Cerro La Ventana type area at Río Seco de los Leones suggesting a common origin. Pb–Pb

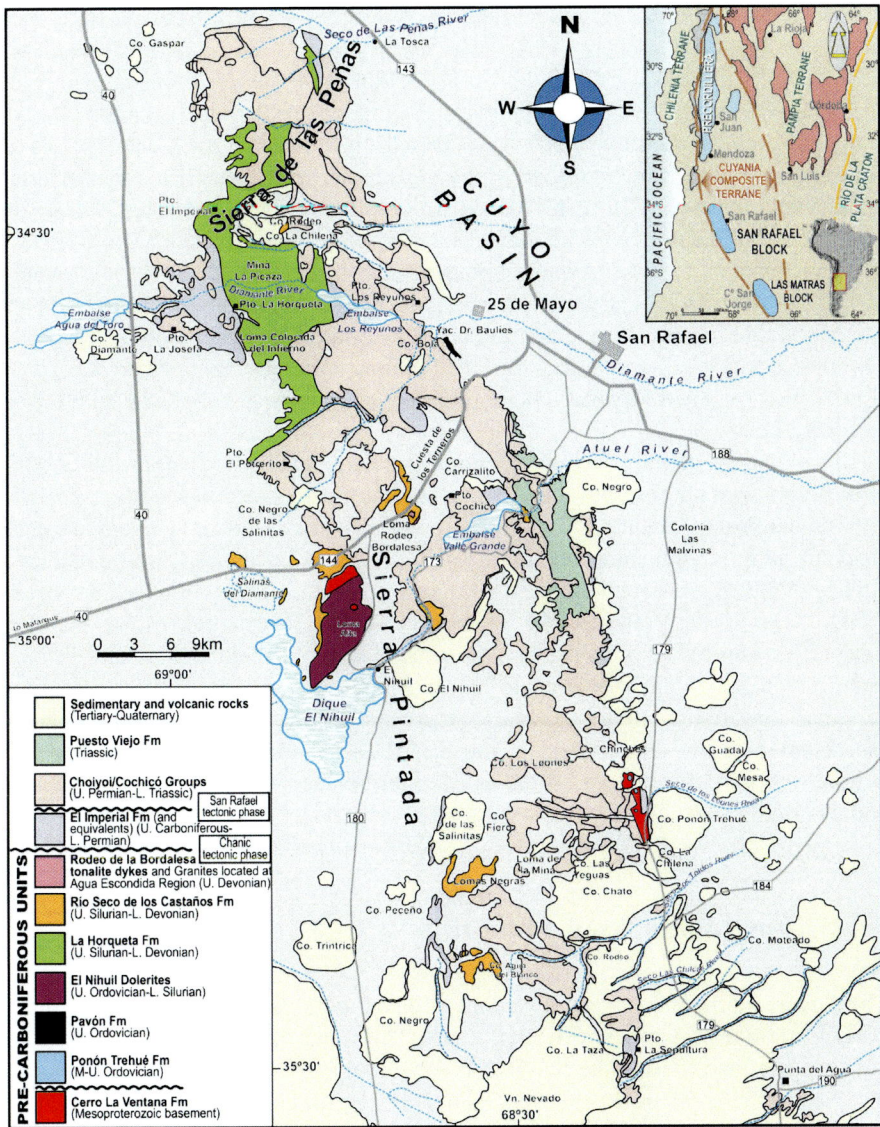

Fig. 1 Geological sketch map of the San Rafael Block showing the distribution of the pre-Carboniferous units (based on Dessanti 1956; Nuñez 1976, 1979; González Díaz 1972). The *inset* shows the location of the San Rafael Block within the Cuyania terrane (Ramos et al. 1986; Ramos 2004)

isotope data from three samples of Cerro La Ventana Formation (Abre et al. 2011) plot in the field of depleted signature, near the field of crustal xenoliths of Tertiary volcanic rocks from Precordillera and Pie de Palo Grenvillian-age rocks (Abbruzzi et al. 1993; Mahlburg Kay et al. 1996; Rapela et al. 2010).

Two additional small basement outcrops south of Cerro La Ventana Formation were described by Holmberg (1973), who named them Cerro Las Pacas Formation. New isotopic information (Cingolani et al. 2014a, b) on the mica schists, which are characterized by their vertical foliation and intruded by Permian granites, records younger Paleozoic detrital zircon ages. Based on these data the Cerro Las Pacas Formation is not considered as part of the Mesoproterozoic basement exposed along the Cuyania terrane. As a result of oil exploration carried out before 1970 in the Triassic Cuyo Basin, some "basement" rocks were correlated with the Cerro La Ventana Formation at the Alvear subbasin, east of the San Rafael Block (Ramos 2004). The well IV-D cut garnet–hornblende–biotite schists and recorded a K–Ar age of 605 Ma (Criado Roqué and Ibañez 1979). Another exploration well near the Corral de Lorca locality intersected a "metasedimentary basement" where Cingolani et al. (2012) obtained Paleozoic detrital zircon ages and was correlated with the La Horqueta Formation.

The signature of the basement rocks of the San Rafael Block point out a common origin with the present eastern part of Laurentia, preserved outboard of the Grenvillian Front (Mosher 1993). Those rocks as the El Llano and Cuyania terranes share in common juvenile signatures, and as whole have been interpreted as a collage of island arc terranes (Fig. 2a) amalgamated during the Grenvillian orogeny (Nelis et al. 1989; Walker 1992; Ramos 2004). The formation of the southern Iapetus Sea during late Neoproterozoic produced the detachment of the Cuyania terrane, which was the lower plate of the Laurentia–Cuyania conjugate margins (Thomas and Astini 1999). Although there are not yet known synrift deposits associated with this detachment in the San Rafael Block, further north (Fig. 2b) they have been described and dated as latest Proterozoic–earliest Cambrian (Astini and Thomas 1999).

3 Drifting Away from Laurentia

The Cuyania terrane has been drifted away developing a carbonate platform during Early Cambrian to Early Ordovician times. The Cambrian carbonate platform has not yet been identified in the San Rafael Block, but is well known further north in the Precordillera. There in the San Juan province, thick sequences of carbonates bearing the typical *Olenellus* fauna, characterize the passive margin facies (Borrello 1971; Bordonaro 2003). The Ordovician (Darriwilian–Sandbian) carbonates that in the San Rafael Block are unconformably deposited over the Mesoproterozoic basement may indicate that near the present eastern slope of the Cuyania terrane, the Cambrian carbonates has been eroded away (see Fig. 3a, b). These sedimentary rocks are included in the Ponón Trehué Formation, which is the only early Paleozoic fossiliferous sedimentary sequence (Nuñez 1962; Keller 1999; Heredia 2006 and references therein) known to record a primary contact with the Grenvillian-age basement of the Cuyania terrane. The lower member of this unit has

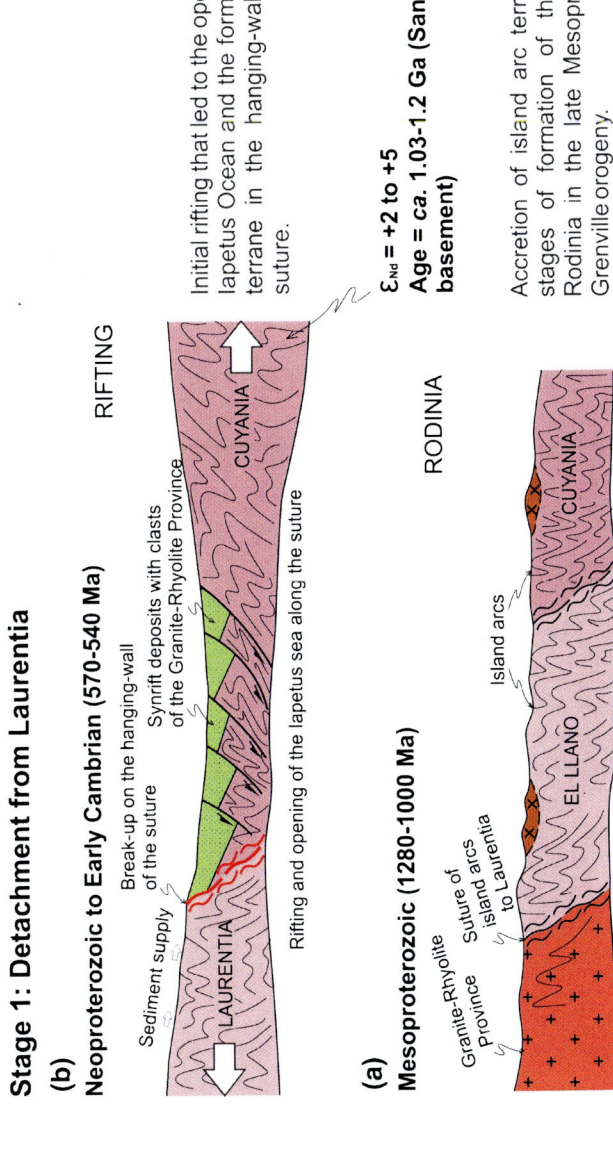

Fig. 2 Meso- to Neoproterozoic evolution of the Cuyania terrane (1280–540 Ma)

carbonates and siliciclastic deposits bearing shelly fauna and conodonts, which biozones are similar to the Precordillera platform carbonates (Heredia and Mestre, this volume). The petrographic data and the geochemical proxies for the Ponón Trehué Formation indicate contributions from dominantly upper continental crustal and subordinated depleted components derived from the basement.

This is confirmed by the in situ U–Pb detrital zircon ages, which cluster around values of 1.2 Ga (inset of Fig. 3), indicating a main derivation from the local basement source of Cerro La Ventana Formation (Abre et al. 2011; this volume).

East of the Cuyania terrane there are some isolated outcrops of orthogneisses and other metamorphic rocks, poorly exposed in the plains of La Pampa province. These rocks have been studied by Chernicoff et al. (2010), who based on geochemical and isotopic grounds showed that they belong to an Ordovician magmatic arc. They established the crystallization ages of these metaigneous rocks by U–Pb SHRIMP in zircons between 475.7 ± 2.3 Ma (meta-quartz diorites) and 465.6 ± 3.9 Ma (metagabbros). The Hf-isotope composition of the dated zircons of the metagabbros indicate model ages around 1.7 Ga, which are results of a mixture of a mafic magma of *ca.* 466 Ma with a much older crustal component, probably older than 2 Ga. This suggests that part of the underlying basement of the southernmost Pampia terrane is at least that old (Chernicoff et al. 2010). The sedimentary cover identified in the adjacent Cuyania terrane along the San Rafael Block has detrital zircons that show a provenance derived from these igneous sources emplaced in the Pampia terrane.

4 The Famatinian Collision Between Cuyania and Gondwana

The timing of the collision of the Cuyania microcontinent with Gondwana was established along the contact with the Pampia terrane, which is the basement of western Sierras Pampeanas. The Famatinian magmatic arc ceased in this region at about 460 Ma, as result of the collision with Cuyania. The arc is represented by orthogneisses of middle Ordovician age (Pankhurst et al. 1998; Ramos 2004 and cites therein).

The Pavón Formation crops out near Cerro Bola sector, in the central-east region of the San Rafael Block in isolated exposures dismembered by the Andean tectonics. It was formed in a peripheral foreland basin linked to the tectonic loading as a result of the accretion of the Cuyania terrane. These syncollisional deposits were developed in a turbidite sand-rich ramp along the continental margin of Cuyania during the collision. The age of this siliciclastic sequence is established by the Upper Ordovician (Sandbian) graptolite biozones (Cuerda and Cingolani 1998; Cingolani et al. 2003a). The illite crystallinity index suggests anchimetamorphic conditions obtained during the Chanic tectonic deformation characterized by east vergence folding. Geochemical provenance proxies display detrital compositions

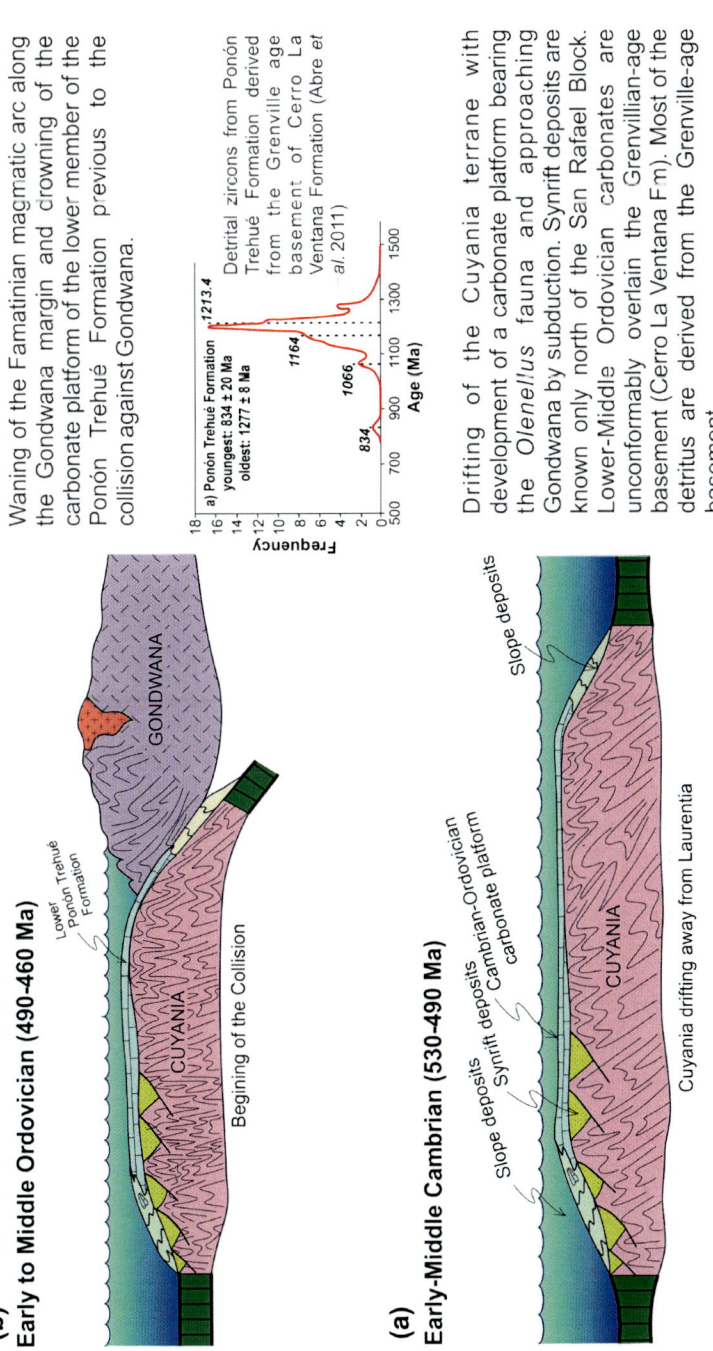

Fig. 3 Evolution of the Cuyania terrane as a microcontinent that collided with the proto-Andean margin of Gondwana, where the Famatinian arc was developed up to middle Ordovician times

derived from an average upper continental crust. The presence of detrital chromian spinels indicate that were derived from a source formed from mid-ocean ridge basalts and continental flood basalts (Abre et al. 2009, 2012). Nd model ages of the Pavón sandstones indicate affinities to Grenvillian-age crust.

U–Pb in situ detrital zircon ages with peaks at 1.1 and 1.4 Ga confirmed the main Mesoproterozoic source (Cingolani et al. 2003a; Abre et al. 2011). The complete provenance dataset suggest that Cerro La Ventana Formation was the main source of detritus, as result of the uplift by thrusting the Mesoproterozoic crust to the east as depicted in Fig. 4a. Paleomagnetic data during the Sandbian (ca. 455 Ma) indicates that the sediments of the Pavón Formation were deposited at latitude of 25.7° ± 2.9°S. This Upper Ordovician paleomagnetic data correspond to the first Cuyania terrane paleopole (Rapalini and Cingolani 2004). It is worth mentioning the El Nihuil Mafic Unit, which is composed by porphyritic dolerites and gabbroic rocks. These rocks were considered as the extension of the Famatinian ophiolite belt on the western side of Cuyania terrane (Haller and Ramos 1993). The undeformed dolerites have a tholeiitic ocean floor basalt geochemical signature (Cingolani et al. 2000) and were interpreted as an Upper Ordovician–Lower Silurian age igneous extensional event based on K–Ar ages. The Nd isotopes and model ages (T_{DM}) show characteristics of mantle derived rocks. The trace and REE pattern is similar to average E-MORB. The gabbroic and tonalitic rocks as part of the basement of Cerro La Ventana Formation are present in several small outcrops within the El Nihuil Mafic Unit and are overprinted by strong ductile deformation (see Fig. 4a).

The orogenic sedimentation depicted in Fig. 5a is represented by the late Silurian–early Devonian sequences of La Horqueta and Río Seco de los Castaños formations, originally described by Dessanti (1956) as a single unit and separated by González Díaz (1981). The La Horqueta is a sandy-dominated metasedimentary marine sequence restricted to the exposures between Río Seco de las Peñas to the north and Arroyo Agua de la Piedra to the south, where the best sections are exposed along the Diamante River. This unit was folded, cleaved, and faulted by a deformational event associated with low-grade metamorphism and medium p-t conditions during the Chanic tectonic phase and are unconformably overlaid by Upper Paleozoic rocks. The Rb–Sr whole-rock age obtained from two different outcrops yielded 371 ± 62 and 379 ± 15 Ma (Tickyj and Cingolani, 2000; Tickyj et al. 2001) interpreted as the age of a metamorphic event. The U–Pb (LA-ICP-MS) detrital zircon ages show main clusters from Mesoproterozoic, Neoproterozoic–Ordovician, and Silurian–Lower Devonian (Cingolani et al. 2008; Tickyj et al. this volume). These data constrain the maximum sedimentation age to the Lower Devonian, and clearly show a provenance derived from the eastern basement outcrops. The main source of detritus was the recently accreted Pampia terrane to the Gondwana margin. Similar outcrops of Devonian rocks were described by Chernicoff et al. (2008) further to the east in La Pampa province.

The Río Seco de los Castaños Formation of Upper Silurian–Lower Devonian age, is a marine-siliciclastic unit (González Díaz 1981) and was interpreted as a

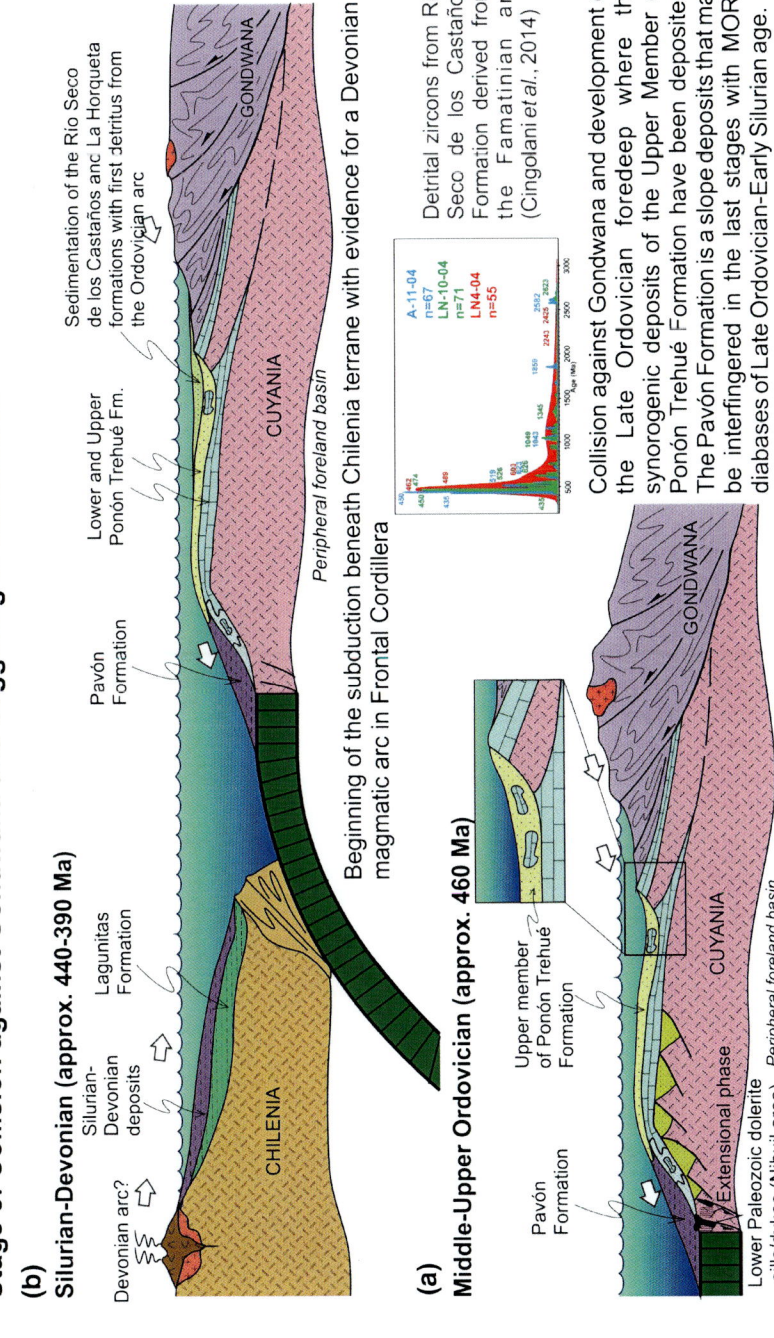

Fig. 4 Collision of the Cuyania terrane against Gondwana, cease of the magmatic arc, widespread deformation and beginning of subduction beneath Chilenia

Stage 4: Collision against Chilenia and beginning of present subduction

(b) Early to Middle Carboniferous (360-320 Ma)

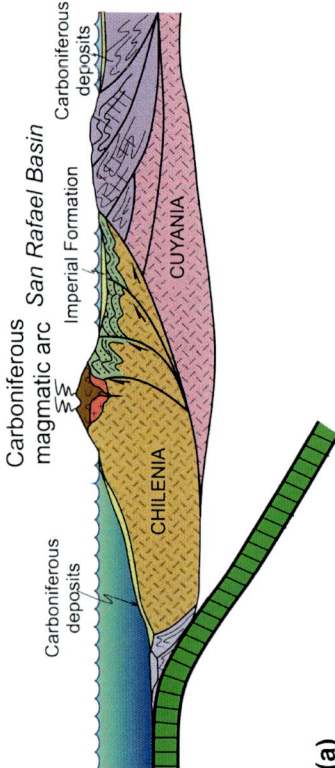

Subduction with east polarity begun in the Early Carboniferous after collision in the late stages of Chanic deformation. A new magmatic arc is associated with synorogenic deposits of the El Imperial Formation and younger units. Strong angular unconformity is developed between Río Seco de los Castaños and El Imperial formations.

(a) Middle Devonian (approx. 390-380 Ma)

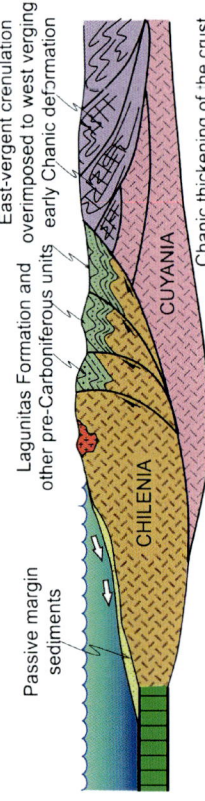

Chanic collision between Chilenia and Cuyania along the Gondwana margin in Middle to Late Devonian times. Previous units are affected by an early west vergent deformation, superimposed by late east vergent structures.

Fig. 5 Collision of Chilenia during the Chanic orogeny and beginning of a new subduction with east polarity beneath the western continental margin of Gondwana

distal to proximal silty platform-deltaic system (Manassero et al. 2009). The dominant sedimentary processes were wave and storm action and the source areas were located to the east. Some acritarchs, lycophyte plants, and several ichnogenera were registered. The lithofacies are mainly immature arkosic sandstones showing both recycled orogen and continental block provenances. Sedimentological characteristics of conglomerate-filled channels and an organic-matter-rich bed were described (Manassero et al. 2009; Cingolani et al. this volume). A very low-grade metamorphism of anchizone overprinted these rocks during the Early Carboniferous (336 ± 23 Ma; Rb–Sr method; Cingolani and Varela 2008). T_{DM} ages and εNd are within the range of the Mesoproterozoic basement and Paleozoic supracrustal rocks from the Cuyania terrane.

The approximation of the Chilenia terrane to the Gondwana margin represented at that time by the basement of the Cuyania terrane produced the beginning of a new subduction beneath the Chilenia terrane (see Fig. 4b). Devonian rocks as the Rodeo de la Bordalesa tonalite dykes and the granitoids crops out south of Cerro Nevado in the "Agua Escondida Mining District" constitute the magmatic arc. The Rodeo de la Bordalesa tonalite dykes, emplaced in the Río Seco de los Castaños Formation, are characterized by high to medium K contents, with metaluminous composition and I-type calc-alkaline signature (Cingolani et al. 2003b). It was dated at c. 400 Ma by U–Pb on zircons and by biotite K–Ar method. Nd model ages (T_{DM}) range between 1 and 1.6 Ga and the negative εNd is characteristic of crustal sources. The age corresponds of an Early Devonian time and the field relationships suggest that correspond to a late-orogenic event.

More recent studies on the La Menta Granite (36°03′S–68°26′W) and on the Borborán Granite (36°01′S–68°24′W) at the Agua Escondida region were associated with the Piedra de Afilar Formation by Tickyj et al. (2015), previously considered as part of an Early Carboniferous suite by González Díaz (1972). However, new LA-ICP-MS U–Pb ages in zircons yielded an age of 388.4 ± 3.0 Ma (Givetian-Eifelian) and 376.6 ± 1.1 Ma (Frasnian), which restricts the igneous activity to the middle to late Devonian for both granites. The chemical characteristics indicate a magmatic arc signature (Tickyj et al. 2015). These rocks as depicted in Fig. 4b, compose the magmatic arc developed on the Chilenia terrane during Devonian times previously to the accretion of this terrane to the Gondwana proto-Andean margin.

5 The Collision of Chilenia Against the Gondwana proto-Andean Margin

The orogenic sedimentary rocks deposited during U. Silurian–L. Devonian times were deformed during the Chanic tectonic phase related to the collision and accretion of the Chilenia terrane against the proto-Andean margin of Gondwana

(Fig. 5a). As a result of that an east vergent cleavage is developed on the previous deformed rocks.

The time of collision was established by Willner et al. (2011) in 390 ± 2 Ma, based on the study of the Guarguaraz Metamorphic Complex at the Frontal Cordillera. These authors recognized the peak of metamorphism by Lu–Hf mineral isochrones from metapelite and metabasite samples, which was interpreted as the time of collision between Chilenia and the proto-Andean margin of Gondwana. The prograded metamorphic conditions of 1.2 GPa, 470 °C and 1.4 GPa, 530 °C were obtained during collision.

Peak pressure conditions were followed by a decompression path with slight heating at 0.5 GPa, 560 °C, dated by a $^{40}Ar/^{39}Ar$ plateau age from white mica at 353 ± 1 Ma, which correspond to cooling below 350–400 °C and pressures of 0.2–0.3 GPa (Willner et al. 2011). This collision is associated with the strong angular unconformity between the La Horqueta/Río Seco de los Castaños Formations (Fig. 5b) and the El Imperial Formation (Dessanti 1956). This last postorogenic unit was deposited in the San Rafael Upper Paleozoic Basin (Henry et al. 2014) that records a succession of sedimentary environments during the latest Mississippian to earliest Permian. That span before, during, and after the glaciation of west central Argentina. The strata of the El Imperial Formation are correlated to similar deposits of other basins of western Argentina.

6 The Gondwanian Cycle and the San Rafael Orogenic Phase

The amalgamation of the Chilenia terrane turned out a new tectonic cycle characterized by normal subduction with East polarity, as depicted in Fig. 5b. As a result of that a typical accretionary prism is developed along the Panthalassa margin. The LA-ICP-MS U–Pb zircon ages of the deposits of the coastal prism between 29° and 36°S identified as the youngest age a cluster of 294–346 Ma (Willner et al. 2008). The maximum depositional ages of most accreted metasediments at the San Rafael Block latitudes, are Middle to Upper Carboniferous. This fact shows that subduction started during Early to Middle Carboniferous after the collision of Chilenia.

The new subduction event that characterized the beginning of the Gondwanian cycle at these latitudes produced a magmatic arc along the Main Andes, which is well preserved in the Chilean and Argentine sides. The activity of this magmatic arc was interrupted by the San Rafael phase that produced intense deformation of all previous rocks. The San Rafael orogenic phase is conspicuously preserved in the San Rafael Block (Fig. 6). The pattern of detrital zircon ages recorded by Rocha Campos et al. (2011) in the upper section of El Imperial Formation indicates a maximum sedimentation age of 297.2 ± 5.3 Ma. These data are compared to those obtained by Mescua et al. (2015) in the Río Salado Upper Paleozoic unit. This

Fig. 6 Angular unconformity between deposits of El Imperial Formation (U. Paleozoic) and the volcaniclastic of the Choiyoi Group (Permian–Triassic) showed at Los Reyunos section, taken to the south

suggests that the San Rafael tectonic phase, at this latitude, could be restricted between 290 and 260 Ma.

The unconformity in the San Rafael Block was related to the beginning of a period of shallow subduction by Ramos and Folguera (2009). This Early Permian tectonic phase is associated with unique processes that resemble more those flat slab episodes of Andean tectonics than those occurred in Paleozoic times at these latitudes. The Early Permian arc abnormally expanded to extend through the entire region. Extensional processes followed the main phase of orogenic building and intraplate rhyolitic sequences of the Choiyoi Group were erupted through the area. These facts point to a flat subduction cycle in Early Permian times, followed by slab steepening and consequent orogenic collapse in the Late Permian to Early Triassic (Ramos and Folguera, 2009).

These outstanding episodes mark the end of the Paleozoic evolution of the San Rafael Block.

Acknowledgements This work was supported by CONICET (grants PIPs 647, 199 and International Cooperation CONICET-FAPESP). We are grateful to Mario Campaña and Norberto Uriz for drafting the figures. We thank to P. Farias and J. García-Sansegundo (University of Oviedo, Spain) for comments and field discussions.

References

Abbruzzi JM, Kay SM, Bickford ME (1993) Implications for the nature of the Precordilleran basement from the geochemistry and age of Precambrian xenoliths in Miocene volcanic rocks, San Juan province. XII Congreso Geológico Argentino, Actas 3:331–339

Abre P, Cingolani CA, Zimmermann U, Cairncross B (2009) Detrital chromian spinels from Upper Ordovician deposits in the Precordillera terrane, Argentina: a mafic crust input. J S Am Earth Sci 28:407–418

Abre P, Cingolani CA, Zimmermann U, Cairncross B, Chemale F Jr (2011) Provenance of Ordovician clastic sequences of the San Rafael Block (Central Argentina), with emphasis on the Ponón Trehué formation. Gondwana Res 19(1):275–290

Abre P, Cingolani C, Cairncross B, F Chemale Jr (2012) Siliciclastic Ordovician to Silurian units of the Argentine Precordillera: constraints on provenance and tectonic setting in the proto-Andean margin of Gondwana. J S Am Earth Sci 40:1–22

Abre P, Cingolani CA, Uriz NJ, Siccardi A (this volume) Sedimentary provenance analysis of the Ordovician Ponón Trehué Formation, San Rafael Block, Mendoza-Argentina. In: Cingolani C (ed) Pre-Carboniferous evolution of the San Rafael Block, Argentina. Implications in the SW Gondwana margin. Springer, Berlin

Astini RA, Thomas WA (1999) Origin and evolution of the Precordillera terrane of western Argentina: a drifted Laurentian orphan. In: Ramos VA, Keppie JD (eds) Laurentia-Gondwana connections before Pangea. Geol Soc Am Spec Paper 336:1–20

Astini R, Benedetto JL, Vaccari NE (1995) The early Paleozoic evolution of the Argentina Precordillera as a Laurentian rifted, drifted, and collided terrane: a geodynamic model. Geol Soc Am Bull 107:253–273

Bordonaro O (2003) Evolución paleoambiental y paleogeográfica de la cuenca cámbrica de la Precordillera argentina. Rev Asoc Geol Argentina 58(3):329–346 (Buenos Aires)

Borrello AV (1971) The Cambrian of the South America. In: Holland CH (ed) Cambrian of the new world. Lower Paleozoic rocks of the world 1, Wiley Interscience, New York, pp 385–438

Chernicoff CJ, Zappettini EO, Santos JOS, Beyer E, McNaughton NJ (2008) Foreland basin deposits associated with Cuyania terrane accretion in La Pampa province, Argentina. Gondwana Res 13(2):189–203

Chernicoff CJ, Zappettini EO, Santos JOS, Allchurch S, McNaughton NJ (2010) The southern segment of the Famatinian magmatic arc, La Pampa province, Argentina. Gondwana Res 17:662–675

Cingolani CA, Varela R (1999) The San Rafael block, Mendoza (Argentina): Rb–Sr isotopic age of basement rocks. II South American Symposium on Isotope Geology, SEGEMAR Anales, vol 24:23–26 (Córdoba)

Cingolani CA, Varela R (2008) The Rb–Sr low-grade metamorphism age of the Paleozoic Río Seco de los Castaños Formation, San Rafael block, Mendoza, Argentina. In: South American Symposium on Isotope Geology, San Carlos de Bariloche, Argentina, pp 1–4

Cingolani CA, Llambías EJ, Ortiz LR (2000) Magmatismo básico pre-Carbónico del Nihuil, Bloque de San Rafael, Provincia de Mendoza, Argentina. IX Congr Geol Chileno, Puerto Varas 2:717–721

Cingolani CA, Manassero M, Abre P (2003a). Composition, provenance and tectonic setting of Ordovician siliciclastic rocks in the San Rafael Block: Southern extension of the Precordillera crustal fragment, Argentina. J South Am Ear Sci 16:91–106

Cingolani CA, Basei MAS, Llambías EJ, Varela R, Chemale JrF, Siga JrO, Abre P (2003b) The Rodeo Bordalesa Tonalite, San Rafael Block (Argentina): Geochemical and isotopic age constraints. 10° Congr Geol Chileno, Concepción, Octubre 2003. 10 p. Versión CD Rom

Cingolani CA, Llambías EJ, Basei MAS, Varela R, Chemale F Jr, Abre P (2005). Grenvillian and Famatinian-age igneous events in the San Rafael Block, Mendoza Province, Argentina: geochemical and isotopic constrains: In: Pankhurst RJ, Veiga GD (eds) Gondwna 12. Academia Nacional de Ciencias, Mendoza, p 103

Cingolani CA, Tickyj H, Chemale Jr F (2008) Procedencia sedimentaria de la Formación La Horqueta, Bloque de San Rafael, Mendoza (Argentina): primeras edades U–Pb en circones detríticos. XVII Congr Geol Argent, Actas, Tomo III, San Salvador de Jujuy, Argentina, pp 998–999

Cingolani C, Uriz N, Marques J, Pimentel M (2012) The Mesoproterozoic U–Pb (LA-ICP-MS) age of the Loma Alta Gneissic rocks: basement remnant of the San Rafael Block, Cuyania Terrane, Argentina. In: Proceedings of 8° South American Symposium on Isotope Geology. CD-ROM version. Medellín, Colombia. Abstract, p. 140

Cingolani CA, Manassero MJ, Uriz NJ, Basei MAS (2014a). Provenance insights of the Silurian-Devonian Rio Seco de los Castaños unit, San Rafael Block, Mendoza: U–Pb zircon ages. XIV Congreso Geológico Argentino, CD-ROM, 2 p. Córdoba

Cingolani CA, Uriz NJ, Manassero MJ, Bassei MAS (2014b) La Formación Cerro Las Pacas al Sur del Cerro Nevado, Mendoza: Basamento Precámbrico o parte de la cuenca devónica de San Rafael? XIV Congreso Geológico Argentino, CD-ROM, S21–11, 2p. Córdoba

Criado Roqué P, Ibañez G (1979). Provincia geológica Sanrafaelino-Pampeana. In: Turner JC (ed) Segundo Simposio de Geología Regional Argentina. Academia Nacional de Ciencias, Córdoba, I:837–869

Cuerda AJ, Cingolani CA (1998) El Ordovícico de la región del Cerro Bola en el Bloque de San Rafael, Mendoza: Sus faunas graptolíticas. Ameghiniana 35(4):427–448

Dessanti R (1945) Informe geológico preliminar sobre la Sierra Pintada, Departamento San Rafael, provincia de Mendoza. Direcc Nac Geol y Min. Carpeta 28, Buenos Aires

Dessanti R (1956) Descripción geológica de la Hoja 27c, Cerro Diamante (provincia de Mendoza). Dirección Nacional de Minería, Boletín 85:1–79 (Buenos Aires)

González Díaz EF (1972) Descripción geológica de la Hoja 27d-San Rafael, provincia de Mendoza. Servicio Nacional Minero Geológico, Boletín 132:1–127 (Buenos Aires)

González Díaz EF (1981) Nuevos argumentos a favor del desdoblamiento de la denominada "Serie de la Horqueta", del Bloque de San Rafael, provincia de Mendoza. Actas 8° Congreso Geológico Argentino 3:241–256 (San Luis)

Haller MJ, Ramos VA (1993) Las ofiolitas y otras rocas afines. In: Ramos VA (ed) Geología y Recursos Naturales de Mendoza, Relatorio 12° Congreso Geológico Argentino, pp. 31–39

Henry LC, Isbell JL, Limarino CO (2014) The late Paleozoic El imperial formation, western Argentina: glacial to post-glacial transition and stratigraphic correlations with arc-related basins in southwestern Gondwana. Gondwana Res 25(2014):1380–1395

Heredia S (2006) Revisión estratigráfica de la Formación Ponón Trehué (Ordovícico), Bloque de San Rafael, Mendoza. INSUGEO Serie Correlación Geológica 21:59–74

Heredia S, Mestre A (this volume) Ordovician conodont biostratigraphy of the Ponón-Trehué formation, San Rafael Block, Mendoza, Argentina

Holmberg E (1973) Descripción Geológica de la Hoja 29d, Cerro Nevado. Servicio Geológico-Minero Argentino, Boletín 144, Buenos Aires, pp 71

Keller M (1999) In: The Argentine Precordillera-sedimentary and plate tectonic history of a Laurentian crustal fragment in South America, vol 341. Geological Society of America, Special Publication, 239 p

Mahlburg Kay S, Orrell S, Abbruzzi JM (1996) Zircon and whole rock Nd–Pb isotopic evidence for a Grenville age and a Laurentian origin for the basement of the Precordillera in Argentina. J Geol 104:637–648

Manassero MJ, Cingolani CA, Abre P (2009) A Silurian-Devonian marine platform-deltaic system in the San Rafael Block, Argentine Precordillera-Cuyania terrane: lithofacies and provenance. In: Königshof P (ed) Devonian change: case studies in palaeogeography and palaeoecology, vol 314. Geological Socity of London (Special Publications), pp 215–240

Mescua JF, Naipauer M, Tapia F, Farias M, Giambiagi L, Ramos VA (2015) Edad U/Pb y correlaciones del Paleozoico de las nacientes del río Salado, y la ocurrencia de la fase Sanrafaélica en la Cordillera Principal de Mendoza. 16° Reunión de Tectónica, Universidad Nacional de Río Negro. Resúmenes: 48–49. General Roca

Mosher, S (1993) Exposed Proterozoic rocks of Texas (part of Proterozoic rocks east and southeast of the Grenville Front). In: Reed JC et al (eds) Precambrian: Counterminous U.S., vol C-2. Geological Socity of America, The Geology of North America, pp 366–378

Nelis MK, Mosher S, Carlson WD (1989) Grenville-age orogeny in the Llano uplift of central Texas: deformation and metamorphism of the Rough Ridge Formation. Geol Soc Am Bull 96:746–754

Nuñez E (1962) Sobre la presencia del Paleozoico inferior fosilífero en el Bloque de San Rafael. Primeras Jornadas Geológicas Argentinas, II:185–189 (Buenos Aires)

Nuñez E (1976) Descripción geológica de la Hoja 28-C "Nihuil", Provincia de Mendoza. Servicio Geológico Nacional. Buenos Aires (unpublished report)

Nuñez E (1979) Descripción geológica de la Hoja 28d, Estación Soitué, Provincia de Mendoza. Servicio Geológico Nacional, Boletín 166:1–67

Pankhurst RJ, Rapela CW, Saavedra J, Baldo E, Dahlquist J, Pascua I (1998) The Famatinian magmatic arc in the central Sierras Pampeanas: an Early to Mid-Ordovician continental arc on the Gondwana margin. In: Pankhurst RJ, Rapela CW (eds) The proto-Andean margin of Gondwana, vol 142. Geological Society of London (Special Publications), pp 343–368

Ramos VA (2004) Cuyania, an exotic block to Gondwana: review of a historical success and the present problems. Gondwana Res 7:1009–1026

Ramos VA, Folguera, A (2009) Andean flat slab subduction through time. In: Murphy B (ed) Ancient orogens and modern analogues, vol 327. Geological Society (Special Publication), London, pp 31–54

Ramos VA, Jordan T, Allmendinger R, Mpodozis C, Kay S, Cortés J, Palma M (1986) Paleozoic terranes of the central Argentine-Chilean Andes. Tectonics 5:855–888

Ramos VA, Litvak V, Folguera A, Spagnuolo M (2014) An Andean tectonic cycle: from crustal thickening to extension in a thin crust (34°–37°SL). Geosci Front 5:351–367

Rapalini AE, Cingolani CA (2004) First Late Ordovician Paleomagnetic pole for the Cuyania (Precordillera) terrane of western Argentina: a microcontinent or a Laurentian plateau? Gondwana Res 7:1089–1104

Rapela CW, Pankhurst RJ, Casquet C, Baldo EG, Galindo C, Fanning CM, Dahlquist J (2010) The Western Sierras Pampeanas: protracted Grenville-age history (1330–1030 Ma) of intra-oceanic arcs, subduction-accretion at continental edge and AMCG intraplate magmatism. J S Am Earth Sci 29:105–127

Rocha Campos AC, Basei MAS, Nutman AP, Kleiman LE, Varela R, Llambías EJ, Canile FM, Da Rosa OCR (2011) 30 Million years of Permian volcanism recorded in the Choiyoi igneous province (W Argentina) and their source for younger ash fall deposits in the Paraná Basin: SHRIMP U–Pb zircon geochronology evidence. Gondwana Res 19(2011):509–523

Sato AM, Tickyj H, Llambías EJ, Sato K (2000) The Las Matras tonalitic-trondhjemitic pluton, Central Argentina: Grenvillian age constraints, geochemical characteristics, and regional implications. J South Am Earth Sci 13:587–610

Thomas WA, Astini RA (1999) Simple-shear conjugate rift margins of the Argentine Precordillera and the Ouachita embayment of Laurentia. Geol Soc Am Bull 111:1069–1079

Tickyj H, Cingolani CA (2000) Metamorfismo de muy bajo grado de la Formación La Horqueta (Proterozoico-Paleozoico Inferior), Bloque de San Rafael (Mendoza), Argentina. IX Congreso Geológico Chileno, Actas 2:539–544 (Pucón, Chile)

Tickyj H, Cingolani CA, Varela R, Chemale F Jr (2001) Rb–Sr ages from La Horqueta Formation, San Rafael Block, Argentina. In: III South American Symposium on Isotope Geology. Extended Abstracts, pp 628–631. Pucón. Chile

Tickyj H, Tomezzoli RN, Basei MA, Fernández MA, Blatter JM, Rodriguez N, Gallo LC (2015) Geología de la Formación Piedras de Afilar, basamento granítico del Distrito Minero Agua Escondida, Mendoza. III Simposio Petrología y Metalogénesis Asociada, General Roca, Río Negro, Abstract. 2p

Tickyj H, Cingolani CA, Varela R, Chemale F Jr (this volume) Low-grade metamorphic conditions and isotopic age constraints of the La Horqueta pre-Carboniferous sequence,

Argentinian San Rafael Block. In: Cingolani C (ed) Pre-Carboniferous evolution of the San Rafael Block, Argentina. Implications in the SW Gondwana margin. Springer, Berlin

Varela R, Basei MAS, González PD, Sato AM, Naipauer M, Campos Neto M, Cingolani CA, Meira VT (2011) Accretion of Grenvillian terranes to the southwestern border of the Río de la Plata craton, western Argentina. Int J Ear Sci 100:243–272

Walker N (1992) Middle Proterozoic geologic evolution of El Llano uplift, Texas: evidence form U-Pb zircon geochronometry. Geol Soc Am Bull 104:494–504

Willner AP, Gerdes A, Massonne HJ (2008) History of crustal growth and recycling at the Pacific convergent margin of South America at latitudes 29°–36°S revealed by a U–Pb and Lu–Hf isotope study of detrital zircon from late Paleozoic accretionary systems. Chem Geol 253:114–129

Willner AP, Gerdes A, Massonne HJ, Schmidt A, Sudo M, Thomson SN, Vujovich G (2011) The geodynamics of collision of a microplate (Chilenia) in Devonian times deduced by the pressure-temperature-time evolution within part of a collisional belt (Guarguaraz Complex, W-Argentina). Contrib Miner Petrol 162(2):303–327

San Rafael Block Geological Map Compilation

Carlos A. Cingolani

Abstract In order to ending the presentation of the Book the geological map compilation of the San Rafael Block is offered. It is separated in three main sectors, with the location of the outcrops of the pre-Carboniferous units described within the Book. From North to South are: Sierra de las Peñas, Sierra Pintada, and Agua Escondida areas. The complete geological compilation is exposed as Supplementary Material that it was based on several bibliographic references.

Keywords Geological map · San Rafael Block · Pre-Carboniferous units

1 Sierra de las Peñas Area

The northern sector extends (Fig. 1A) from the Seco de las Peñas River to the south of Diamante Volcano-Cerro Bola. It comprises a longstanding destruction plain dissected by Quaternary erosion, where the highest relief forms are the modern basaltic volcano El Rodeo (1858 m b.s.l.) and the Agua de la Chilena (1775 m b.s.l.). The plateau-like relief of this region is crossed by a dense network of dry streams. The San Rafael city and the town of 25 de Mayo are both located within this area; the drainage system flows into the Diamante River, which through their canyons (of up to 300 m deep) it is possible to access numerous outcrops of the La Horqueta unit. This permanent river gets the snow melting from the Andes. A series of hydroelectric power plants can be found in this place. The Cuyo extensional basin (Triassic) is located immediately north and northeast. The Cerro Diamante

Electronic supplementary material The online version of this chapter (doi:10.1007/978-3-319-50153-6_14) contains supplementary material, which is available to authorized users.

C.A. Cingolani (✉)
División Científica de Geología, CIG-Centro de Investigaciones Geológicas (CONICET-UNLP), Museo de La Plata, Paseo del Bosque, 1900 La Plata, Argentina
e-mail: ccingola@cig.museo.unlp.edu.ar; carloscingolani@yahoo.com

(2354 m b.s.l.) stands out from a plateau; it was volcanically active until the Pleistocene remaining as a geographic reference within the area. The group of mines of La Picaza district (or Diamante distric) is situated at the southern edge of the Diamante River and is emplaced within the La Horqueta Formation; they had been exploited around 1877 for Pb-sulphides, arsenopyrite, pyrite, sphalerite. The ore origin is linked to Permian-Triassic igneous activity. The most important uranium deposit of the area can also be found here ('Dr. Baulíes' open mine from National Atomic Energy Commission), which is genetically related to Permian volcaniclastic units. Outcrops of the units described in the following chapters of this Book are developed within this area: Chapters "The Pavón Formation as the Upper Ordovician Unit Developed in a Turbidite Sand-Rich Ramp. San Rafael Block, Mendoza, Argentina" (Abre et al.), "Low-Grade Metamorphic Conditions and Isotopic Age Constraints of the La Horqueta Pre-carboniferous Sequence, Argentinian San Rafael Block" and "La Horqueta Formation: Geochemistry, Isotopic Data and Provenance Analysis" (Tickyj et al. and Abre et al.). Outcrops of the Upper Paleozoic sedimentary El Imperial Formation as well as the Permian-Triassic volcaniclastic Cochicó and Choiyoi Groups and the continental Triassic Puesto Viejo Formation dominate the area.

2 Sierra Pintada Area

This area (Figs. 1B and 2B) extends from the Atuel Canyon-Colonia Las Malvinas to the Plateado volcano. Important localities are the villages of El Nihuil and Punta del Agua. Outcrops of pre-Carboniferous units described in the following chapters are exposed in this sector: Chapter "The Mesoproterozoic Basement at the San Rafael Block, Mendoza Province (Argentina): Geochemical and Isotopic Age Constraints" (Cingolani et al.), Chapters "Sedimentary Provenance Analysis of the Ordovician Ponón Trehué Formation, San Rafael Block, Mendoza-Argentina" and "Ordovician Conodont Biostratigraphy of the Ponón Trehué Formation, San Rafael Block, Mendoza, Argentina" (Abre et al. and Heredia and Mestre), Chapters "Lower Paleozoic 'El Nihuil Dolerites': Geochemical and Isotopic Constraints of Mafic Magmatism in an Extensional Setting of the San Rafael Block, Mendoza, Argentina" and "Magnetic Fabrics and Paleomagnetism of the El Nihuil Mafic Unit, San Rafael Block, Mendoza, Argentina" (Cingolani et al. and Rapalini et al.) and Chapters "Silurian-Devonian Land-Sea Interaction Within the San Rafael Block, Argentina: Provenance of the Rio Seco de los Castaños Formation" and "Primitive Vascular Plants and Microfossils from the Río Seco de los Castaños Formation, San Rafael Block, Mendoza Province, Argentina" (Cingolani et al. and Morel et al.) and "The Rodeo de la Bordalesa Tonalite Dykes as a Lower Devonian Magmatic Event:

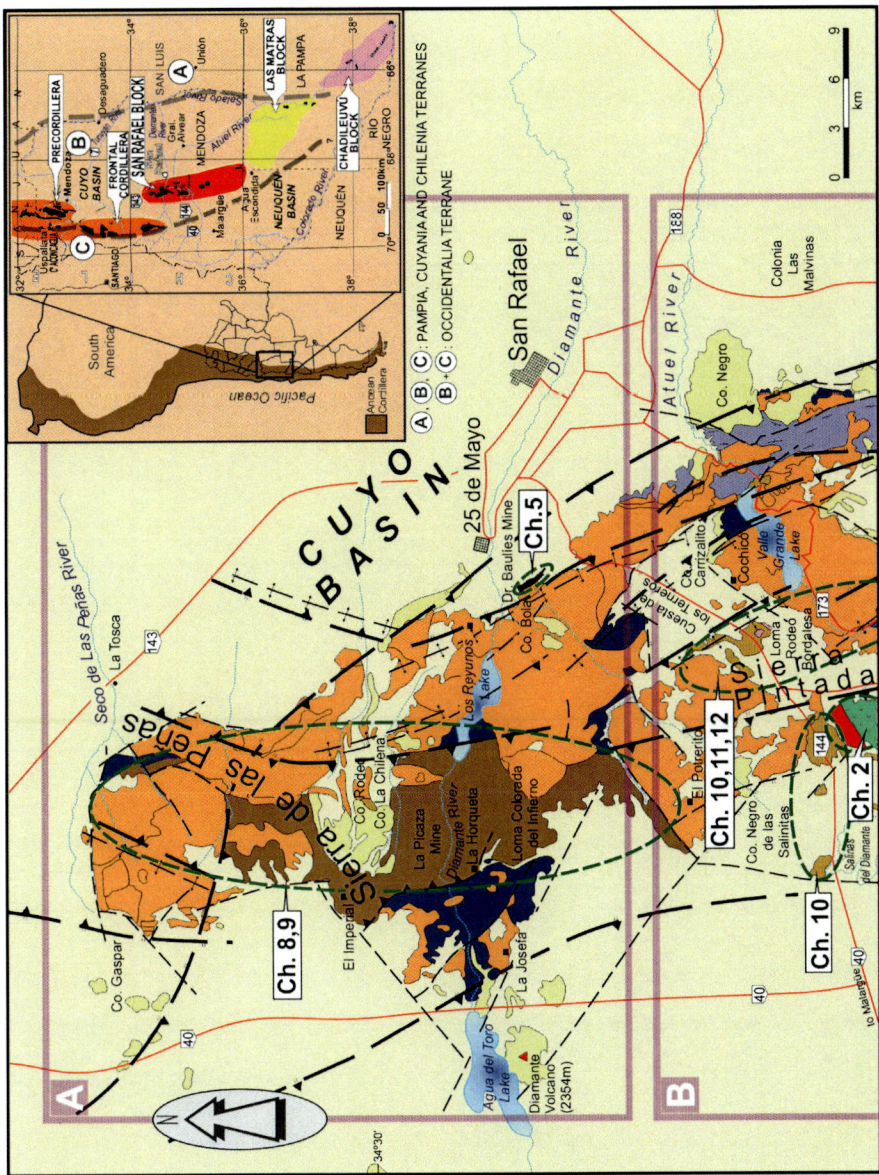

Fig. 1 Northern sector of the San Rafael Block. Location of the outcrops described in Chapters "The Pavón Formation as the Upper Ordovician Unit Developed in a Turbidite Sand-Rich Ramp. San Rafael Block, Mendoza, Argentina", "Low-Grade Metamorphic Conditions and Isotopic Age Constraints of the La Horqueta Pre-Carboniferous Sequence, Argentinian San Rafael Block" and "La Horqueta Formation: Geochemistry, Isotopic Data and Provenance Analysis", are remarked. Geological references in Fig. 3

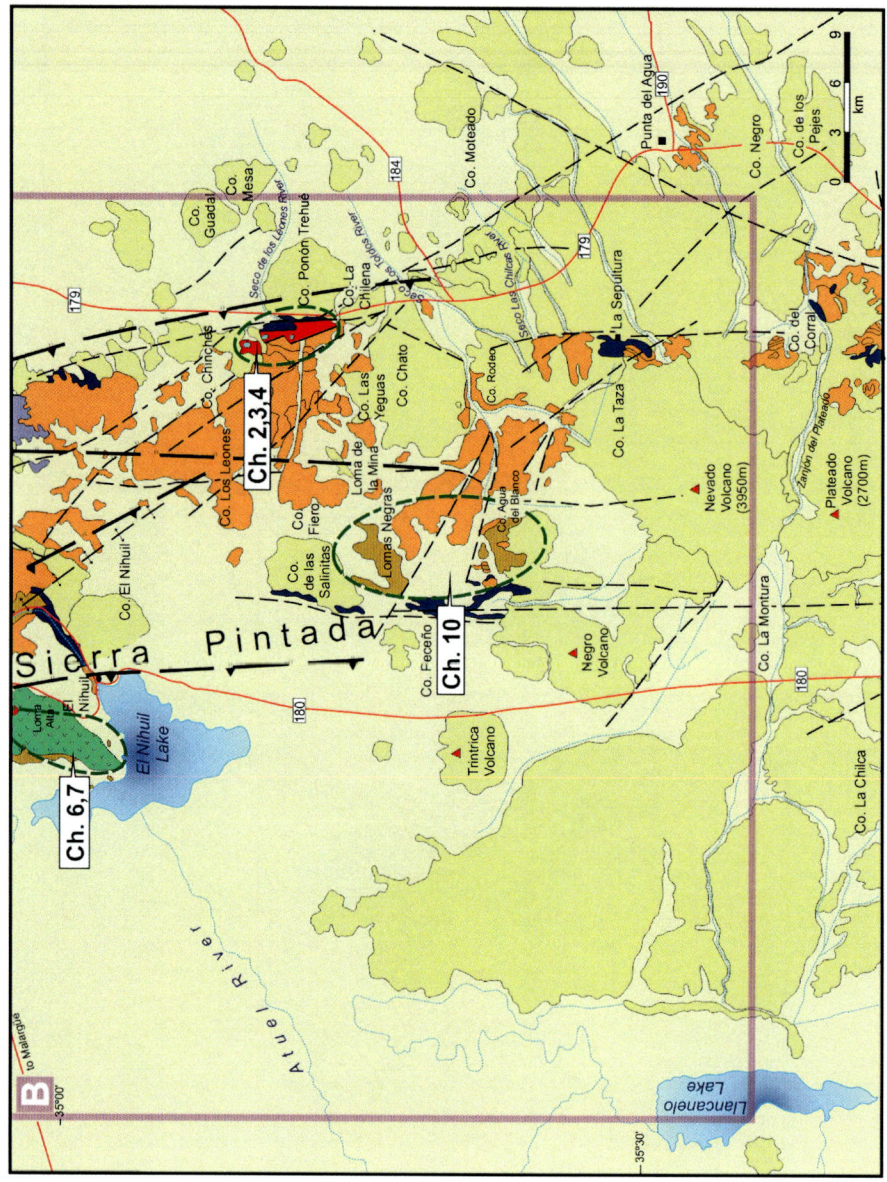

Geochemical and Isotopic Age Constraints". The El Nevado (3950 m b.s.l.) and Plateado Hills (m b.s.l.) record the ingression of the Neogene volcanic arc more than 500 km towards the continental crust (Fig. 2B).

◄**Fig. 2** Central sector of the San Rafael Block geological map. Location of the outcrops described in Chapters "The Mesoproterozoic Basement at the San Rafael Block, Mendoza Province (Argentina): Geochemical and Isotopic Age Constraints", "Sedimentary Provenance Analysis of the Ordovician Ponón Trehué Formation, San Rafael Block, Mendoza-Argentina" and Ordovician Conodont Biostratigraphy of the Ponón Trehué Formation, San Rafael Block, Mendoza, Argentina , "Lower Paleozoic 'El Nihuil Dolerites': Geochemical and Isotopic Constraints of Mafic Magmatism in an Extensional Setting of the San Rafael block, Mendoza, Argentina" and "Magnetic Fabrics and Paleomagnetism of the El Nihuil Mafic Unit, San Rafael Block, Mendoza, Argentina", "Silurian-Devonian Land-Sea Interaction Within the San Rafael Block, Argentina: Provenance of the Rio Seco de los Castaños Formation" and "Primitive Vascular Plants and Microfossils from the Río Seco de los Castaños Formation, San Rafael Block, Mendoza Province, Argentina" and "The Rodeo de la Bordalesa Tonalite Dykes as a Lower Devonian Magmatic Event: Geochemical and Isotopic Age Constraints" are remarked. Geological references in Fig. 3

3 Agua Escondida Area

It is a transitional zone developed in between the San Rafael Block, the Neuquén basin (located towards SW) and the Las Matras region (located to the south-east), comprising Mendoza and a part of the La Pampa provinces (Fig. 3C). It extends from the El Plateado Volcano towards the Agua Escondida town that outstands as the most important locality of this region. The hills are up to 1700 m b.s.l. The Agua Escondida River flows toward the homonymous town. The region comprises a mining district of relative importance in the past, such as the mines Haydee and Santa Cruz (Mn linked to hydrothermal–epithermal origin), Liana (fluorite), and smaller deposits of W, Mb, galena and sphalerite (Mines Elsiren, Potosí among others). The mineralizations are linked to the Piedras de Afilar (Devonian) and Choiyoi (Permian-Triassic) magmatic units. Outcrops of the Upper Paleozoic (El Imperial Formation and equivalents) as well as the Permian-Triassic Cochicó and Choiyoi Groups are widely spread. Some outcrops of Devonian granitoids were recognized in the Piedras de Afilar Formation and the metasedimentary rocks around the Cerro de las Pacas are mentioned in Chapter "The Mesoproterozoic Basement at the San Rafael Block, Mendoza Province (Argentina): Geochemical and Isotopic Age Constraints" (by Cingolani et al.) and Chapter "The Rodeo de la Bordalesa Tonalite Dykes as a Lower Devonian Magmatic Event: Geochemical and Isotopic Age Constraints" (by Cingolani et al.) of this Book.

The complete geological compilation of the San Rafael Block is offer as Fig. 4 (Supplementary Material), it was based on: Groeber (1939); Padula (1949); Dessanti (1956); Agua y Energía Eléctrica (1960); Polanski (1964); Rolleri and Criado Roqué (1970); González Díaz (1972a, b); Holmberg (1973); Nuñez (1976); Nuñez (1979); Moreno Peral and Salvarredi (1984); Delpino (1988a, b); Sepúlveda et al. (2001, 2007); Manassero et al. (2009); Cingolani et al. (see this volume chapters "The Mesoproterozoic Basement at the San Rafael Block, Mendoza Province (Argentina):

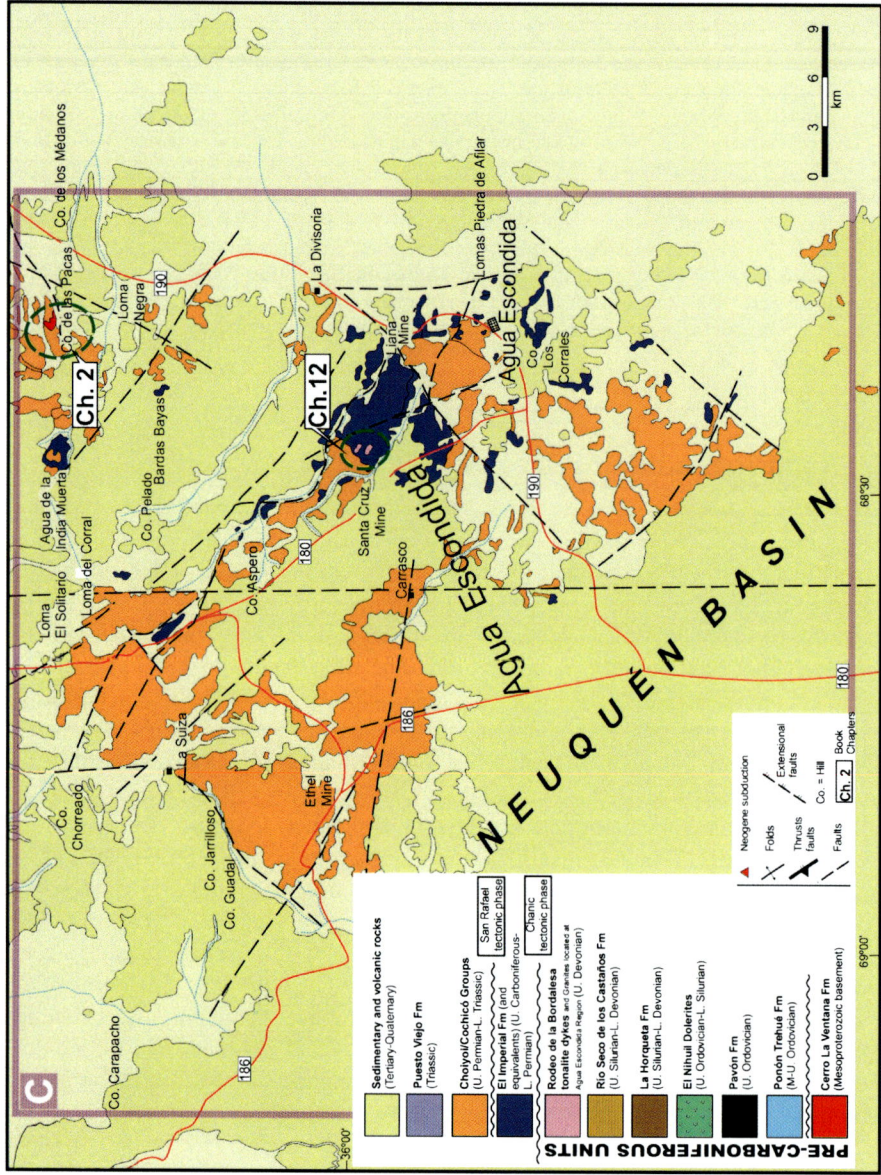

Fig. 3 Southern sector of the San Rafael Block geological map as a transition zone to the Las Matras Block and Neuquén Basin. Location of the outcrops mentioned in Chapters "The Mesoproterozoic Basement at the San Rafael Block, Mendoza Province (Argentina): Geochemical and Isotopic Age Constraints" and "The Rodeo de la Bordalesa Tonalite Dykes as a Lower Devonian Magmatic Event: Geochemical and Isotopic Age Constraints" are remarked

Geochemical and Isotopic Age Constraints" and "Lower Paleozoic 'El Nihuil Dolerites': Geochemical and Isotopic Constraints of Mafic Magmatism in an Extensional Setting of the San Rafael Block, Mendoza, Argentina").

Acknowledgements I am grateful to Eduardo J. Llambías, Víctor Ramos, Norberto Uriz and Paulina Abre for their comments, suggestions and stimulated interactions in several aspects of the map compilation. Thank to Mario Campaña for his technical assistance during final compilation. I am also keen to acknowledge my appreciation to editors of Springer, for their efficiency, patience, and professional help to bring this book to completion.

References

Agua y Energía Eléctrica (1960) Mapa geológico del río Diamante, Mendoza. Unpublished Report, Buenos Aires

Cingolani CA, Llambías EJ, Chemale Jr F, Abre P, Uriz NJ (this volume) Lower Paleozoic 'El Nihuil Dolerites': Geochemical and isotopic constraints of mafic magmatism in an extensional setting of the San Rafael Block, Mendoza, Argentina. In: Cingolani C (ed) Pre-Carboniferous evolution of the San Rafael Block, Argentina. Implications in the SW Gondwana margin. Springer

Delpino DH (1988a) Mapa Geológico del Bloque de San Rafael. Unpublished Report

Delpino DH (1988b) Hoja 29c "Laguna Llancanelo". Dirección Nacional de Geología y Minería, Buenos Aires

Dessanti R (1956) Descripción geológica de la Hoja 27c "Cerro Diamante". Dirección Nacional de Minería, Boletín 85. Buenos Aires

González Díaz EF (1972a) Descripción Geológica de la Hoja 30e "Agua Escondida", provincias de Mendoza y La Pampa. Servicio Nacional Minero Geológico, Boletín 135, 79 p., Buenos Aires

González Díaz EF (1972b) Descripción geológica de la Hoja 27 d San Rafael. Servicio Nacional Minero Geológico, Bol. 132, 127 p. Buenos Aires

Groeber P (1939) Mapa geológico de Mendoza. 2da. Reunión de Ciencias Naturales (Mendoza). Physis 14(46):171–220, Buenos Aires

Holmberg E (1973) Descripción geológica de la Hoja 29d "Cerro Nevado", Provincia de Mendoza. Dirección Nacional de Geología y Minería. Boletín 144, Buenos Aires

Padula E (1949) Descripción geológica de la Hoja 28c "El Nihuil". YPF (unpublished report), Provincia de Mendoza

Polanski J (1964) Descripción geológica de la Hoja 26c "La Tosca", Provincia de Mendoza. Dirección Nacional de Minería, Bol. 101. Buenos Aires 1964

Rolleri EO, Criado Roqué P (1970) Geología de la provincia de Mendoza. 4° Jornadas Geológicas Argentinas (Mendoza, 1969). Actas 2:1–60, Buenos Aires

Manassero MJ, Cingolani CA, Abre P (2009) A Silurian-Devonian marine platform-deltaic system in the San Rafael Block, Argentine Precordillera-Cuyania terrane: lithofacies and provenance. In: Königshof P (ed). Devonian Change: Case studies in Palaeogeography and Palaeoecology. The Geological Society, London, Special Publications, 314:215–240

Moreno Peral CA, Salvarredi JA (1984) Interpretación del origen de las estructuras anticlinales del Pérmico inferior en el Bloque de San Rafael, provincia de Mendoza. 9° Congreso Geológico Argentino, Actas 2:396–413, San Carlos de Bariloche

Nuñez E (1976) Descripción geológica de la Hoja 28-c "Nihuil". Buenos Aires (unpublished report), Provincia de Mendoza. Servicio Geológico Nacional

Sepúlveda E, Carpio F, Regairaz M, Zanettini J, Zárate M (2001) Hoja Geológica 3569-II, San Rafael, Provincia de Mendoza. Servicio Geológico Minero Argentino, Instituto de Geología y Recursos Minerales, Boletín 321:1–77

Sepúlveda EG, Bermúdez A, Bordonaro O, Delpino D (2007) Hoja Geológica 3569-IV, Embalse El Nihuil, Provincia de Mendoza. Servicio Geológico Minero Argentino, Instituto de Geología y Recursos Minerales, Boletín 268:1–52

Index

A
Acritarch, 13, 184
Agua de la Chilena, 141
Agua del Blanco, 189
Agua Escondida, 4
Alvear sub-basin, 25
Amphibolites, 32
Andean Belt, 2
Atuel River, 186

B
b-parameter, 145

C
Calc-alkaline, 40
Cathodoluminescence images, 69
Cenozoic volcanism, 5
Cerro Bola, 8
Cerro de la Chilena, 4
Cerro Las Pacas Formation, 25
Cerro La Ventana Formation, 22, 51, 69, 242
Cerro Nevado, 4
Chanic, 193
Chanic Orogenic phase, 155
Chanic tectonic phase, 13, 138, 246
Channels, 193
Charcoal bed, 191
Chilenia, 150, 199
Chilenia terrane, 13, 249
Choiyoi group, 251
Chromian spinels, 67, 246
Clay minerals, 194
Climacograptus bicornis Biozone, 12
Collision, 14, 244
Conglomerates, 193
Conodont fauna, 11
Conodonts, 75, 79, 81–84
Corral de Lorca, 25
Cratonic, 152

Crustal basement, 53
Crystallinity index, 194
Crystallization age, 49
Cuyania, 2, 22
Cuyania terrane, 70, 88, 92, 94, 97, 99–101, 123, 138, 161, 173, 176, 177, 179, 242

D
Darriwilian, 71
Darriwilian to Sandbian Ponón, 60
Deformed gabbros, 122
Detrital zircon, 68, 139, 151, 205
Detrital zircon dating, 91, 97, 101
Devonian, 150, 184
Devonian Chanic Orogenic phase, 12
Dictyodora, 191
Diorites, 34
Dolerites, 25
Ductile deformation, 53, 150
Ductile-deformed gabbros, 37

E
Early Carboniferous, 199
East polarity, 250
Ectasian- Stenian, 51
El Imperial Formation, 14, 144, 250
El Nihuil, 51, 186, 258
El Nihuil dolerites, 12, 14
El Nihuil mafic unit, 10
E-MORB, 246
E-MORB-type, 122
Epizone, 145, 146
Extensional event, 14, 123
Extensional history, 25

F
Famatinian, 13, 200
Famatinian cycle, 22, 138, 152
Famatinian ophiolites, 14

Fold vergence, 144
Foliated gabbros, 32
Foreland systems, 205
Fortín San Rafael del Diamante, 8
Frontal Cordillera, 4, 10

G

Gabbro Loma Alta, 113
Gabbros, 26
Garnets, 34
Geochemical analyses, 71
Geochemistry, 91, 93, 94, 161, 165, 173, 176
Geochronological, 222, 225, 234, 235
Geological Map, 14
Gneisses, 26
Gondwana, 54
Gondwana margin, 138
Gondwanian cycle, 14, 250
Granites, 26
Graptolite, 8
Grenville orogeny, 54
Grenvillian age, 26, 246
Grenvillian orogeny, 242

I

Isotope data, 161
Isotope dilution, 49
Isotope geochemistry, 96
Isotope system, 67

J

Juvenile rocks, 46

K

Kübler index, 138

L

La Estrechura, 26
La Horqueta Formation, 12, 25, 138, 161–165, 167, 168, 170, 171, 173–177, 179, 200, 258
LA-ICP-MS, 49
LA-MC-ICPMS, 151
Land plants, 13
La Picaza district, 258
Las Matras, 22
Las Matras pluton, 54
Las Matras TTG, 42
La Tortuga section, 63
Laurentia, 242
Leones, 53
Leones-Ponón Trehué, 10
Leones type section, 25
Limestone clasts, 189
Lithotypes, 189

Loma Alta, 128
Loma del Petiso, 25
Lomas Orientales, 25
Lomitas Negras, 189
Los Gateados, 142
Low anchizone, 194
Lower Devonian, 13, 210, 218, 222, 225, 235
Lower Paleozoic dolerites, 12, 106
Low-metamorphic, 13

M

Mafic rocks, 26, 109
Magmatic arc, 240
Magmatic event, 222, 235
Magnetic fabrics, 129
Magnetic foliation, 132
Mendocino-Pampeano mobile belt, 4
Mendoza, 75, 76, 83, 84, 209, 216
Mendoza province, 240
Mermia, 186
Mesoproterozoic, 10, 51, 152, 197, 200
Mesoproterozoic basement, 14, 240
Meta-conglomerates, 142
Metamorphic foliation, 37
Metasandstones, 142
Microfossil, 186, 213
Mining district, 261

N

Nazca plate, 22
Nd model ages, 116
Nereites, 186
Nonradiogenic signature, 54

O

Olistoliths, 25
Olistostromic, 60, 71
Ophiolitic signature, 120
Ordovician, 60, 75, 76, 78–81, 84, 87, 91, 97, 99–101
Orogenic collapse, 251
Orogenic sedimentation, 246

P

Pájaro Bobo formation, 60
Paleomagnetic, 129
Paleomagnetism, 12, 99
Pampean, 152
Pampean-Brasiliano, 200
Pampia terrane, 244
Pavón Formation, 12
Pb analyses, 49
Pb evolution, 68
Pegmatites, 34

Penetrative foliation, 143
Piedras de Afilar, 261
Platform, 189
Platform-deltaic system, 249
Pleurodyctium, 189
Ponón Trehué Formation, 10, 63
Porphyritic dolerites, 111
Postorogenic, 250
Pre-Andean, 199
Pre-Andean region, 22, 106
Precordillera, 8, 22
Proto-Andean margin, 14
Provenance, 67, 71, 87, 88, 91, 92, 94, 96, 97, 100, 101, 161, 165, 173, 178–180, 185
Provenance analyses, 12
P-T conditions, 138
Puesto Imperial, 146
Punta del Agua, 258

R
Rb–Sr systematic, 44
Rb-Sr whole-rock, 155
Recycled orogen, 194
Río Diamante, 4
Río Leones, 8
Río Seco de los Castaños Formation, 13, 184
Rodeo de la Bordalesa, 13

S
San Rafael Block, 5, 76, 79, 83, 161–163, 173, 174, 176–179, 209, 210–212, 221, 222, 234, 235
San Rafael phase, 250
San Rafael tectonic phase, 14
Shallow section, 123
Shallow water, 186
Sierra de las Peñas, 4
Sierra de las Peñas area, 257
Sierra Pintada, 4

Sierra Pintada area, 258
Sills and dykes, 122
Site lineation, 131
Sm–Nd, 47
Source rocks, 66
Sources, 199
Storm action, 191
Syncollisional deposits, 244

T
Terranes, 138
Tholeiitic, 40
Tonalite dykes, 221, 234, 235
Trondhjemites, 40

U
Unblocking temperatures, 135
Unconformably, 14
Unconformity, 186
Unrecycled crust, 196
U–Pb isotopic data, 51
Uranogenic- and thorogenic-Pb, 68

V
Valle Grande, 186
Vascular plants, 186, 217, 218
Veins of quartz, 142
Volcano-plutonic complex, 53

W
Weathering, 65
Western Precordillera, 120
White mica peak, 145

Z
Zircon fractions, 50
Zircon grains, 51